Penguin University Books

SCIENTIFIC KNOWLEDGE
AND ITS SOCIAL PROBLEMS

Jerome R. Ravetz was born in 1929 in Philadelphia, U.S.A.;
his grandparents were Russian Jewish immigrants, his father a
van driver and trade-union organizer. He attended Central
High School, Philadelphia, went to Swarthmore College on a
Pepsi-Cola Scholarship and then to Cambridge University on a
Fulbright Scholarship. After teaching in America and at
Durham University he came to Leeds University in 1957 as
Research Fellow and subsequently as Lecturer and Senior
Lecturer in the History and Philosophy of Science. In 1964–5
he was visiting Professor of History of the Exact Sciences at
the University of Utrecht in the Netherlands. He is a naturalized
Briton; his wife is a social historian and they have three children.

Scientific Knowledge and its Social Problems

JEROME R. RAVETZ

PENGUIN BOOKS

Penguin Books Ltd, Harmondsworth,
Middlesex, England
Penguin Books Australia Ltd, Ringwood,
Victoria, Australia

First published by Oxford University Press 1971
Published in Penguin University Books 1973

Copyright © Oxford University Press, 1971

Made and printed in Great Britain by
Hazell Watson & Viney Ltd, Aylesbury, Bucks
Set in Monotype Ehrhardt

To the memory of my father

GUS RAVETZ

who taught me to think for myself

PREFACE

THE thoughts that are developed in this book have occupied me for such a large part of my life that to describe their origins, or even to give credit to all those who have influenced my progress, would require an autobiographical volume. My concern for the problems of society and ethics has been with me since childhood; and my curiosity about the nature of scientific knowledge was aroused at high school. On the former problem, I at least had Lincoln Steffens' *Autobiography* as an illuminating introduction; but on science, I found the accepted 'philosophy of science' to be only another university subject, interesting in its own right but irrelevant to the problems that were germinating in my own mind. For some years I had hoped that Marxism would provide a coherent synthesis of all these problems; but in the versions then extant it was very much less a successful doctrine than an extremely general guide to action. I recall that I was aware of a philosophical problem-situation to which I wished eventually to devote myself some twenty years ago when I first came to England from America; about seven years ago I had an opportunity to make a first statement of the problems; and since then my work on this material has become steadily more intense, up to the present. Through all these years, I have been mulling over problems in my mind; and they would gain evidence from experience in a variety of ways, mostly unsystematic and informal. My reading and teaching in the history of science was fundamental for the growth of my ideas; indeed, had I not been able to become an historian rather than a mathematician, it is very unlikely that this book could have been written. Of equal importance, however, were the insights and instances derived from a great multitude of discussions, with friends inside and outside the academic world. Many of these discussions were on such ordinary and practical questions, that neither I nor my friend was aware at the time that something would be 'going into my book'; and when in retrospect I would realize that a particular idea had been suggested or stimulated by a discussion, it was often difficult to remember which one, with which friend, had been the occasion for the crystallizing of a vague notion. As I have written up the text, I have tried to give credit to friends for points which struck me at the time of discussion; I fear that this biases the credits towards recent years, and also omits a goodly number of friends from whom I have gained a generalized wisdom.

It is possible for me to identify those institutions and organizations to which I am indebted, in various ways, for the growth of my work. When I attended Central High School, Philadelphia, there still remained enough of the atmosphere, and the teaching staff, of the *Gymnasium* that it had once been, for me to experience the challenge and discipline of serious academic study. Through the generosity of the Pepsi-Cola Company in awarding me a scholarship, I was able to attend Swarthmore College, where windows opened for me. There I discovered the delights of intellectual exploration; and also found a community where people are honourable and trustworthy in the absence of any propaganda or sanctions, simply because of an atmosphere created by the College's Quaker traditions. Coming to Trinity College, Cambridge on a Fulbright Scholarship, I had the privilege of apprenticeship to a master-craftsman of mathematics, Professor A. S. Besicovitch, F.R.S. At the (then) Durham Colleges in the University of Durham, I found myself the protégé and friend of the Warden, Sir James Fitzjames Duff; and through him learned more of the subtle differences between English and American society. There too I met Mr. Peter Doris, who taught me the importance of finding the fundamental questions and sticking to them.

When I moved to Leeds University to work with Professor Stephen Toulmin on a Leverhulme Fellowship, I could concentrate my energies on the history of science and the philosophical problems suggested by it. The occasion for the first formal presentation of the ideas of this book was provided by the Workers' Educational Association; during 1962–3 I held a short course on 'science and society', whose participants included colleagues and students from my department and from the Department of Social Studies, and secondary school pupils with whom I had worked in the Campaign for Nuclear Disarmament. Just a half-century earlier, W.E.A. teaching in Leeds had been the occasion for the achievement of a most influential book in the philosophy of science, Norman Campbell's *What is Science?* If this book is successful, it will serve as another example of useful ideas being developed outside the context of teaching to a captive audience preparing for degree examinations. For several years, a small but vigorous discussion group held under the auspices of the Leeds University Union Student Christian Movement examined drafts of the basic chapters of this book, and subjected them to a sympathetically ruthless criticism. Another small but vigorous group, a seminar in 'Social Problems of Contemporary Science' at the Harvard University Summer School in 1967, helped me formulate the problems discussed in the latter part of the book. Equally important for this work has been my contact with the peace movement in Britain, and more recently, with the British and American societies for social responsibility in science. Even with all these opportunities and stimuli, I would not have been able to create this book, and give it time to mature, were I not working in the happy and encouraging environment of the University of Leeds. My debt to the University, through successive Heads of my department and Vice-Chancellors, together with the administrative, clerical and teaching staff with whom I have worked, is enormous. The motto of the University is Bacon's 'et augebitur scientia'; I hope that my own work will both explain its deeper meaning, and contribute to the fulfilment of its prophecy.

Among the individuals to whom my obligations are particularly strong, I recall from earlier years the late Professor A. S. Besicovitch, Professor C. Truesdell, Dr. D. T. Whiteside and Dr. (now Professor) I. Lakatos, for insights into the nature of the work and of the knowledge achieved in the exact sciences. When Dr. D. S. L. Cardwell and I were colleagues at Leeds, we had many conversations whose fruits are inadequately acknowledged in the occasional citations in the text. On the basis of those discussions, I was able to appreciate the insights contained in Michael Polanyi's *Personal Knowledge*, particularly in connection with the nature and role of craft skills in scientific work. More recently, conversations with Dr. P. M. Rattansi provided me with the first insights on the 'craftsman's romantic philosophy of nature' and on the category of folk-science; and from J. E. McGuire and C. Webster I learned much concerning the philosophical and social aspects of science during its formative period in the seventeenth century. To R. G. A. Dolby I am indebted for his truly incisive criticisms of crucial sections of the text. Discussions with Dr. N. D. Ellis and Dr. A. Coddington have been essential for the development of my ideas on 'immature sciences'; and N. Parry exposed the obscurities at the foundations of earlier drafts of my argument on ethics in science. Dr. Alison Ravetz has contributed to the work over the years, through sharing her own experience as an archaeologist and historian, and also providing an astringent and penetrating stylistic criticism at several crucial points in the evolution of the text. It was through reading the work of Professor Barry Commoner and his group at St. Louis, that I came to the conception of 'critical science', which gives this whole work what unity it has. I must also thank the two typists who produced a fair text with such speed and enthusiasm, Mrs. Lily Minkin and Miss Margaret Knapp; and also thank the proprietors of Beck Hall, Malham, Yorkshire, whose friendly hospitality enabled me to achieve an intense concentration on the most difficult problems of shaping this book, in ideal conditions. I am grateful finally to the Clarendon Press for their enthusiasm for this difficult book; and to them and the printers for producing it so quickly and well. I hardly need add that the many imperfections of this book, some of them known to me but doubtless more unknown as yet, are entirely my own responsibility.*

July 1970   J.R.R.

---

* Regretfully, due to technical limitations, it has not been possible to make any emendations in this paperback edition.

CONTENTS

INTRODUCTION

THE problem-situation that gave rise to the present work is the recent rapid change in the character of the social activity of disciplined inquiry into the natural world, and the consequent changes in its understanding of itself. The established traditions of research in 'the philosophy of science' and in 'the sociology of science' have been recognized as losing contact with the actual practice and real problems of science in the present period. Quite independently of these academic studies, there has developed a new common-sense understanding of science, derived from the daily experience of working scientists and from the problems of decision and government within science.[1] There has been a rapidly increasing flow of studies of one or another aspect of science, conditioned by this new common-sense, but none as yet that attempts a new synthesis.

It appeared to me that the problems of the character of scientific knowledge, of the sociology and ethics of science, and of the applications of science to technology and to human welfare, are so intimately connected that a proper study of any one of them requires an informed awareness of the others. A mixture of inherited assumptions from the academic studies, with elements of the new common-sense, varying in its contents from one person to the next, is not sufficient for this. Hence I have tried to create a coherent framework of

[1] The most extensive codification of this new common sense of science is in W. O. Hagstrom, *The Scientific Community* (Basic Books, New York and London, 1965). Many of the points made in this book can be found there, with a wealth of illustrative material from interviews and historical examples. But his problem was very different from mine, and the organization of the argument, and the emphasis given to the various points, is quite different. I felt that it would be otiose to provide a concordance of material between the two texts, and so I am mentioning it at this early stage, with a recommendation for its study as a parallel source.

Rather closer to my own concerns are C. Wright Mills, *The Sociological Imagination* (Oxford University Press, 1959; Grove Press, New York, 1961) and D. S. Greenberg, *The Politics of American Science* (Penguin, London, 1969; first published as *The Politics of Pure Science*, New American Library, New York, 1967). I have quoted from them occasionally to provide evidence for some particular points, but each in its own field provides materials and insights complementary to those developed here.

concepts in whose terms each of the problems could be discussed
in relation to the others.

I am well aware that my extended arguments can be difficult for
those who have not already begun to reflect on science in terms of this
new common-sense. In order to ease the way in, I have provided two
introductory chapters. The first shows that there *is* a problem in the
understanding of science. The second picks up several themes that
are current in popular discussions of science, and provides them with
a unified theme: the industrialization of science and its consequences.

In the main body of the text, the order in which topics are discussed is governed mainly by the requirements of clearly defined
terms and established conclusions that are necessary for a proper
analysis at each point. I have therefore proceeded from the aspects
of science that are more abstracted from their social context, towards those of the fullest complexity. Thus, it is possible to analyse
problem-solving in research with only an occasional reference to its
wider aspects, while the problems of quality-control and ethics in
science are difficult to appreciate without a systematic knowledge of
the craft of research and the nature of its products. Lest this ordering
from the simple to the complex give the impression that the earlier
topics are in any way more fundamental, I have referred forward,
informally, to later sections of the book whenever appropriate.

The sequence of major problems is then epistemology, sociology
and ethics, and applications in the technical and human context. On
the first, I am not so much concerned to refute the earlier assumption
that science somehow produces truth by the application of a standard
method, as to see how recognizeable, objective scientific knowledge
can (but need not) eventually emerge from the very personal and
fallible activity of research. On the second set of problems, I analyse
the sorts of social mechanisms and attitudes that are necessary in
order that worthwhile research may be done; and I also study the
conditions under which they are effective, and those under which
they are not. Having established the concepts required for a discussion of both the 'individual' and the 'social' aspects of scientific
work, I can enrich them further and discuss the applications of
'science' and 'the scientific method' both to disciplines concerned
with man rather than with nature, and to problems relating to technology and to human welfare. With all this accomplished, I can
return to the general problems of the present and future condition of
science, and offer a critical discussion of these. Because I build up

my technical vocabulary as I proceed, using ordinary words in special restricted senses, it is not easy to dip into the book halfway through and follow the argument. This difficulty is unavoidable, but the analytical index should help the reader locate the main discussion of the technical terms that I use in a special sense.

There is one basic term that I have deliberately used in several senses, and that is 'science' itself. A glossary of the various meanings of the English word now in use would be lengthy and tedious; and if one extended this to the German *Wissenschaft* and the French *science positive*, both of which made important contributions to our inherited notions of 'science', then any worthwhile study would be a major undertaking in itself. Also, the different parts of my analysis apply to different (though overlapping) senses of 'science', and the context has been sufficient for establishing, in a general way, which sort is being discussed. Thus in Part I, the concern is with the natural sciences, pure and applied; while in Parts II and III, the analysis applies to any sort of disciplined inquiry. In Part IV, the 'social sciences', pure and applied, are to be included as a major object of attention; and in the Conclusion, my concern reverts primarily to the natural sciences. I have tried to indicate these particular focuses of reference by my use of words standing for 'science' in the text, and by the examples discussed in the footnotes. Although this method still leaves some obscurity, it seemed preferable to a pedantic differentiation of terms.

Another ambiguity in usage should be pointed out, as it may eventually prove to be a weak point in my structure of concepts. I have not considered the problem of the different levels of aggregation of problems, methods and results, described loosely by such terms as 'field', 'discipline', and 'science'. There was no place in my analysis where I could perceive any difference being made by the level of aggregation assumed. Hence I have rather promiscuously used these terms as synonyms; and the occurrence of one rather than another should not be taken as having some particular significance.

In almost all cases, I have kept the discussion of particular examples in footnotes separate from the text. Although this may be inconvenient for the reader, and also deprives the examples of some of their force as evidence, there are several reasons for my adopting this style. The consistent alternative, of providing evidence for every thesis by a significant example from history or current practice, thoroughly established as valid in itself and woven into the argument, would have required a German style of scholarship, involving a preparation time and a book both several times longer than the present work. Also since historians of science have themselves only recently come to share the new common-sense of science that I am trying to formalize, reliable and appropriate case-studies are still rare; and to provide comprehensive evidence for each of my theses, I would have needed first to re-create much of the history of science. There are other reasons for the separation of examples from argument, deriving from the character of the work itself. No set of particular examples can provide conclusive evidence for general theses of the sort argued here. Many of my arguments will appear as restatements of the obvious to those involved in the practice of science, and for them the examples will function as light relief. On the other hand, those who find my conclusions repugnant or incomprehensible would be resistant even to the most persuasively argued case-studies; they could well dismiss each of them as unrepresentative instances. Hence my arguments will carry conviction if and only if they strike the reader as offering real solutions to real problems; and for this, examples that illustrate the argument without entangling and lengthening it seem most appropriate.

Although the main intended function of this work is a practical one, to assist in the understanding and solving of the deep problems confronting science now and in the near future, I have consistently attempted to maintain a standard of rigour and consistency appropriate to philosophical discussion. Also, I believe that some of the concepts and distinctions I have introduced in the discussion may be of use in other general discussions of science. The basic concepts fall into two classes, those concerned with knowledge and those concerned with action. For the former, I define 'problem', and give a sketch of its anatomy, distinguishing the different stages of development of its materials, as 'data', 'information', 'evidence', and 'conclusion'. In analysing 'methods', I distinguish among the criteria on which the controlling judgements of scientific inquiry are based: particularly those of 'adequacy' and of 'value'. I make strong use of the concept of 'pitfall', which more than any other explains how apparently rigorous research can lead to disasters. Studying the subsequent history of a solved problem, I distinguish between the 'result' itself, the 'fact' which may evolve from it (and I define this term closely), and 'scientific knowledge'. The condition of 'immaturity' of a field can be defined in terms of the absence of 'facts', resulting from the lack of those appropriate criteria of adequacy which steer the investigation away from pitfalls.

Concerning action, I make strong use of distinctions within the 'final cause' of a task, such as the 'goal' which defines what is to be attempted, the 'function' to be performed by the successfully accomplished task, and the 'purpose' in human satisfaction to which the performed function is a means. The operator on the task has his own purpose (or purposes) to be achieved, not necessarily identical with the fulfilment of the goal; and for brevity I use 'motive' for his reasons for being on that task at all. The application of this battery of final causes to the social activity of science enables an analysis of the problems of quality-control and ethics. Also, with it one can distinguish between the 'scientific problem' of the traditional sort, the 'technical problem' in which the goal is defined by the desired performance of a pre-assigned function, and the 'practical problem' defined by the achievement of given purposes. One can also use these distinctions for an analysis of various pathologies of social activities in general; in their terms I am able to define 'corruption',

Finally, all these concepts can be used to enrich the earlier discussion of important styles of scientific activity. The 'industrialized science' of the present can be distinguished from the 'academic science' that dominates the folk-memory of leading scientists of the older generation, in terms of the capital-intensity of the tools of scientific work and the consequent new social relations within the world of science. We now see emerging a 'critical science', in which science, technology, politics, and ultimately the philosophy of nature are involved, and which may be the most significant development in the science of our age.

Part I

THE VARIETIES OF
SCIENTIFIC EXPERIENCE

INTRODUCTION

THE activity of modern natural science has transformed our knowledge and control of the world about us; but in the process it has also transformed itself; and it has created problems which natural science alone cannot solve. Modern society depends increasingly on industrial production based on the application of scientific results; but the production of these results has itself become a large and expensive industry; and the problems of managing that industry, and of controlling the effects of its products, are urgent and difficult. All this has happened so quickly within the past generation, that the new situation, and its implications, are only imperfectly understood. It opens up new possibilities for science and for human life, but it also presents new problems and dangers. For science itself, the analogies between the industrial production of material goods and that of scientific results have their uses, and also their hazards. As a product of a socially organized activity, scientific knowledge is very different from soap; and those who plan for science will neglect that difference at their peril. Also, the understanding and control of the effects of our science-based technology present problems for which neither the academic science of the past, nor the industrialized science of the present, possesses techniques or attitudes appropriate to their solution. The illusion that there is a natural science standing pure and separate from all involvement with society is disappearing rapidly; but it tends to be replaced by the vulgar reduction of science to a branch of commercial or military industry. Unless science itself is to be debased and corrupted, and its results used in a headlong rush to social and ecological catastrophe, there must be a renewed understanding of the very special sort of work, so delicate and so powerful, of scientific inquiry.

If we are to achieve the benefits of industrialized science, and avert its dangers, then both the common sense understanding of science and the disciplined philosophy of science will need to be modified and enriched. As they exist now, both have come down from periods when the conditions of work in science, and the practical and ideological problems encountered by its proponents, were quite different from those of the present day. Science is no longer a marginal pursuit of little practical use carried on by a handful of enthusiasts; and it no longer needs to justify itself by a direct answer to the challenge of other fields of knowledge claiming exclusive access to truth. As the world of science has grown in size and in power, its deepest problems have changed from the epistemological to the social. Although the character of the knowledge embodied in a particular scientific result is largely independent of the social context of its first achievement, the increase and improvement of scientific knowledge is a very specialized and delicate social process, whose continued health and vitality under new conditions is by no means to be taken for granted. Moreover, science has grown to its present size and importance through its application to the solution of other sorts of problems, and these extensions react back on science and become part of it. For an understanding of this extended and enriched 'science', we must consider those sorts of disciplined inquiry whose goals include power as well as knowledge.

To achieve an understanding of the new social character and problems of science, it is useful first to review the deficiencies of the various prevalent common sense and traditional images of science; and then to show by example that the industrialized science of the present has social problems which arise from the technical and social conditions of its work. This discussion opens the way to a new philosophy of science which, instead of asking 'What sort of truth is embodied in perfected scientific knowledge?', proceeds by asking 'By what activities and judgements, individual and social, can genuine scientific knowledge come to be?' The second part of this book is concerned with this question, and the concepts developed there are then applied to some important problems in the social activity of science. In the third part I examine the practices and attitudes which are necessary for the maintenance of the health of scientific endeavour, and to prevent its decline into stagnation or corruption. In the fourth part, I consider the special features of the application of science to technical and practical problems; and in this connection I analyse the peculiarly difficult problems of immature sciences. Finally, I discuss the moral and political problems of industrialized science, and show that it is already giving rise to its opposite, 'critical science', which may offer a means to the resolution of some of the problems which disturb and perplex all those who recognize the condition of science in our modern world.

I

'WHAT IS SCIENCE?'

THE SUCCESS and vigour of scientific research, and the effectiveness of its technological application, are generally accepted as indicators of the quality of a nation's life. Science is so important, and expensive, that the major policy decisions concerning its development are increasingly being taken by the State, rather than being left to the judgement of the scientists and their private patrons. Accordingly, the progress of science becomes a matter of politics; everyone in the community is involved in the consequences of decisions on 'science policy', and every citizen is responsible, however indirectly, for the formation of those decisions. This social involvement of science will necessarily increase over the decades to come. An increasing number of practical problems will need to be solved through planned programmes of scientific and technological research, and the rising cost of the many-sided work of science will call for the most direct involvement of the State in its planning.

In such a situation, it is most important that there should be a common knowledge of what it is that is being discussed and planned, and what are the special features that distinguish it from other objects of public concern. Such a common understanding should eventually develop, but we are as yet a long way from it. At present, we find a variety of partial views on the nature of science. Each of these derives from some special experience of a very complex activity, and from traditional views of science that were developed in response to problems that have since vanished. Such a variety is natural; but the present situation has the potentially dangerous feature that there is no general awareness of this variety. Each view of 'science' seems obvious common sense to its proponent; and in discussions of science, one can find the participants using the same word to refer to radically different things.

Thus, the common sense of science actually includes several

varieties of scientific experience. The understanding of science is truly fragmented; but each of these fragments reflects an important aspect of the total thing. It would be impossible to fit these fragments together into a coherent whole, as if they were pieces of a jigsaw puzzle; but any attempt at unified understanding of science must take account of the experience (of the past and of the present) from which these views take their origin.

Views of science

Since natural science depends on the general public for its support, in this period as never before, the public understanding of science is crucial, in the long run, for the continued health of the community of science. In many ways, the public appreciation of science leaves nothing to be desired. We are all grateful for the comfort and security of life that is achieved by modern technology, and prepared to accept the claim that all these good things are the by-products of scientific research. Moreover, in the common-sense understanding of man's relations with his natural environment, and even with his fellows, 'science' reigns supreme. Supernatural explanations of natural events, even of great disasters, are no longer taken seriously. The areas of ordinary life where inherited craft-wisdom is valued more highly than the judgements of scientific experts, are shrinking down to the vanishing point. This arises partly from the increasing artificiality of our material environment, and from the rapid changes in it as well as in our social environment. A sign of the triumph of 'science' is the reliance on textbooks for such personal crafts as rearing children and even achieving a happy married life. This tendency is more marked in the United States than anywhere else; there the prevailing assumption is that every problem, personal and social as well as natural and technical, should be amenable to solution by the application of the appropriate science.

Without such a general appreciation of science, it would be impossible for the scientific community to continue to grow, and to lay claim to an ever-increasing share of the national wealth. But the limitations of this appreciation must be understood if the support is to continue on a genuine and healthy basis. First, what the public appreciates is not the same as that for which the support is required. In all the varied activities of science (including popularization, teaching at all levels, technological application, and administration and decision-making) it is research that lies at the nerve-centre.

Without research—the continuous, disciplined advance from the known into the unknown—these other activities would either lose their meaning, or become stale, sterile and eventually corrupted. Yet scientific research is most difficult, if not impossible, for a layman to comprehend. The work itself cannot be appreciated without some prior familiarity with the very specialized activity of investigating problems in the context of an abstract, technical discipline; while it has some similarities to the running of a business, or even to the driving of a bus, its special character can be communicated only to a layman with a well-developed imagination. And the object of the research work, the new scientific result, is almost always so technical, so dependent for its meaning and significance on the context of the existing information in the special field, that the layman can never get more than the most general idea of what it is all about. In this respect, the 'layman' may equally well be a scientist in another field, for the problem of internal communication within the world of science is certainly severe. But at least one working scientist can recognize the moves in an exposition by another, in spite of not understanding the content; while the man in the street would find a detailed account of a new discovery to be totally incomprehensible.

Hence what the general public appreciates in science is not what the scientists are doing. Rather, it can be classed under two headings: techniques and natural magic. The first is the collection of devices that make life easier to live, or the destruction of life more efficient. The second is the production of strange and wonderful effects without recourse to supernatural agents. Although the term 'natural magic' fell into disuse some time ago, it is quite natural for this interpretation of science to persist. When the layman (child or adult) sees a demonstration of some astonishing effect, can he imagine the human endeavour and intricate technical work which led to its creation? No; it is the effect itself which captures the imagination, and produces wonder and delight: this is natural magic. Of course, the audience for the 'wonders of science' is told that these achievements are simply the application of the laws of Nature, and that the creator of the effect is in no sense a magician. But since the audience cannot understand the laws of Nature that are relevant to the production of the effect, it is strange and wonderful (until it becomes a commonplace part of the environment) and so differs little from natural magic of the old sort. We can see the strength of this attitude from the great popularity of projects that are politely described as science but which are clearly recognizable as nearly pure natural magic; the manned exploration of space is the best example.

In earlier ages, when natural science was pursued by a small band of enthusiasts with the support of a handful of patrons, the ignorance of the general public did not usually matter. But it is now the general public which, indirectly or directly, pays the piper, and which, through its elected representatives, ever more frequently wishes to call the tune. No amount of popularizing effort will overcome the fact that at the present stage of civilization scientific research is a rather esoteric activity. Even in their relations with their patrons in the past scientists have sometimes been tempted to play something of a confidence game, inventing a middle ground between what was important to their patrons and what was worthwhile to them. The temptation is even stronger now, but it must be resisted. If relations between the scientists and the general public are to remain healthy, they must be based on mutual respect between people in different worlds. The scientist must not dismiss the layman as an ignoramus, and the layman must trust the scientist as a man whose work, while incomprehensible, is genuinely worthwhile for its own sake. Such mutual respect cannot be achieved until both sides recognize the difficulties in the way of genuine communication; and of course there is a special responsibility on the scientist to be worthy of the trust placed in him.

Between the general public and the professional scientists lies a rather large group, roughly between five and ten per cent of the population, who have some detailed acquaintance with scientific knowledge, although no experience of research. These are the people who have made a disciplined study of some branch of science; students, present or past. It is widely hoped that a higher proportion of places of responsibility in public life will be taken by such people, so that there will be an intelligent and competent mediation between the scientific community and the general public. In the student's view of science there is assuredly little of magic; but unfortunately it is false to suppose that the experience of learning a selection from the body of existing scientific knowledge confers any insight into the process of its creation.

Up until very recently the overwhelming impression of science imparted to a student by all but a few eccentric and gifted teachers was that of a mass of accomplished, solid facts. The student's task was to assimilate a pre-sorted sample of these and to be capable of reproducing them under examination conditions. The facts could be of many types: special bits of information, techniques of manual or mental manipulation, and general laws and theories. Whatever their sort, they were recognizably hard, and announced to be clear. For a student in such an environment, the frontiers of science would be located on the remote heights of a pyramid of facts, and the work of research would be imagined as the addition of some more hard facts on top.

Much of the teaching of science in the later years of secondary schools and in universities is still conducted within this framework. But in recent years there has been a massive onslaught from the proponents of the many reformed syllabuses for the teaching of science. The difficulties encountered by these reforms may serve to remind us that while this older view of science is dangerously one-sided it is by no means without foundation. The great achievement of the natural science of the past few centuries, unique in the history of civilization, is the wealth of facts (of all the sorts I mentioned above) which it has produced. And these facts are, to an astonishing degree, hard and reliable, and are usually sufficiently clear to serve their purpose. Moreover, scientific research, even of the most inspired and revolutionary sort, is not accomplished by a great man opening his eyes to the world about him, but necessarily grows out of the matrix of a body of highly technical special results. Except in the very rare cases when a whole field is created, it is on the basis of old facts that new ones are made. Finally, to be able to do this work, the scientist must be an accomplished craftsman; he must have undergone a lengthy apprenticeship, learning how to do things without being able to appreciate why they work.

Thus, the problem of imparting a correct view of science to students cannot be solved by throwing out the dry old facts and bringing in the excitement of open-ended research. Some sorts of students may enjoy this much more, and those who go out into the world will have a better appreciation of research than students of earlier generations trained under the former 'discipline. But one cannot be a craftsman unless one can manipulate one's tools; and one cannot appreciate craftsmanship in others, as in judging between solid and spurious research, unless one has been trained up to it one-self. Thus the students of the present day may have either one of two diametrically opposed views of science, depending on whether they have been the subjects of pedagogical experimentation. Coming out of the old school, they will tend to see science as the mountain of facts; and coming out of the new school, they will tend to see it as questions with ends opening in all directions. Some of the new syllabuses manage to strike a balance; but in the difficult and delicate task of training inexperienced and immature minds to appreciate the complex interplay of facts and problems it is too much to expect uniform success within a short time.

It might seem that the group of people whose view of science is the most important and valid are the working research scientists themselves. But although research is at the centre of scientific activity, it remains as one very specialized part of a large and complex whole. The experience of the research worker will, in its own way, be as specialized as that of the student. His task is to achieve new results in a special field; even if he teaches part-time (as at a university) there will usually be little or no connection between the new results he is creating and the established knowledge which he is passing on. Also, unless he is already in a position of seniority and responsibility, he will have little involvement in the work which requires him to see his own efforts in the context of the field as a whole, in making the judgements and decisions which determine the directions of future research.

Over the generations there has developed a view of science which is well suited to produce the intensely committed, completely specialized activity which modern science has required from the majority of its practitioners. For them, 'research' is not merely the centre of scientific activity; it *is* science. Even if this is not stated explicitly, it can be inferred from the attitudes towards those involved in other aspects of scientific work. Those who have tried research, but abandoned it for easier things, are objects of some contempt; those who have never even tried, but are merely teachers, are to be pitied.

This exclusive, sometimes fanatical, concentration on the production of isolated scientific results can be traced back to the German scientific research schools of the last century. It is conducive to the production of solid work, but it is also anarchic and selfish. Its prevalence has discouraged the growth to intellectual maturity of many scientists who would otherwise have been capable of it. It creates personal difficulties for scientists who, in their middle years, would be more happily and productively employed on other sorts of work related to science; but they are reluctant to have it said of them that 'he has given up research'. The conditioning produced by years of concentrated and narrow research has the effect of making it more difficult for the leaders of science to see the work of themselves and their colleagues in its broader context, both within science and in relation to the outside world. And finally, any social activity which depends for its recruits on a supply of single-minded enthusiasts is dangerously vulnerable to changes in intellectual fashion.

Science as the Pursuit of Truth

The question 'What is Science?' supplies the title or the subject-matter of many books on the 'philosophy of science'. In them the question usually takes the form, 'What sort of truth is embodied in completed scientific knowledge?' Ideas developed in the course of an attempt to answer such a question will not be well suited for describing science as a human activity, always changing and never perfect. Treatises on 'Scientific Method' written within such a framework of ideas seem to have little relation to the real work of discovering new knowledge and are frequently scorned by practising scientists who have become amateur philosophers of science.

It would be a serious mistake to dismiss all this effort as irrelevant to the understanding of science. The question itself is one of deep and perennial philosophical problems, considered in the special context of modern science. The essays in 'scientific method' are an attempt to reveal the anatomy of scientific knowledge, analogous to the way that Aristotle's logic of syllogisms functions for correct verbal arguments. Moreover, many of the greatest 'philosophers of science' have been scientists of distinction who were able to reflect in a disciplined way on what they were doing. Even if this approach to the question seems to provide few of the answers we are now seeking, it will help our understanding if we appreciate why it once seemed the best way, and also why it no longer does so.

The 'scientific revolution' of the seventeenth century, although better described as a reformation in natural philosophy, had some of the essential characteristics of what we now recognize as revolutions. In particular, it was inaugurated by a small group of prophets, fully conscious of their role in attempting to destroy an existing order and to replace it by something better. Their purpose was the achievement of truth, and they were committed to the view that this could be achieved only through a certain approach to the study of the natural world. A classic expression of this ideology can be found in Galileo:

> If this point of which we dispute were some point of law, or other part of the studies called the humanities, wherein there is neither truth nor falsehood, we might give sufficient credit to the acuteness of wit, readiness of answers, and the greater accomplishment of writers, and hope that he who is most proficient in these will make his reason more probable and plausible. But the conclusions of natural science are true and necessary, and the judgement of man has nothing to do with them.[1]

Although the men who consolidated and extended the scientific revolution were generally neither so certain about achieving truth, nor so exclusive in their claims for natural science, this Galilean commitment remained in the tradition, available for renewed emphasis in times of institutional and ideological struggle on behalf of natural science. This occurred during the nineteenth century; and a hint of its consequences for the self-consciousness of science, and the philosophy of science, is given in the preface to the first edition of Karl Pearson's classic *Grammar of Science* of 1892. There he mentioned the fact that science previously had to carry on a 'difficult warfare with metaphysics and dogma'; and the exigencies of this struggle explained, for him, the deliberate concealment of the 'obscurity which envelops the *principia* of science'.[2] Pearson saw that this 'warfare' had distorted the self-consciousness of science and had harmed its teaching; and considering the battles as won, he thought it appropriate to exhibit the structure of scientific knowledge, in its weaknesses as well as its strengths.

The 'warfare' to which Pearson referred was documented by other writers of his time;[3] and those who had fought for science were not necessarily anti-philosophical or irreligious. But in Germany there had been a bitter institutional struggle for the establishment of

---

[1] G. Galileo, *Dialogue on The Great World Systems*, in The Salisbury Translation, ed. G. de Santillana (University of Chicago Press, 1953), p. 63. The word 'judgement' in the last sentence is a translation of 'arbitrio' and, like all key words, it is loaded with meanings. The other modern English translation, that of S. Drake, *Dialogue Concerning The Two Chief World Systems* (University of California Press, 1953), uses 'will' (p. 53). The latin root has the ambiguities from which the English derives both 'arbitrate' and 'arbitrary'; 'judgement' seems best suited to this context.

[2] Karl Pearson, *The Grammar of Science* (1st edn., 1892; Everyman, London, 1937), p. 3.

[3] A. D. White, *A History of the Warfare of Science with Theology in Christendom*, 2 vols. (1896; reprinted Dover Publications Inc., 1960), is the classic in this literature. White was President of Cornell University and American Ambassador to Russia.

natural science as a discipline worthy of recognition as different from, and independent of, the academic philosophy of the earlier nineteenth century. And in England there had been a frequently hysterical public debate over the theory of the evolution of the human species, with ministers of religion denouncing the Darwinists as the new Antichrist. What was at stake in these struggles was the autonomy of the goals and methods of inquiry in natural science; the freedom of scientists to draw conclusions from the evidence they considered relevant, and to ignore that which they considered irrelevant (such as traditional Christian teaching, or accepted metaphysical doctrine). In the last resort the struggle was over the social functions and social responsibilities of natural science. Some saw it as subordinate in its functions to some other sphere of learning or work, in particular religion; and they considered that scientists had a responsibility to work in such a way that both their methods and their conclusions would reinforce the authority of accepted religion rather than endanger it. The best defence against this was, as usual, a good offensive; and the ideology of truth in science, a truth independent from and perhaps superior to that of philosophy or religion, performed a useful function in the struggle for the freedom of science.

The struggle flared up briefly in the 1930s, when a group of eminent British scientists (some native-born and others Continental immigrants) saw a new threat to the freedom of science coming not only from the racist persecutions of the Nazis but, more insidiously, from the pressures on science in the cause of social responsibility and Marxist philosophy, enforced in the Soviet Union and urged by British supporters of Soviet Communism.[4] The later episode of the capture of Soviet genetics by T. D. Lysenko, who destroyed a brilliant school in the name of Marxism, seemed to confirm all their fears. In reaction to this disaster, a distinguished American historian of science used the Galileo text in a condemnation of the Marxist approach to science and its history:

To suppose otherwise is to give the game away. It is to suppose that the papal court was justified when it decided to condemn Galileo on the ground that the Copernican system, however convenient as a mathematical device, tended to unsettle men's minds and weaken authority and order. It is to agree that the Soviet state was justified when it imposed

[4] For a brief history of the 'Society for Freedom in Science' see Sir Solly Zuckerman *Scientists and War* (Hamish Hamilton, London, 1966), pp. 141 ff.

Lysenkoism on the science of genetics as a theory comfortable to Marxist dogma. . . .[5]

Thus the Galilean commitment to truth in science, seen as a necessary means to the defence of its freedom, has continued down to our own age, and will be expressed whenever the freedom of science seems endangered.

Such a directly political motivation for the conception of science as the pursuit of truth is only a part of a long and complex history. Partly because of its indubitable successes natural science has seemed for many decades to be the paradigm of genuine knowledge. In a variety of ways, the pursuit of science, or a popularized image of scientific work or scientific knowledge, has functioned as a religion in substitution for, or in opposition to, the accepted religion of a society. At the more practical level, the belief in the possibility of attaining some truth that will live on after one's personal death, has furnished a motive for selfless and dedicated work by many scientists; and some schoolmasters, forced by circumstances to transmit what is mainly a manipulative craft, derive self-respect and peace of mind from the conviction that they are imparting clear and distinct truths to their young charges. Since the pursuit of truth can appear to be the most noble and harmless of human activities, this conception of science helps to justify their work to scientists themselves; and it also helps them argue for public support, especially when public relations work is done on behalf of expensive experimental fields of research.

The Technocratic Conception of Science

The obsolescence of the conception of science as the pursuit of truth results from several changes in the social activity of science. First, the heavy warfare with 'theology and metaphysics' is over. Although a few sharp skirmishes still occur, the attacks on the freedom of science from this quarter are no longer significant. This is not so much because of the undoubted victory of science over its ancient contenders as for the deeper reason that the conclusions of natural science are no longer ideologically sensitive. What people, either the masses or the educated, believe about the inanimate universe or the biological aspects of humanity is not relevant to the

[5] C. C. Gillispie, letter to *American Scientist*, 45 (September 1957), 266A–74A, answering criticisms of his review of J. Needham *Science and Civilization in China*, vol. ii.

stability of society, as it was once thought to be. The focus of sensitivity is now in the social sciences; and the techniques of control by those in authority vary in subtlety in accordance with local requirements and traditions. Hence the leaders of the community of natural science no longer need to hold and proclaim this sort of ideological commitment as a rallying slogan for their followers and potential recruits. Also, the experience of modern scientists in their work, seeing the rapid rate of obsolescence of scientific results, makes the vision of the pursuit of truth not so much wrong as irrelevant. But, more important, the attention of the general educated public has shifted away from the problem of the nature of pure science and its relation to philosophy and religion. It is now concentrated on the visible triumph of technology based on applied science. Applied science has now become the basic means of production in a modern economy.[6] The prosperity, and economic independence, of a firm or of a nation does not rest so much in its existing factories as in the 'research and development' laboratories, where the industry of the future is being created and the competition of the future is being met. Thus, industry has been penetrated by science.

All this has not only happened but it has recently been seen to happen by those who plan the future of our societies. The 'technocratic' view of science is that of a basic factor of production, needing ever-increasing supplies of highly-trained 'scientific manpower'. This view of science is a descendant, in a simplified and vulgarized form, of a tradition extending from Francis Bacon down through Karl Marx. Bacon gave the aphorism 'knowledge and power meet in one'; and Marxist historians have attempted to show that the major advances in science have come as a response (however indirect) to the particular needs of production at the time. Even in the nineteenth century, 'science' was given much credit for the advances in technology which so dramatically transformed life in the advanced nations; and so it is the widespread adoption of this view, rather than the insight itself, which is characteristic of the present age.

But the 'technocratic' view of science is dangerously one-sided. It assumes 'science' to be an independent, self-contained factor in the situation, needing only financial support and administrative planning

[6] Science can thus be considered as one sector of the 'knowledge industry'. One study of this (F. Machlup, *The Production and Distribution of Knowledge in the United States*, Princeton, 1962) estimates its product at 27 per cent of the G.N.P.; although his definition of 'knowledge' would be better termed 'software'.

to provide unlimited blessings for us all. Until very recently, there was no systematic appreciation of the fact that as science grows and penetrates industry it becomes industrialized. This is partly a matter of scale; as the investment in scientific research increases, both as a whole and in the cost of individual projects, there arises a need for a formal administrative machinery for decision-making and executive action. Moreover, many practically useful scientific results can be treated as a sort of commodity, to be produced under contract, tailored to the special needs of the purchaser. The bulk of the public funds which are given to 'science' are in fact contracts for such specialized products; and it is only natural for the older forms of patronage of science to be displaced by contractual agreements for the production of results, whether directly useful or not. Hence, in any large and vigorous field, the social atmosphere becomes increasingly 'industrial' where a large organization, with labour force directed to specialized tasks, produces the sorts of results for which the directors have been able to obtain contracts from agencies which invest in such production.

Without some such organization it would be impossible for the scientific community of the present, and of the future, to operate. But the assimilation of the production of scientific results to the production of material goods can be dangerous, and indeed destructive of science itself. For producing worthwhile scientific knowledge is quite different from producing an acceptable marketable commodity, like soap. Scientific knowledge cannot be mass-produced by machines tended by semi-skilled labour. Research is a craft activity, of a very specialized and delicate sort. The minimum standards of accuracy and reliability for worthwhile scientific results are extremely high. But there are no automatic, external tests of its quality, neither gauges to check its agreement with specifications, nor a market mechanism for the public rejection of inferior products. Two separate factors are necessary for the achievement of worthwhile scientific results: a community of scholars with a shared knowledge of the standards of quality appropriate for their work and a shared commitment to enforce those standards by the informal sanctions the community possesses; and individuals whose personal integrity sets standards at least as high as those required by their community. If either of these conditions is lacking—if there is a field which is either too disorganized or too demoralized to enforce the appropriate standards, or a group of scientists nominally within the field who are content to publish substandard work in substandard journals—then bad work will be produced. This is but one of the ways in which 'morale' is an important component of scientific activity; and any view of science which fails to recognize the special conditions necessary for the maintenance of morale in science is bound to make disastrous blunders in the planning of science.

It may appear that this concern is somewhat beside the point. It is applied science, or technology, that enters into the system of industrial production, and every national plan for science ensures that 'pure science' or 'basic research' should not be neglected. Unfortunately, it is impossible to make a neat line of division between the two sectors, allowing one to serve Truth and the other Caesar. It is more than the simple fact that no piece of scientific knowledge can be guaranteed 'pure', or free of application. In addition, much of 'pure' or 'basic' research involves capital outlay on an industrial and occasionally gigantic scale, and those who undertake and organize such projects must necessarily have some of the qualities of an industrial manager.

No sector of science is immune from the problems of industrialization; and, conversely, modern sophisticated technology is sufficiently like pure science to share its delicacy, and is also prone to special diseases of its own. For technological innovation is also removed from the immediate controls of the market. Every important new device or process takes several years of development before it can be produced, and hence a decision to invest in development is necessarily partly a speculative assessment of prices, demand, and competition in the future. In military technology, the criteria for decision are even more complex and speculative. Mistakes can be made here, as in science; but here they cannot be quietly buried. There is then a danger that a firm (or more commonly a government) will try to redress an error by making it bigger, and all considerations of what is economically viable (not to say socially desirable) will be cast aside. Supersonic airliners are a case in point. Hence scientific technology, as much as industrialized science, can suffer from a new and dangerous form of corruption.

The '*Humanist*' *Critique*

Now that 'dogma and metaphysics' no longer do battle with science, the traditional opposition views are collected together under the label 'the Arts'. This is usually considered to be an outsider's view, seeing science as a new Leviathan; a body of knowledge which is esoteric, inhuman, and increasingly dominant. Those who hate and fear this new science are even more mixed than the Oxford classicists and Chelsea poets who provided the initial evidence for C. P. Snow's famous lecture.[7] Indeed, such critics are even to be found within the camp of science itself, especially among students who have been discreetly press-ganged into a scientific career by their schoolmasters, or research workers in the position of helpless proletarians on the production-line of some modern research establishment.

The debate between the 'arts' critics and the defenders of science is necessarily confused since it depends on a partial understanding on both sides, and also involves a mixture of personal philosophy, politics, and quite directly professional concerns (as among the teachers of upper forms of English grammar schools). Also, the debate oscillates between two levels: that of scientific knowledge itself and that of its technological applications. This further confuses the debate, for while the results of scientific inquiry are, in their statement, remote from specifically human concerns and values, the industrial applications of those results are transforming the material and social conditions of human life. A fully consistent 'humanist' critic must not merely point out the sterility (in his terms) of knowledge of the natural world but he must also deny the human value of modern scientific technology. This latter task can be done, but only by fabricating a legend of some bygone 'good old days' when men were poor but happy, or by denying the human importance of the majority of the human race. The fully consistent defenders of science also have their difficulties. It is easy enough for a certain sort of intellect to find only triviality in the works of literature and philosophy, and to claim that natural science has the only path to truth.[8] But the social responsibilities of scientists, in the face of

[7] C. P. Snow, *The Two Cultures and the Scientific Revolution*, Rede Lecture, 1959 (Cambridge University Press, 1959).

[8] Thus, I. I. Rabi, 'Science and Technology', in *The Impact of Science on Technology*, eds. Warner, Morse, and Eichner (Columbia University Press, 1965), pp. 9–36, can say 'The arts are certainly central to life. Yet they are not the kind of thing that will inspire men to push on to new heights ... Even the works of Shakespeare, which are essentially an exploration of human character, are really wonderful, glorified gossip. I am not degrading gossip, because we live by it, but it does not take us outside ourselves, outside the human race' (p. 20). In fairness to Professor Rabi, it should be added that his comments on social aspects of science, in his essay and in the subsequent discussion, are a valuable source of information and insights.

destructive and evil applications of their work, cannot be brushed
aside lightly. A tempting way out is to treat the scientist's search for
truth about nature as an absolute good, and then to disclaim all
responsibility for the application of this knowledge at the hands of
the technologists. In an age when industrialized science and
science-based industry are so closely related this neat bifurcation
may strike an unsympathetic audience as shallow, if not indeed
hypocritical.[9]

Between these extreme attitudes lies a rich mixture of sincere but
worried criticisms and defences. The critics of natural science
implicitly (or explicitly) contrast it to their idealized image of Arts
studies as explorations in the philosophy of man. The teaching of
science is claimed to be irrelevant to true education, and indeed
destructive of the values of education. It is authoritarian in style,
giving students no opportunity to develop their powers of criticism
and judgement. Dealing with an artificial and abstract subject-
matter, it gives the student no materials for developing himself as a
mature person in human society. Moreover, run the criticisms,
scientific knowledge cannot be part of the intellectual equipment of a
cultivated person. Either one is a highly trained specialist, or one can
learn of science only at third-hand through popularizations. The
specialists themselves are accused of having a narrow vision of the
world since their training and professional practice removes them
from contact with genuinely human concerns. The increasing power
of natural science thus threatens the destruction of a humane
understanding among educated people as the humane studies are
increasingly deprived of prestige, of time in university teaching
programmes, and of resources for research. As a result, our thinking
about ourselves and the world around us becomes grossly material
and quantitative; the higher sensibilities and values are crushed
beneath the machine.

This comprehensive indictment would have more force if the
aggrieved humanists had previously achieved reliable methods for
teaching wisdom through their various approaches. But as Plato

[9] L. K. Nash, in *The Nature of the Natural Sciences* (Little Brown & Co., Boston, 1963),
distinguishes between the moral responsibility of the *scientist*, when he 'elects to take
part in the technologic exploitation of science for destructive purposes', and *scientific
knowledge*, 'ethically as neutral as iron', capable of becoming swords or ploughshares.
He concludes, 'Ambivalence attaches to the works of science simply because their
technologic exploitations rest in the hands of men' (p. 113), but he does not discuss
whether a similar ambivalence could be justifiably attached to scientists.

showed long ago, the task is not an easy one. Descartes' savage indictment of his own humanistic education at one of the very best schools of Europe should be read by all those who believe that the virtues of 'humanism' are so obvious as to need no defending.[10] Moreover, most of the present criticisms of natural science can apply to any developed discipline, be it classical philology or Shakespearean criticism. For illumination on the great, eternal questions can be achieved and transmitted only by those rare individuals whose intellect and vision are equal to the task. The vast majority of scholars must do the humdrum work of investigating narrow, technical problems, and teachers must try to impart some competence to immature minds. When people trained as research technicians are forced to be philosophers in their teaching, the results can be disastrous. The deepest and noblest ideas of great minds, when retailed by lesser men, can easily become banal commonplaces, mockeries of the original inspiration. It is also worth remembering that on occasion some 'humane' disciplines have suffered from precisely that dehumanization of which natural science is now accused; it was most marked in classics and history in Germany at the end of the last century.[11]

In fact, the differences between 'Arts' and 'Science' disciplines are more of degree than of kind. Rivalry and conflict between different spheres of learning is as old as learning itself. The perceived lines of division, and the sharpness of conflict, are the result of temporary and local circumstances. It can be very entertaining to proceed from impressionist judgements on a very partial experience, by undisciplined generalization to erect abstract entities like 'Arts' and 'Science', and then let this pair of straw men do battle with each other. But without a recognition of the many-sidedness of the work of the advancement of learning in any sphere, and of the systematic difficulties common to all, little new understanding will be achieved.

---

[10] In the *Discours de la Methode*, 1st part, Descartes first describes his enthusiasm for the various subjects in his humanistic curriculum, and then disposes of them one by one, showing that they either confuse or mislead the mind, or actually corrupt our morals, or (as theology) are condemned to sterility by their own assertions. Only mathematics survived his destructive analysis.

[11] See F. Lilge, *The Abuse of Learning; the Failure of the German University* (MacMillan, New York, 1948), ch. 4: 'Criticism and Satire of Academic Culture: Nietzsche'. This unfortunate philosopher was not alone in his critique; the great Swiss historian Jacob Burckhardt also stood aside from the industry of narrow factological historical scholarship (pp. 100–4).

It is worth recalling that the lecture which gave rise to the pro-
longed discussion of the 'Two Cultures' was really directed towards
quite a different problem. An earlier title was 'The Rich and the
Poor', and the purpose of the lecture was to condemn the ignorance
in both 'cultures' of a phenomenon which will determine our
future: the power of the new scientific technology for transforming
the lives of the men, women and children now helping to emerge
from what Snow calls 'the great anonymous sludge of history'. But
Snow's English audience was not prepared to comprehend the
challenge laid down in that thesis, and so the tea-cups rattled
agreeably while the revolution went on outside.

Science as Dirty Work

Quite recently there has developed another very critical image of
science: not so much that it is destructive of human values, but that
science, through its military and industrial applications, may well
destroy the human race. There has long been a popular tradition of
imagining 'the scientist' as a white-coated magician, concocting
weird and potent substances in his fuming test-tubes; the comic-
book literature provides evidence for this. But this image has been
far from dominant, as we can see from the very limited overt popular
reaction to the early nuclear weapons. However, as the political
campaigns against the use and testing of nuclear weapons gained an
audience, and the world found itself on the brink of nuclear annihila-
tion over Cuba, the sorcery of military science became a part of
everyday common sense. Even this new image did not react directly
on the activity of science, except perhaps in contributing to the
beginning of the decline in recruitment. But shortly thereafter the
problems of the environment, or 'biosphere', began to come into
prominence. At the same time as the public was learning of ever new
and more subtle ways in which their technology is poisoning the
environment and themselves, there arose several large movements
against the society of which this technology is an integral part. These
ranged from protests against the Vietnam war and its particularly
sophisticated horrors, through 'student' revolts against various
manifestations of bureaucracy, over to the 'hippie' or 'drop-out'
communities which have abandoned our civilization entirely.

The tarnishing of the image of science is very real, and is likely to
be one of the main reasons for the international 'swing from science'

among schoolchildren.[12] It is certain to persist, especially since distinguished scientists now voice their perplexity and disquiet over the work they have been engaged in.[13] Yet to conclude that all science is 'dirty work', and solely responsible for our present ecological problems, is both naïve and fruitless. It is true that our runaway technology is bringing us towards the brink of disaster; but old-fashioned myopic engineering was quite sufficient to do the same (on a smaller scale) in earlier times. The salination of irrigated lands and the deforestation of the countryside have contributed to the decay of civilizations through all of human history. Moreover, the problems of the biosphere will not be solved by protests alone. Even to determine what sort of damage is being caused, and by which agents, is a sophisticated scientific problem; and to devise the means of rectification and control calls for a variety of talents. If the study of science is abandoned by all except those who are content to be manpower-units, then within a generation there will be no one competent and concerned to analyse and expose the ecological blunders produced by our bureaucratic technology. In that case, the hippies will have been proved correct, and we can anticipate at least some centuries of barbarism before the next try at civilization.

Towards a New Understanding

Although none of the partial views of science are adequate for an understanding of the industrialized science of the present, three of them have great significance for the future course of science. The ideal of the pursuit of truth has functioned for generations to provide science with a defence against its external enemies and with a basis for the morale and commitment necessary for the production of worthwhile work. It will persist for some time to come, but its base in experience is eroding very rapidly. The clue to its demise lies in

[12] There is evidence that the image of science as 'dirty work' has now become diffused among younger schoolchildren; and to the extent that this is so, the problems of recruitment to science cannot be expected to be solved for nearly a decade. In a survey of children of ages 12 to 13, using relatively unstructured discussions, and picking up repeated key words as clues to stereotypes, the most often repeated theme (18 per cent of the total against 10 per cent for any other) was along these lines: 'They may invent good things like new drugs ... well, other things I can't name but also things which are not very good (18 per cent) like H-bombs and other weapons, giving diseases to animals; and the thousands of scientists breeding germs.' See C. Selmes, 'Attitudes to Science and Scientists: the Attitudes of 12–13 Year-Old Pupils', *School Science Review*, 51 (1969), 7–14.

[13] For an example of this new self-questioning, see S. Silver, 'Science and Society', *Science Journal* (October, 1969), pp. 39–44.

the concept of 'scientific manpower'; and a warning about that term, issued some years ago by Sir Eric Ashby, can now be seen as the epitaph of academic science and with it the ideal of truth.

I do not deny that for certain very narrow purposes it is convenient to do sums in which scientists and technologists are considered as so many units of scientific or technological man-power. The danger is that people who habitually do such arithmetic come to think of scientists and technologists as nothing but man-power units, and the places which produce them as assembly-lines for man-power production. It is the unforgiveable sin to introduce the concept of man-power into education. For centuries universities have been concerned with individual men, not man-power. No one, except as a joke, talks about classical man-power or philosophical man-power. Universities must be firm, even with their friends, in rejecting any implications that they are concerned with man-power. They are not. They are concerned with individual men.[14]

This was written in 1958, and since then the calculations with units of scientific manpower have increased steadily. Indeed, the term 'scientist' itself is giving way to the acronym Q.S.E. (Qualified Scientist or Engineer) as a grading for a particular class of skilled manpower. If Sir Eric Ashby were a backward-looking romantic then his worries could be dismissed; but he is a scientist of distinction, with a broad experience of the world of science and universities. That his 'unforgiveable sin' has now become standard practice is evidence of the complete victory of the technocratic conception of science.

Science lives not by manpower units alone; and without some ideal to replace that of the pursuit of truth it could soon degenerate. Moreover, to the extent that science becomes organized around the service of commercial and military industry, it will be subject to the criticisms of being dirty work. Attempts by leaders of science to conceal this connection from an audience of the lay public and potential recruits, while using it to obtain support from industry and the State, will accomplish nothing but a further decline in the prestige and morale of science.

There is certainly no easy way out of this impasse; and to the extent that industrialized science depends on large-scale investment from industry and the State, it must be responsible to those institutions. But it is possible that the reaction against runaway technology will not take a purely negative form. Many leading scientists have

[14] Eric Ashby, *Technology and the Academics* (MacMillan, London, 1958), p. 94.

already begun to criticize what is done in the name of science; and there are now emerging focal points of 'critical science' where research is organized around the identification and exposure of the damage being done to the biosphere. As yet this movement is small; and, even if it grows, it will necessarily exist on the margin of industrialized science. Also, it must expect incomprehension and hostility when it tries to establish an institutional base for its teaching; for by its programme it implicitly rejects our inherited approach to science, with its faith in Facts and in Progress, as a relic of the Victorian age. But as it gains strength and coherence it will develop a new philosophy of science, and a new philosophy of nature and of man's place in it. For this, it can draw on a suppressed tradition within natural science itself, which saw beyond the accumulation of facts, and beyond the domination of nature, to the welfare of humanity living in harmony with itself and its neighbours. Whether such a philosophy could flourish within the context of our industrial civilization, and whether the new science based on such a philosophy could gain influence in time to avert the destruction of civilization, are unanswerable questions. But it is certain that without a new understanding of science our present problems, within science and in its applications, will multiply to the point of no return.

2

SOCIAL PROBLEMS OF
INDUSTRIALIZED SCIENCE

IN recent years, the vision of 'science' as the pursuit of the Good and the True has become seriously clouded, and social and ethical problems have accumulated from all directions. Each particular one of these can be discussed in isolation, and solutions sought as if it were the only or the main perturbing force in an otherwise tranquil scene. But there are systematic connections between these various problems, which can be organized under the title 'industrialization of science'. This means, in the first place, the dominance of capital-intensive research, and its social consequences in the concentration of power in a small section of the community. It also involves the interpenetration of science and industry, with the loss of boundaries which enabled different styles of work, with their appropriate codes of behaviour and ideals, to coexist. Further, it implies a large size, both in particular units and in the aggregate, with the consequent loss of networks of informal, personal contacts binding a community. Finally, it brings into science the instability and sense of rapid but uncontrolled change, characteristic of the world of industry and trade in our civilization.

The social and ethical problems consequent upon the industrialization of science are the deepest problems in the understanding of the science of our period. In response to these problems, both the informal common-sense understanding of science, and the scholarly philosophy of science, have been shifting their perspectives and modifying their assumptions about what is 'real' in science. The present work is intended as a contribution towards the understanding of these problems; and before going into a lengthy discussion of the foundations, it will be useful for us to review the problems themselves. Much of what I say will be harsh and unpleasant, and to some will appear to be muckraking for its own sake. But these things can and should be said, not only because they reflect the judgements of eminent scientists, but also because there are many men who have devoted their lives to science, whose efforts to solve these problems will be helped by their identification and explanation.

Paradise Lost

A very important feature of the social problems of contemporary science is their novelty.. Although the gross quantitative indices of scientific work show a fairly steady growth-rate over many decades,[1] the self-awareness of the scientific community, as reflected in the pronouncements of leading men in leading fields and in the sorts of studies made of science, has experienced a sudden change. Whether the situation of science as a whole has really altered so drastically in the very recent period is not of crucial importance in this context. For the belief in the rapidity of change means that the present leadership of science considers itself as lacking a stock of inherited attitudes and social skills for coping with the problems created by the present condition of science. As we shall see, the new awareness of social and ethical problems of great complexity, together with a sense of changes beyond the control of the scientific community itself, and of inexperience in handling them, are themselves factors in a complex and unstable situation.

The sense of loss of a very precious quality of pre-war science is conveyed clearly in the reflections of the distinguished Soviet physicist Kapitsa:

The year that Rutherford died (1938) there disappeared forever the happy days of free scientific work which gave us such delight in our youth. Science has lost her freedom. Science has become a productive force. She has become rich but she has become enslaved and part of her is veiled in secrecy. I do not know whether Rutherford would continue nowadays to joke and laugh as he used to do.[2]

The loss of innocence of science is even deeper than Kapitsa indicated. Science is not merely harnessed to industry; it can be

[1] The pioneering work in establishing these statistical regularities and in appreciating their significance was done by D. J. de S. Price. See his *Little Science, Big Science* (Columbia University Press, 1963). I was informed by the author that the basic quantitative regularities were discovered by the techniques of 'little science': working from data obtained by sampling standard reference works. The computers were brought in later.

[2] Quoted from V. K. McElheny, 'Kapitsa's Visit to England', *Science*, 153 (1966), 725-7.

applied to produce effects which for some centuries, since the decline of belief in magic and miracles, had been thought impossible. Norbert Wiener described this revival of ancient moral problems under the heading of simony: the perversion of the magic of the Host to other ends than the Greater Glory of God.

Dante indeed places it among the worst of sins, and consigns to the bottom of his Hell some of the most notorious practitioners of simony of his own times. However, simony was a besetting sin of the highly ecclesiastical world in which Dante lived, and is of course extinct in the more rationalistic and rational world of the present day.

It is extinct! It is extinct. It is extinct? Perhaps the powers of the age of the machine are not truly supernatural, but at least they seem beyond the ordinary course of nature to the man in the street. Perhaps we no longer interpret our duty as obliging us to devote these great powers to the greater glory of God, but it still seems improper to us to devote them to vain or selfish purposes. There is a sin, which consists of using the magic of modern automatization to further personal profit or let loose the apocalyptic terrors of nuclear warfare. If this sin is to have a name, let it be Simony or Sorcery.[3]

Still more serious is the changed perception of the character of scientific work, and of the people who engage in it. For in the traditions of propaganda on behalf of science, as well as in the practice of the social activity of scientific inquiry, the work requires codes of behaviour which are more sophisticated and more enlightened than those adequate in many other spheres, and hence the scientific community requires members of sufficient moral calibre to adhere to them. This ethical aspect of science has long been an important part of its self-awareness. Since science cannot honestly be defended purely on its contribution to industry, and its products are generally too esoteric to contribute to general culture even at the highest level, the work of scientific inquiry as a worthwhile thing in itself must be an important component of any justification of the total activity of science.

Traditionally, the image of science has been of a work which is demanding, and also productive, of the highest standards of morality. The work is arduous, involving the sacrifice of leisure and convenience and lacking any of the rewards of wealth and power which are supposed to spur ordinary mortals on in their endeavours. In addition, scientists offer the products of their work openly and

[3] N. Wiener, *God & Golem, Inc.* (M.I.T. Press, 1964), pp. 51-2.

without charge to all colleagues, ignoring the barriers of politics, nationality, race, or class. Moreover, the inevitable debates over scientific results are conducted within a set of rules which seem inhumanly strict, and yet which are adhered to by consensus. The scientist cannot hire an advocate to make his case sound better than it is; he must not resort to personal abuse or denigration of opponents, or use his political influence (within the community or with lay society) to determine an issue; and any resort to dishonest reporting of work is sufficient to ruin a man's reputation. The willing adoption of such rigorous standards of practice, and their operation in the absence of any formal institutions for enforcement, seems to require individuals of a genuinely superior moral character, who willingly submit to such a discipline in the service of a noble end: namely, the advance of knowledge and power on behalf of humanity.

There is no doubt that this elevated view of the ethical aspects of science has had a strong foundation in reality, in the social practice of science as well as in its public self-awareness, during the period of 'academic science' which began to be displaced during the Second World War. The initial contact of scientists raised in pure academic science, with men of affairs, often produced reactions like the one expressed by the physicist Szilard:

> When a scientist says something, his colleagues must ask themselves only whether it is true. When a politician says something, his colleagues must first of all ask, 'Why does he say it?'; later on they may or may not get around to asking whether it happens to be true.[4]

The context of this quotation is a story called 'The Voice of the Dolphins', in which an international organization of scientists has found the simple secret of gaining the adherence of any politician to any cause, irrespective of his predispositions: find his price, and buy him. Szilard's estimated round figure at which any American politician's soul could be purchased was a guaranteed lifetime annual income of $200,000 (at 1960 prices).[5]

In recent years, the traditional portrait of the noble scientist has receded from the centre of the image of science. There has not been a denial of its accuracy for that part of scientific work which it describes; but there has been a growing realization that there is more to the activity of science than the achievement and assessment of the

---

[4] See L. Szilard, *The Voice of the Dolphins and Other Stories* (Simon & Schuster, New York, 1961), pp. 25-6.
[5] Ibid., p. 41, where the point is illustrated by example.

results of research. And the tangible rewards of a successful career in high-prestige pure science, and in industrialized science, are common knowledge. Szilard himself recognized that when scientists engage in work other than straightforward research they are subject to the same hazards as anyone else involved in practical affairs. In another short story, he sketched a plan for achieving a desired deceleration of scientific progress in the future through the creation of an administrative machinery that would turn scientists into bureaucrats. In this way, talented men would be diverted from research, and mediocrity would be systematically fostered. His model for this stifling bureaucracy was, very unfairly, the American National Science Foundation.[6]

Nostalgia for the bygone simplicities of particular fields of science can exaggerate the depth and novelty of the social problems of contemporary science, and thereby hinder their solution. Also, the confusion among lay audiences between 'pure' and 'applied' science and 'technology' obscures real differences which must be appreciated if the different aspects of scientific work are to be understood and maintained in a healthy relation. Yet this confusion is not altogether unjustified; the creation of nuclear weapons was the achievement of men who had been trained in the most pure and philosophical field of science, atomic physics. And even 'pure science' has come to require gigantic projects such as high-energy particle accelerators, involving vast funds and also managerial and political skills for their achievement.

In these new conditions, there has emerged a new literature about scientists, in which the scientist is seen primarily as a man of affairs, and the painstaking, disciplined work of research is only a part, perhaps a minor part, of the story. There have been protests at this practice of ignoring the science when talking about scientists;[7] but

[6] op. cit., pp. 89–102, 'The Mark Gable Foundation', especially pp. 100–1. The plan was to have the best men in each field appointed as high-salaried chairmen, and to have about twenty annual prizes of $100,000 for the best papers of the year. 'As a matter of fact, any of the National Science Foundation bills which were introduced in the 79th and 80th Congresses would perfectly well serve as a model.' The effects would be achieved by taking the best scientists out of their laboratories, and inducing the others to follow fashions.

[7] See N. Reingold's review of *Organization for Economic Co-operation and Development* (O.E.C.D., Paris, 1968), and of E. B. Skolnikoff, *Science, Technology and American Foreign Policy* (M.I.T. Press, 1967), and of D. S. Greenberg, *The Politics of Pure Science* (New American Library, New York, 1967), *Isis*, 59 (1968), 216–20. Against the first two, he argues that 'one cannot understand either (social and economic) output or process without detailed knowledge and understanding of the 'input', which consists of

it appears that for the discussion of the social problems of science at a reasonable depth, it can be done. Also, men of administrative experience have (incorrectly, in my view) dismissed the ethical component of scientific research as a purely prudential strategy by scientists with application only to an isolated part of their work.[8]

Even more corrosive of the traditional image of the scientist than all this explicit discussion are the implicit assumptions that underlie all debates on 'science policy'. For it is assumed that, when decisions are made at the highest level on priorities between projects and entire fields, the work for which support is offered will attract scientists to it, regardless of their prior interests or views on the choice. Thus, the bulk of the scientific community can be and is treated as manpower, which will flow as directed between tasks for which it has the requisite skills. As we shall see, such policy decisions, and the resultant flows, are necessary in the conditions of industrialized science; and the enterprise could not survive if the majority of competent scientists were extreme individualists who would work well only on problems of their own choosing. But such manpower-units cannot be considered as scholars, and any propaganda that projects the image of the typical scientist as an independent searcher, following his own path in the exploration of Nature, is now worse than false: it is a bore.

Whether there has been a Fall of Scientific Man over the last generation is a topic on which any debate will produce more heat than light. The world of science is variegated in the extreme, and our knowledge of the previous period depends largely on a folk-history of science which naturally stressed certain aspects of science and ignored others, and created a simplified, idealized picture out of disparate and contradictory elements. But we have already seen that the industrialization of science brings with it certain strong tenden-

specific scientists working in particular scientific contexts'. Greenberg's book he considers to be 'not about pure science or even about pure scientists', but rather about power, international hostility, industrial competition, and regional conflicts; the real focus of the 'politics of pure science' is not in the scientific establishment but in the State and industrial agencies that wield real power. In general, Reingold observes that the one-sidedness of such books reflects the division and weakness within the history of science, where 'internalist' and 'externalist' approaches, although no longer tied to hostile ideologies, make hardly any contact.

[8] D. S. Greenberg quotes the plain-spoken Robert M. Hutchins, 'There have been very few scientific frauds. This is because a scientist would be a fool to commit a scientific fraud when he can commit frauds every day on his wife, his associates, the president of the university, and the grocer. . . .' (*The Politics of Pure Science*, pp. 20–1.)

cies to change, in the realities of the social practice of science as well as in its self-awareness. In this chapter I will describe some of these new problems, and briefly indicate some of their causes. Because many of the phenomena themselves are ruled out as inconceivable, let alone inexplicable, in the terms of the traditional philosophies of science (in particular, the circumstances in which scientific inquiry proves abortive), it will be necessary first to establish the elements of a new philosophy of science for their systematic description and analysis. On that basis we will be able to return to a deeper study of the social problems of industrialized science, and consider the ways in which they might possibly be resolved.

The World of Academic Science

It will be useful for our purposes to sketch some aspects of the social activity of science in the period preceding the present one.[9] The institutions and attitudes of science and scientists at the present time are largely inherited from that period, and the memory of that era is a point of reference for all analyses of the present except for those which see science simply as a factor of production. This period may be considered as having its terminal points in the French Revolution and in the atomic bomb. The event that marks its inception is the establishment of the École Polytechnique in Paris, where scholarship students chosen by ability received a scientific education from the great masters. Although the school was designed to provide engineers, and only a small proportion of its graduates were able to pursue a scientific career, it provided a model and inspiration for later scientific education, as distinct from practical training in medicine or engineering. Within a few decades, first in Germany and then more gradually in England, science achieved an institutional base in teaching posts in universities, which themselves were reformed from their previous stagnation. All the other social institutions internal to science were either created during this period, or changed radically from their earlier forms. Generally, dilettantes and a lay audience were gradually excluded. The work became concentrated and specialized, with an apparatus of journals and specialist societies, and an increasing dependence on an occupational base in teaching and research. The work of science itself progressed at an

[9] A most useful collection of papers on the institutional aspects of contemporary science, seen in the perspective of the recent past, is *Science and Society*, ed. N. Kaplan (Rand McNally, Chicago, 1965).

accelerating pace; a succession of new fields were opened up for effective disciplined inquiry, and the increase in factual knowledge was stupendous. By the later part of the century the centre of organized research was in the German universities, and there the system of academic science may be said to have achieved its greatest moments.

By the end of the nineteenth century there were two developments that in retrospect can be seen to mark the inception of the latter phase of the cycle of development of academic science as a social institution existing in a particular, historically conditioned, context. First, the German universities ceased to expand at their previous rate; and second, the systematic institutionalized connection of an important science with industry was established. This was chemistry, which had always had a very close relation with practice, but which in later nineteenth century Germany became the basis for a sophisticated, capital-intensive industry on the pattern which is now dominant. The example and ideals of academic science, on a model derived from Germany, remained vigorous well into the current century. 'Science' has been understood as pure, university-based science, in spite of an involvement of science in the First World War so deep that it has been called 'the chemists' war'. Indeed, not until the Second World War had produced a scientific-technological effort of a new order of magnitude, culminating in the atomic bomb, did the interpenetration of industry and science, and the resulting industrialization of science, destroy the claim of the 'academic' image to represent the essential nature of science.

Our description of the social practice and self-awareness of science in this past era will be under four headings: the assurance of a diffuse social benefit; the ethic of the search for truth; work in the context of an autonomous community of gentlemen; and the particularly refined sort of personal property achieved in the work. For the first two themes, we can do no better than to quote from the classic pronouncement of Helmholtz:

In fact, men of science form, as it were, an organized army labouring on behalf of the whole nation, and generally under its direction and at its expense, to augment the stock of such knowledge as may serve to promote industrial enterprise, to adorn life, to improve political and social relations, and to further the moral development of individual citizens. After the immediate practical results of their work we forbear to inquire; that we leave to the uninstructed. We are convinced that whatever contributes to the knowledge of the forces of nature or the powers of the human mind is worth cherishing, and may, in its own due time, bear practical fruit, very often where we should least have expected it. Who, when Galvani touched the muscles of a frog with different metals, and noticed their contraction, could have dreamt that eighty years afterwards, in virtue of the self-same process, whose earliest manifestations attracted his attention in his anatomical researches, all Europe would be traversed with wires, flashing intelligence from Madrid to St. Petersburg with the speed of lightening? In the hands of Galvani, and at first even in Volta's, electrical currents were phenomena capable of exerting only the feeblest forces, and could not be detected except by the most delicate apparatus. Had they been neglected, on the ground that the investigation of them promised no immediate practical result, we should now be ignorant of the most important and most interesting of the links between the various forces of nature. . . .

Whoever, in the pursuit of science, seeks after immediate practical utility, may generally rest assured that he will seek in vain. All that science can achieve is a perfect knowledge and a perfect understanding of the action of natural and moral forces. Each individual student must be content to find his reward in rejoicing over new discoveries, as over new victories of mind over reluctant matter, or in enjoying the aesthetic beauty of a well-ordered field of knowledge, where the connection and the filiation of every detail is clear to the mind, he must rest satisfied with the consciousness that he too has contributed something to the increasing fund of knowledge on which the dominion of man over all the forces hostile to intelligence reposes.

Helmholtz then remarked on the likelihood that a scientist will not receive due recognition for his achievement, but noticed an improvement in recent times. After giving many examples of the necessary interconnections between the various disciplines of natural and human science, he ended with this exhortation:

In conclusion, I would say, let each of us think of himself, not as a man seeking to gratify his own thirst for knowledge, or to promote his own private advantage, or to shine by his own abilities, but rather as a fellow-labourer in one great common work bearing upon the highest interests of humanity. Then assuredly we shall not fail of our reward in the approval of our own consciences and the esteem of our fellow-citizens. To keep up these relations between all searchers after truth and all branches of knowledge, to animate them all to vigorous co-operation towards their common end, in the great office of the Universities. Therefore it is necessary that the four faculties should ever go hand in hand, and in this conviction will we strive, so far as in us lies, to press onward to the fulfilment of our great mission.[10]

The particular form of Helmholtz's statement is conditioned by the particular traditions in which he, and his audience, participated. One was that of the German universities, in which the cultivation of knowledge was considered as a good in itself, and also a contribution to the greatness of the nation. The other was that deriving from the revolution in natural philosophy of the seventeenth century, with its twin goals of knowledge and power for mankind.

In other national traditions, the formulation of the themes of diffuse social benefit and of the search for truth, would be slightly different; but the underlying similarities are sufficiently great, at least in those fields which were not deeply involved in technical work, to justify speaking of a single 'ideology' of academic science; by 'ideology' we mean a definition of reality and a set of values which serve to guide and to justify the practice of a particular group. A remarkable feature of academic science is that for at least a century this ideology could be dominant, even to the point of becoming tacit common sense, without being modified by any internal strains or by contradiction with real experience.

This happy situation could persist because of the nature and social context of the activity of scientific inquiry at that time. Throughout the nineteenth century, science was a very small-scale affair. In any given field, there would be only a few small schools of master and pupils. And opportunities for employment in science were so few as to offer no attractive prospects to careerists and time-servers; those who survived the personal and economic hardships of initiation into the work were likely to be highly competent and committed. Hence there was a small set of overlapping communities, whose members were bound together not merely by accident or convenience, but by the personal ties of shared endeavour and a common loyalty to an ideal. They had the excitement and gratification of opening up one field after another to disciplined scientific investigation. Because of the nature of their work, and its investigation, its institutional context, and the social basis of recruitment, most of the members of such communities would be gentlemen, who

[10] H. von Helmoltz, 'On the Relation of Natural Science to Science in General', in *Popular Lectures on Scientific Subjects*, 1st series (London and New York, 1893), pp. 1–28; extracts from pp. 24–8.

could be motivated to act by goals more refined than the mere acquisition of wealth or power.

Because of their size and their membership, these communities of scientists could manage their affairs with the very minimum of formality. The tasks of obtaining and administering external funds for research were few, because of the small size both of the community and of the funds involved. The management of novelty in science, which inevitably involves conflict, was facilitated in Germany by the decentralized system of universities, and in England by the independent status of individual scientists. And the maintenance of the quality of work done could be accomplished quite informally, through the close personal ties which linked effective members of a field with the leadership.

Idealized pictures of this community of academic science can make it appear as a primitive-communist lay priesthood. It was not that; not only were there great rivalries and debates, but each scientist was necessarily involved in the society of his time. Each man had a career to build, and personal interests requiring protection. This communal function was performed by the system of publication in journals, where each published paper, certified by a referee or the editor as good, was a piece of intellectual property of the author. Others might use it freely, but they were obliged to cite it so that the value of the borrowed property would be publicly recognized. The stock of published papers, with their citations in later work by others, was the scientist's return on the investment of his time and energies. Although it could not be simply exchanged for cash, it had a material value of fixed capital investment through its function as evidence of a man's worth when he was being considered for advancement.

Of course, the protection of property was not the sole or even the main function of the system of publication through journals. The system served primarily as a means to the attainment of the general goals which defined the community: the advancement of knowledge. And it did this through the rapid diffusion, to the widest possible audience, of those results which had been certified for their quality. In the social conditions of academic science, the two functions harmonized remarkably well. A man would not want to risk the disgrace of having offered faulty goods to his colleagues, and so he would not mind the delay caused by the checking of his results by a colleague before they were released. By this test, the community was protected from the annoyance and waste of attempting to use unreliable results, and also from the temporary honouring of false claims to property. And for really urgent news, a private letter to one or two friends would suffice.

In practice, this form of property is a very subtle thing; for example, the formalized technique of citation cannot always encompass the manifold relations which a result might have to a predecessor. Hence if the system is to work effectively in both of its functions, and not to break down through evasions, there must be a rigid etiquette for the treatment of this property. This cannot be enforced unless the members of the community acquiesce in it; and for this they must subscribe to a working ethic, itself based on the underlying ideology of the activity. To be effective, this needed to cohere with the daily practice of the scientists, and also with their general view of the world. Thus the metaphysical belief that atoms of truth exist and can be discovered in isolated investigation, had its function in the maintenance of the health of the community of academic science.

I will discuss these different aspects of the social activity of science in greater detail later, but at this point one important conclusion can be drawn. A healthy and vigorous state of scientific work is not at all a 'natural' condition, but requires a leadership capable of providing enlightened direction and imparting morale and commitment to the community. The persistence of excellence in academic science over several generations in any one locale, and its diffusion to new centres, shows that these qualities of leadership could be passed down from master to pupil. This cannot be by formal precept alone, but only by everyday practice. It is for this reason that we can be sure that the ideology of academic science, and its associated ethic, were really effective, and not merely propaganda and retrospective folk-history.[11]

This idealized account abstracts from the many variations, over time, between centres, and between fields, which occurred during the development of academic science. Also, many branches of science were never in this autonomous and balanced state; chemistry always had relations with industry, and biology with medicine; and

[11] I am indebted to Mr. N. Parry for forcing me to realize the necessity of this particular argument.

astronomy was always capital-intensive and related to the State.[12] But the simplified picture on which my analysis is based does have a certain historical reality: it was the dominant self-consciousness of academic physical science in Germany in the latter part of the nineteenth century, whence it spread to other fields and other nations.[13] As such it is important, for it is such an ideology that provides the only existing foundation for an ethics of science which is distinct from that of technological development or commerce.[14] The danger is that this ideology will be kept in a fossilized state for particular public-relations functions, while becoming less and less relevant to the experience of those who live in the world of industrialized science. Then the inherited ethical principles of scientific work will

[12] The basic historical studies of the social and institutional aspects of European academic science have been made by Joseph Ben-David. His earlier papers are listed in *Science and Society*, ed. N. Kaplan (Rand McNally, Chicago, 1965), pp. 581–3. A pioneering effort to characterize the difference between 'academic science' of the nineteenth century and the earlier social forms, is his 'Scientific Growth: A Sociological View', *Minerva*, 2 (1964), 455–77.

[13] In 'The Scientific Worker' (Ph.D. thesis, University of Leeds, 1969), N. D. Ellis has explored some of the ambiguities in the concept of the 'autonomy' of academic science. He suggests that the legend of 'pure science' may be a conflation of two quite different situations: the German academic, 'pure' with respect to applications, but unless a full Professor, *not* autonomous in his choice of problems; and the British amateur, free to follow his choice, but not at all averse to involvement in commercial ventures. More detailed historical studies are required before this insight can be considered as fully established.

[14] Two examples, widely separated in context, indicate the great variety of principles that can be invoked for the preservation of the integrity of science. An extreme case is cited by D. S. Greenberg, *The Politics of American Science* (p. 93), concerning a group of distinguished biologists in the early twentieth century, who preferred penury for themselves and for their work, rather than accepting grants from the Carnegie Foundation and from the University of Chicago which had 'unacceptable' conditions attached. A striking contrast is provided by a comment on a priority dispute between Charles Wheatstone, F.R.S., and then Professor at King's College, London, and an obscure mechanic, Alexander Bain. In earlier nineteenth-century London, it was not thought wrong for a man of science to concern himself with practical (and profitable) matters; but the distinction between the two sorts of work was clear to this public. In this case, the dispute was over a patent, and so the reviewer of a book on the controversy could write: 'The Professor, in the case before us, has laid aside his gown—he has stepped from his chair into the shop of the artisan. His discoveries there are not the treasures presented to the scientific world, they are riches heaped up for his own private emolument . . . when a philosopher avails himself of the protection of the Patent Laws, he *loses caste* as a philosopher, and descends to an equal rank with all others who seek the protection of the same banner.' The quotation is from a review of J. Finlaisson, *An Account of Some Remarkable Applications of the Electric Fluid to the Useful Arts by Mr. Alexander Bain . . .*, in the *Electrical Magazine* (October 1843), pp. 139–42. I am indebted to Mr. R. A. Muir for this reference. An account of Wheatstone's career, with a sympathetic history of his involvement in another priority dispute, is G. Hubbard, *Cooke and Wheatstone and the Invention of the Electric Telegraph* (Routledge, London, 1965).

become increasingly divorced from reality; and under the pressures of present conditions, they could not long survive as effective controls on action.

The Industrialization of Production in Science

For a comparison of the present age with the one preceding it, we can best start with the changed technical character of the work of scientific research; for from this follow the changes in its social institutions, and social practices.[15] The basic difference is a simple one: research is now capital-intensive. Any significant piece of work is almost certain to cost far more than an individual scientist can afford out of his own pocket; it will generally cost much more than his annual income. Hence he is no longer an independent agent, free to investigate whatever problem he thinks best. Nor is he likely to have personal contact with a private patron who will provide for all his needs. Rather, in order to do any research at all, he must first apply to the institutions or agencies that distribute funds for this purpose; and only if one of them considers the project worth the investment can he proceed.

This change is as radical as that which occurred in the productive economy when independent artisan producers were displaced by capital-intensive factory production employing hired labour. The social consequences of the Industrial Revolution were very deep, and those of the present change in science, while not comparable in detail, will be equally so. With his loss of independence, the scientist falls into one of three roles: either an employee, working under the control of a superior; or an individual outworker for investing agencies, existing on a succession of small grants; or he may be a contractor, managing a unit or an establishment which produces research on a large scale by contract with agencies. Of course, he may have other tasks and responsibilities, as refereeing for journals,

[15] I am particularly indebted to Professor D. Humphrey, of Oregon State University, for helping me develop the ideas of this chapter. I should also say that even before I began to think seriously on these problems, they had been indicated by John Ziman in a broadcast talk, 'Scientists, Gentlemen or Players?', *The Listener*, 68 (1960), 599–607. For surveys of the history of the process whereby academic science evolved into industrial science, see D. S. L. Cardwell, *The Organization of Science in England* (Heinemann, London, 1957), and H. and S. Rose, *Science and Society* (Allen Lane, London, 1969). It is interesting that throughout their historical study, the authors of the latter book show impatience with the slow growth of industrialization over the past century; but in their final chapter they find themselves making a radical criticism of the state of affairs that has finally been achieved.

advising investment agencies, and so on; but in relation to his means of production and the decisions which determine his work, his position must be one of these three.

Along with this differentiation of the positions of individual scientists, there comes a concentration of the power to make decisions, and the development of a formal administrative system for this function. The dispersal of large sums of money, and even more the decisions between competing demands, are matters which require proper procedures of information and control. A completely informal consensus of a large community is not sufficiently precise or reliable to be the basis for such work; and the investing agencies must work from the evaluations and judgements of a group of advisers. With this concentration of powers of decision and control, the free market place of scientific results, whose value is established after they are offered and by an informal consensus, is replaced by an oligopoly of investing agencies, whose prior decisions determine what will eventually come on to the market.

This is an inevitable consequence of the costs of research; even if individual universities were given large block grants out of which to finance all the research done by their members, the same sorts of decisions would be necessary, and a similar set of structures would develop.[16] Also, the investing agencies try to maintain some of the old conditions of consensus, by choosing their advisers from among the recognized leaders of each field. But the incoming generation of leaders, and all those following, have built their careers and reputations under these new conditions, and will be subject to its influences in ways which I shall soon describe.

The most basic effect of these technical and structural changes is a tendency to a change in the location of the intellectual property of a scientist, possession of which is desired for the achievement of his personal purposes. We recall that under the old system it was fundamentally the published research report that constituted his property; on the basis of the informal evaluations of it by his

[16] In his broadcast talk on 'Planned Science' in 1948, Michael Polanyi argued cogently that 'no committee could forecast the routine progress of science except for the routine extension of the existing system'. Hence, 'The function of public authorities is not to plan science, but only to provide opportunities for its pursuit. All that they have to do is to provide facilities for every good scientist to follow his interests in science.' Polanyi may have been correct in believing that the Marxist impulse to the planning of science had evaporated; but the impossibility of carrying out his suggested policy soon made planning inevitable. See *The Logic of Liberty* (Routledge, London, 1951), pp. 86–90.

colleagues, he expected appropriate rewards in his career, and the personal satisfaction produced by public recognition of his work. In the present situation, the research contract is not merely a prerequisite for the future possession of the property embodied in a published paper; it also brings immediate benefits in itself, in the way of prestige and possible material conveniences. Moreover, with the concentration of decision-making power to the investment agencies and their few advisers in each field, their estimate of a man is of more practical significance for his career, than that of some future diffuse consensus. Hence the location of a successful scientist's property tends to shift from his published results to his existing research contracts, and the personal contacts that will ensure their continuation. With this shift in the location of the scientist's property, there is a tendency to a corresponding shift in his conception of a successful career. Especially for someone who enjoys the role of a contractor, with its incidental benefits, the goals of a career in science can change from being a series of successful research projects made possible by a parallel series of adequate contracts, to being a series of successful research contracts made possible by a parallel series of adequate projects.[17] When this happens, the man is better described as a 'scientific entrepreneur' than as a 'scientist'.

Under such conditions of division of labour, concentration of decision-making power, and tendency to the shifting of goals, it is impossible for a 'community' to survive in its old form. The mere expansion in size of every field, speciality, and subspeciality, would make it difficult in any case. But under the present conditions, differences in wealth (taken here as a measure of the scientist's intellectual property) produce such extreme differences in prestige, power, and material benefit, that we can truly speak of classes in a society of science, rather than of more and less eminent colleagues in a community. Since most scientists engaged on 'pure' research are employed by a university (or by one of the few State-supported research laboratories where conditions are equivalent to those at universities), they will tend to settle at institutions of a prestige status comparable with their own. The leading men, forming a very perceptible 'invisible college', will congregate at the great univer-

[17] This distinction is made implicitly by D. S. Greenberg when he contrasts 'grantsmanship', the (ethically neutral) practice of extracting funds for research, and 'chiselling' (*The Politics of American Science*, pp. 351–3).

sities, enjoying favourable conditions of employment there, and with their intimate connections with the investing agencies (a prerequisite for the existence of such a group) they will pursue the researches they please in comfort. And since in most countries the high-prestige universities cluster in a special geographical region, those who are left, or cast, out of the charmed circle of the successful men may be isolated in every way. Deprived of the personal contacts whereby one keeps up with new work, and whereby one makes an impression on those who advise the investing agencies, they are left to do derivative or second-class research on less generous grants, or none at all.

Adulteration of the Products of Research

We can now consider some of the changes in social practice consequent on the changes in production in science; and in so doing, identify and explain some recognized social problems. The first of these has been the subject of anxious discussion for some years: the 'information crisis'.[18] At first this appears to be a purely quantitative problem; the number of journals in each field is already so large, and is increasing so rapidly, that a scientist cannot 'keep up' with the literature without great difficulty.[19] Worse, because of the difficulty of finding any particular item, even with the existing abstracting services, scientists frequently suffer the waste and disappointment of duplicating research already done elsewhere. Several schemes have been developed for more effective information retrieval systems; but it has to come to be seen that the complexity of the contents of a given research report makes purely mechanical identification less valuable than was at first thought.[20]

On closer examination, it appears that the problem is not a purely quantitative one. For along with the expansion of the traditional

[18] An illuminating survey of problems of communication in science, including some striking examples of blunders in research that appear in print, is given by H. V. Wyatt in his essay 'Communication in Obscurity' in *The Use of Biological Literature*, eds. R. T. Bottle and H. V. Wyatt (Butterworth's, London, 1966).

[19] See Bentley Glass, 'Information Crisis in Biology', *Bulletin of the Atomic Scientists*, 18 (October 1962), 6–12. He reports a survey of fifty scientists, in both physical and biological sciences, on their use of the literature; the conclusions suggest a 'growing isolation in which much scientific research is carried on'. He also deplores the decline of genuine review articles, as opposed to mere annotated bibliographies; this can be explained by their not counting as 'research' by the author.

[20] See P. Cranefield, 'Retrieving the Irretrievable; or the Editor, the Author and the Machine', *Bulletin of the Medical Library Association*, 55 (1967), 129–34.

channels of publication, there has been a rapid development of less formal channels. Through mailing-list distributions, scientists will circulate not only reprints, but also preprints, duplicated preliminary research reports, conference abstracts, and informal 'newsletters' of people, events, and results. These other types of publication are not merely supplements to the official channel of communication, making results available more quickly than can the printed journals. Rather, they are complementary to that channel; and their function is to provide publication of a sort which is not subject to the hazards, as well as the delays, of the scrutiny of referees. For the same sort of function is performed by the raw collections of conference papers, published in hard covers and sold on the market, but appropriately called 'non-books'. The fact that this channel is a dilution of the other is recognized in discussions of the problem of deciding whether such materials can be cited in a paper published traditionally. For the materials so cited should not only be public, they should also be of tested quality; otherwise a research report which has been certified by publication will be dependent in part on a component which is uncertified. Of course, there is no easy answer to this problem; for whatever the criterion of distinction between orthodox and unorthodox publication, some clever fellow will find a technique for straddling it.[21]

Thus the problems created by this diluted channel are far greater than the creation of additional headaches for those who do the work of abstracting and information retrieval. For they show that a large number of scientists are quite happy to evade the traditional rules for publication, and to be seen doing so. This phenomenon can easily be explained in terms of the shift in the location of intellectual property. Through the adulterated channels, one can hope that one's results, dressed up in their most attractive form, will catch the eye of someone important, where another routine paper in a crowded

[21] In 1960, the editor of *Physical Review Letters* took a firm stand against initial publication of results in the daily press, remarking that 'Scientific discoveries are not the proper subject for newspaper scoops', and warning that previously released results would not be accepted by his journal. See D. S. Greenberg, *The Politics of American Science*, p. 74.) A decade later, and in a field in which pressures might well be stronger, the barrier was surrendered. Dr. Morris Fishbein, the eminent leader of the American Medical Association, wrote: 'The distinguished editors of the clinical journals would do a disservice to the medical profession by demanding that medical investigators withhold information about their observations and conclusions until sufficient time has passed to allow the medical journals of record to publish their work in its totality. Nor should they expect investigators to heed such directives.' (Quoted in *Science*, 167 (1970), 148.)

journal, or a bare title in an abstract, would be passed over without notice. But to engage in this practice, a man must be willing to put untested products on offer to his community, knowing that some of the work might be substandard.[22] ✓

That there is no shortage of such people, becomes clear when we discover that one of the most serious aspects of the 'information crisis' is a phenomenon known as 'pointless publication'. But since every act of publication has some purpose, I shall refer to it as 'shoddy science'. For it appears that the majority of journals in many fields are full of papers which are never cited by an author other than their own, and which, on examination, are seen to be utterly dull or just bad.[23] Now, the existence of bad scientific research is something of which every scientist (or at least every good one) must be aware by his experience of attempting to use other published results. But hitherto it has not been a topic for polite discussion in print; it is mentioned neither in the philosophy of science nor in the formal teaching in science. Research students learn of it, frequently by very painful experience, as something of a dirty secret. Indeed, not very long ago an elder statesman of science could assert that such a thing is nonexistent.

Would it be too much to say that in the natural sciences today the given social environment has made it very easy for even an emotionally unstable person to be exact and impartial in his laboratory? The traditions he inherits, his instruments, the high degree of specialization, the crowd of witnesses that surrounds him, so to speak (if he publishes his results)— these all exert pressures that make impartiality on matters of *his* science almost automatic. Let him deviate from the rigorous role of impartial experimenter or observer at his peril; he knows all too well what a fool So-and-so made of himself by blindly sticking to a set of observations

[22] The American Psychological Association has recently found itself in conflict over a project designed to provide an institutional machinery for the publication of untested materials. An 'early dissemination' scheme, which forms one part of a comprehensive plan for improving communications, would distribute manuscripts to special-interest subscription lists shortly after submission. This has been criticized as 'a vast sewer carrying garbage from one scientist to another'. Disquiet over the very principles of the system, which was designed to cope with a genuine information crisis in the subject, is mixed with concern that the bureaucracy necessary for running it will dominate the Association. See P. M. Boffey, 'Psychology: Apprehension over a New Communications System', *Science*, 167 (1970), 1228–30.

[23] D. J. de S. Price, *Big Science, Little Science*, reports a paper by D. J. Urquhart on loans of journals in the stock of the Science Library in London; about half were not used at all, and a quarter only once in the year of survey (pp. 75–6).

or a theory now clearly recognized to be in error. But once he closes the laboratory door behind him, he can indulge his fancy all he pleases....[24]

The most remarkable thing about this argument is that it assumes no special virtue on the part of the scientist; but it does claim that the system of quality control in science is perfect. The claim may have been plausible at some time in the past, but it is no longer so. Shoddy work exists, and in large quantity. References to it can be gleaned from published discussions of the state of particular fields. And it is a truly pathological symptom of the social condition of industrialized science.

For a paper to be published, it is sufficient that the author, an editor, and a publisher all find some purpose served by its publication. From the side of the publisher, it is a matter of economics: given the guaranteed library subscriptions and the economics of journal publication, it is possible to make a profit even on a obscure journal.[25] The editor may receive an honorarium from the publisher, and will certainly derive prestige at his own university by virtue of his position. The author may need to have another title in his record, as a demonstration (for his employer) of his continued competence in research, or as another point to be included in his aggregate score of publications when he applies for a grant from a large and impersonal investing agency.

Of course, the publication of shoddy work would exclude a man from membership in a community devoted to the advancement of science; hence those who do this are either in no community at all, or in one with different goals. In such a community, the traditional ethic of the disciplined search for truth is either forgotten, or is a sick joke.[26] If the publication of shoddy science were restricted to the dim and obscure men in remote provinces, and their local journals, then the situation could be explained partly at least by the natural differentiation by quality of scientists. But the participation in this abuse by men of high prestige requires another explanation.

[24] James B. Conant, *On Understanding Science* (Yale University Press, New York, 1947; Mentor Books, 1951), p. 23 of Mentor edition.

[25] See 'How Many More New Journals?', *Nature*, 186 (2 April 1960), 18–19.

[26] For a penetrating analysis of the problem of shoddy science, see Howard A. Mayerhoff, 'Useless Science', *Bulletin of the Atomic Scientists*, 17 (March 1961), 92–4. This article is a reprint of a speech originally given in 1954 by the author, a former editor of *Science*. He describes the various pressures on scientists to accumulate publication-points, and gives useful case studies on how insignificant research can be made fruitful and multiplied.

The Penetration of Science by Industry

It is well known that science, even the most pure science, is important for the work of industry. Any firm engaged in modern industrial production, and by extension an advanced industrial nation, secures its future existence by the work done in the 'research and development' laboratories. Such work does not merely use scientific results; in its more sophisticated forms, it is continuous with scientific inquiry. Problems and discoveries thrown up in such technological inquiry can stimulate important scientific research; and conversely, any scientific result, even one deriving from a problem which was investigated with no thought of application, may find a use in industry. All this should be familiar; and in this unification of science and technology lies the hope for the realization of the age-old dream of material plenty for all of mankind.

What is less familiar is that industry has, in many ways, made its own penetration into science. The industrialization of scientific research is one manifestation of this. By itself it involves an increase in scale, and in formal organization; but we can be sure that the vast expense of industrialized science would not have been incurred by the State unless some tangible benefit through a close association of the two sides was expected. How close the association can be was indicated in a report on the ethical problems of physicists in America a few years ago.[27] There it was found a man could be simultaneously filling nine roles: at his university, to be teacher, administrator, and research scientist; with various State agencies, to be a contractor for research, an assessor of research proposals, an official adviser on existing projects, and a private consultant on particular technical problems; and with commercial industry, to be a private consultant to firms, as well as a businessman manufacturing equipment of his own invention. It is hardly surprising that little financial abuses should occasionally occur, nor that 'conflict of interest' should be a matter of concern. What might be asked is whether such a man could change his attitudes and values every time he picks up a different piece of correspondence, or whether he operates in a world where all distinctions are blurred, and it's all business.

The penetration of science by industry proceeds by yet another

[27] See 'Hazards of Sponsored Research', *Physics Today*, 18 (March 1965), 98, 100; a report of a statement issued the previous December jointly by the American Council of Education and the Council of the American Association of University Professors on the prevention of conflicts of interest in government-sponsored research at universities.

path, created by an inevitable ambiguity in the specification of every research project. This is in the possible function of the result to be achieved. Even if the scientist is personally interested only in its significance for advancing the field, he can well imagine a possible relevance of the result to some sophisticated and unsolved technical problem. An emphasis on this relevance will help the project to be considered as 'mission-oriented research', which is naturally better endowed with funds than the useless sort. Now, it is possible for the scientist or the investing agency to keep the different aspects of a project, and the different values they represent, in separate compartments, considering it accidental and unsought good luck that interest in the problem should be so diversified as to bring it financial support. But such a state of affairs cannot persist indefinitely for a whole group of scientists. The natural tendency, to ensure good relations and continued support, is to give serious attention to those whose interest is essential and for the research to shift in their direction.

Of course, the community of science has always had a very difficult problem in justifying itself in terms comprehensible to the lay world; and in its generalized claims to be the basis for progress in industry it was to some extent playing a gentle confidence game with its public. But it did not matter too much, for the resources devoted to science were very small, and of them only a fraction came from the public purse.[28] Also, the claims on behalf of the practical importance of science were necessarily diffuse or retrospective; for it was only rarely until the end of the nineteenth century that anything like systematic scientific inquiry could be successfully applied to the solution of industrial problems. But in the present period the sums are significant, and the claims are true enough for those who advance the cash to feel entitled to see some return. Hence the indi-

[28] On occasion, the conflict of values between excessively pure-minded scientists and a particularly vulgar governmental patron could yield disasters. See G. D. Nash, 'The Conflict Between Pure and Applied Science in Nineteenth-Century Public Policy: the California State Geological Survey, 1860–1874', *Isis*, 54 (1963), 217–28. Norman Kaplan uses the term 'bounty' to describe the grants formerly made by European governments to individual scientists for the pursuit of their research; these would be on a small scale, and recognized as contributing to the research work of a university for the enhancement of its, and the nation's prestige. See N. Kaplan, 'The Western European Scientific Establishment in Transition', *American Behavioural Scientist* (December 1962), 17–21, reprinted in *Science and Society*, ed. N. Kaplan, pp. 352–64. It should be observed that a very important part of this 'bounty', at least between the two wars, came from American foundations, notably the Rockefeller.

vidual scientist, and even more those leaders of the scientific community who plead for public funds for particular projects as well as for general purposes, must be able to talk the language of economic (or military) benefit at least as well as that of the search for knowledge.

Influences from Runaway Technology

The relations of science with industry will not be uniformly close in all fields; the connections will be strongest with the most modern, rapidly developing technologies, where innovation depends entirely on large-scale, sophisticated 'research and development'. It is these areas, such as aerospace, electronics, and parts of biological engineering, where the pace of development is so rapid, and the ecological and social effects so unpredictable and dangerous, that have been the focus of public concern in the menace of 'science' to humanity. Those who take the decisions to plunge into ever greater 'progress' in this work are not afflicted by any special wickedness or even irresponsibility. They are merely continuing the attitudes and practices inherited from the industry of the past, which might be called 'myopic engineering'. Provided that a particular development was technically viable and not at risk of penalties under the law, then so long as it seemed likely to make a profit, it would be adopted with no further thought of its consequences. Hitherto, the effects of such a policy, however disastrous, were localized to the region where they were put into practice; thus the rural South of England knew little, and generally cared less, of what was being done to the Midlands and North by the Industrial Revolution. And it is undeniable that the generally short-sighted and ruthless men who created the industry of the nineteenth century laid the material foundations for the prosperity of the present.

But the engine of innovation and production which we now possess is qualitatively different in many respects. Its effects are so pervasive that there is no place to hide from them. Also, the work of innovation in the advanced technologies is now a large industry in its own right. Its projects have some special features, which make them very different from the work of inventors and scientific consultants in the past. First, the investigation of any technical problem of development requires the prior commitment of enormous resources, both in funds and skilled manpower. Also, such problems are necessarily speculative, in several ways. It cannot be guaranteed in advance that the research will produce a device which works at all. Moreover, even if it works, there is the risk that during its years of development, either a change in the technical or commercial context, or the appearance of a better suited product from a competitor, will deprive it of its market. Perhaps most significant of all, the concept of 'profit' has been transformed. For much of the sophisticated technological work is done for the State, for use outside the market sector, as for war. Although the prospect of foreign sales of a device are part of the calculation of 'benefit', the basic component of benefit is assessed through a scientific study of its potential uses. Even where a key industry is nominally in the private sector, the State will take responsibility for its continued prosperity, through research contracts, guaranteed purchases, and other techniques. Thus a particular innovation may be recognized as risky from the technical point of view, dubious from the commercial point of view, of very slight use to anyone at all, even the State, and a potentially serious nuisance to the public and source of legal and political difficulties, and yet still receive enormous sums from the State because of its contribution to national prestige and its importance for maintaining employment and morale in a key industry. The Anglo-French supersonic transport is a perfect case in point of this phenomenon. At the extreme, national prestige may become so involved in a project that all considerations of cost, benefit, and profit (except, of course, to the private firms doing research and manufacture) are cast to the winds, and a glamorous technical project, such as the moon-race, absorbs resources on a gargantuan scale, all in the name of 'science'.

Although this new industry of 'R. and D.' employs many scientists (indeed, the bulk of graduates in science and technology go there rather than into teaching or university research), its working ethics are descended from industry, private and state-supported, rather than from academic science. In America, the enormous defence and aerospace industries carry on in the time-honoured American tradition of 'boondoggling' on Government funds; the most effective path to super-profits being to keep the relevant Government agencies for cost-accounting and quality-control either remote, or weak, or complaisant.[29]

[29] The first extended discussion of this problem is in H. L. Nieburg, *In the Name of Science* (Quadrangle Books, Chicago, 1966). A summary appears as 'R and D in the Contract State: Throwing away the Yardstick', *Bulletin of the Atomic Scientists*, 22

Thus we can speak of this new technology as 'runaway' in several respects. In calculating cost and benefit, it ignores all those costs of a project for which it cannot legally be called to account: in particular, the degradation of the natural and human environment. Since the combined effect of the present and future technological developments is likely to be catastrophic, this rush onwards can truly be considered as out of control. And in its internal workings, the absence of that traditional discipline, crude and frequently distorted but in the last resort effective, of the test of a commercial market, makes the category of 'profit' an artificial one, to be determined by the judgement of men in State agencies, in co-operation with the promoters themselves.

It is in the borderland between science and this sort of technology that we find some significant pathological phenomena. The first occurs when a contractor (individual or institutional) develops a really big enterprise, which is most likely to be on some mission-oriented research in a field where money is plentiful and not too many questions are asked.[30] There then develops a research business, making its profit by the production of results in the fulfilment of contracts. The director of such an establishment is then truly an entrepreneur, who juggles with a portfolio of contracts, prospective, existing, extendable, renewable or convertible, from various offices in one or several agencies. The business is precarious, of course, for his only capital is in his friendly contacts with those who decide on the allocation of funds. In such a research factory, conditions are not usually conducive to the slow, painstaking, and self-critical work which is necessary for the production of really good scientific results. Hence much, most, or even all the work can be shoddy; but

---

(March 1966), 20–4. A recent, more general survey of American military procurement and spending is W. Proxmire (Senator), *Report from Wasteland* (Praeger, New York, 1970). For the report of the Lang Committee on the English Bloodhound missiles contract see *The Times*, 10 February 1965.

[30] D. S. Greenberg has invented a prototype entrepreneurial scientist, 'Dr. Grant Swinger'. In 'Grant Swinger: Reflections on Six Years of Progress', *Science*, 154 (16 December 1966), 1424–5, Dr. Swinger describes the exploits of the 'National Animal Speech Agency', which 'had its origins in the President's challenge to the nation "to teach an animal to speak in this decade"'. Previously, Dr. Swinger had advocated a new model of high-energy particle accelerator which would extend from Palo Alto, California to Cambridge, Massachusetts. It would be shaped so as to pick up maximum local political support. An alternative was a vertical accelerator at the unique intersection of four States, in the West; the design yielded a very attractive bev/dollar/vote analysis. See '1965: Herewith, a Conversation with the Mythical Grant Swinger, Head of Breakthrough Institute', *Science*, 151 (1 January 1965), 29–30.

the entrepreneur does not operate in the traditional market of independent artisan producers who evaluate work by consensus. So long as he can keep his contacts happy, or at least believing that they personally have more to lose by exposing themselves through the cancellation or non-renewal of contracts than by allowing them to continue, his business will flourish.[31]

It is in such circumstances that a man of high prestige and real talent will produce a stream of shoddy work. Too busy to do any thinking himself, and yet requiring a steady stream of publications as a proof of his continued competence, he will toss off pieces, either alone or with associates, which will produce a list of titles of the necessary length. Although large-scale science is more exposed to the risk of invasion by entrepreneurs, size is not the determining factor. Whenever a research contractor, however modest his plant, sets the goals of his establishment to be the renewal and extension of contracts rather than the achievement of worthwhile results, he is an entrepreneur.

Even when scientific work of good quality is being done, the style of runaway technology can infect a field; the old, diffuse ideal of material benefit gives way to something more sharply defined and intoxicating: the possibility of the creation of new technical powers. The patent dangers of some of these powers, in the present state of civilization, have been brushed aside as of little consequence, or as the responsibility of someone else. It is so many generations since people in our civilization believed that there are 'secrets too powerful to be revealed', that a scientist of our age cannot conceive himself as being in the position of the sorcerer; and yet he is. Thus 'reckless science', as a special product of the technical and social conditions of scientific inquiry in our time, must be identified and controlled, for the safety of humanity in the long run and for the preservation of science in the short run.[32]

[31] For a satirical but none the less penetrating account of the techniques of entrepreneurial science, see H. Miner, 'Researchmanship: the Feedback of Expertise', *Human Organization*, 19 (1960), 1–3. The reference comes from J. Barzun, *Science the Glorious Entertainment* (Secker & Warburg, London, 1964), where this and many other targets are put under heavy fire.

[32] The leading candidate for the status of 'reckless science' at the time of writing is molecular genetics, which can enable the controlled manipulation of human genetic material. See S. L. Luria, 'Modern Biology: a Terrifying Power', *The Nation* (20 October 1968), 406–9. One of Luria's most promising students, James Shapiro, has taken the message and left science. He gave three reasons, of which the first is his fear that his results will be 'put to evil uses by the men who control science'. See J. K. Glassman, 'Harvard Genetics Researcher quits Science for Politics', *Science*, 167 (1970), 963–4.

Finally, the demands of military technology in particular provide opportunities for employment of scientists on research projects whose intended application lies beyond the pale of civilized practice and morality. The weapons called 'ABC'—atomic, biological, and chemical—are each, in their own ways, morally tainted. Research and development of such weapons can be plausibly justified in terms of defence and deterrence; but the experience of the scientists on the original atomic bomb project shows that once the weapon is available, the tender consciences of the scientists who created it will not have much influence on the decisions on its use.

These four abuses, shoddy science, entrepreneurial science, reckless science, and dirty science, are distinct in their natures, but there will be tendencies for them to overlap in practice.[33] Also, each of them arises from conditions inherent to the situation of contemporary science; and there is no clear line of demarcation, in the results of the work or in the attitudes of the scientist, whereby one can condemn one man and exonerate the next. But because they are more closely related to the demands of modern technology, and sometimes more easily popularized as exciting than traditional research, they will tend to attract a lion's share of the available funds, thereby providing the most attractive career prospects for recruits, and drawing into their ambit those who could not otherwise carry on their research. This effect can be seen in the United States, where the total budget for 'science' is enormous, but where all save a small fraction is allocated to military R. and D. and the space-race. Even there, it can be argued that Congress would never allocate more than it does for the direct support of the esoteric and peculiar activity of pure research; but the result is that the 'scientist' is seen as costing a lot of money to the taxpayers, and if he wants to use some of it, he is under pressure to make his accommodation with those who control these branches of runaway technology.

Problems of Morale, and of Morals

Up to now, our discussion has been of problems which could be shown to arise naturally out of the changed technical and social conditions of science, and whose existence can be established by representative instances. The problems we shall now discuss are

[33] A case allegedly illustrating such tendencies is cited in S. M. Hersh's *Chemical and Biological Warfare; America's Hidden Arsenal* (MacGibbon & Kee, London, 1968), pp. 141-2. It was from the station mentioned there that a cloud of nerve gas killed six thousand sheep in 1969.

more subtle, although equally important; and for evidence I shall need to use certain recognizable symptoms rather than hard cases.

The need for good morale is never mentioned in general discussions of science directed to a lay audience; and this is evidence that hitherto its presence could be taken for granted. For doing good scientific work is strenuous and demanding, and the quality of the work done in any field of science is dependent, to a great extent, on the integrity and commitment of the community of scientists involved. In other spheres of social activity where the success of the enterprise cannot be assured by the imposition of systems of discipline and control, the factor of morale is given due recognition. Military organizations are the outstanding case in point; but voluntary associations of any sort are known to require good morale; and nearer home to science, the performance of students at a university is also seen to depend on their morale. The maintenance of morale is a most subtle task; it is easiest in a voluntary community of equals doing successful work; and where there are gradations of status and power, the leaders must show themselves as standing for the whole group, in its problems, its interests, and its difficulties.[34]

Whether there has been a decline in morale affecting all of science over the last generation is impossible to say. Even if one were to conduct a social survey of a large sample of scientists, the evidence for the past would have to come from the personal memories of older men, and these are not reliable as historical testimony. But there is one recognized phenomenon, which is probably a symptom of declining morale, and certainly a cause of its further decline. This is the disappointment in the recruitment to science, measured by the number of students electing to study science at university and also continuing into research. The cause, or causes, of this phenomenon are not at all understood. It may simply be that the proportion of young people who, in the present conditions of our culture and educational system, have the inclination and talent to pursue science as a career, is strictly limited; and that merely making more places

[34] An excellent analysis of the importance of good morale for scientific work is given in *Biochemistry*, *'Molecular Biology' and Biological Sciences* (The Biochemical Society, London, 1969), 26–7. The argument is extended from science to the universities, as a source of a necessary idealism in modern society, and for which excellence, and 'in the last resort, excellence of morale' is of the utmost importance. This emphasis on morale in science and in universities can be explained by the circumstances of the production of the report as an answer to the 'Kendrew Report', *Report of the Working Group on Molecular Biology*, Cmnd. 3675 (H.M.S.O. London, 1968), whose recommendations could be interpreted as favouring a contraction of research to a few centres.

available at universities does not, in the short run, increase the supply of worthwhile candidates. It may also be that many young people of the classes which provide the recruitment to universities do not, as did their parents, have a keen awareness of a need to obtain a qualification leading to a secure job. So if their personal interests lie in the arts or in society, they will be less inclined to sacrifice them during their university studies.

Another possible factor is the decline in the quality of science teaching in the schools. A bright but uncommitted pupil can easily be put off such an abstract and demanding discipline if he never encounters a teacher who by his personal example makes it an exciting challenge. And through failing to ensure that well-qualified and well-motivated scientists returned to school-teaching the community of science was, for two decades after the Second World War, eating its seed-corn. Even if the supply of science teachers now improves, there is no guarantee that it will soon yield to science a crop of recruits of the desired quality. Pleasant classroom experiences are not the only factor that influences the choice of the bright pupils, to say nothing of those who are really gifted.

It is possible (one can say no more in the absence of very sophisticated social investigations) that part of the falling-off of recruitment to science is a result of the changed image of science which is projected to the public at large and school-children in particular. Throughout the nineteenth century many of the great men of science shared their enthusiasm with a wide public, through books and lectures. What they conveyed was the sense of excitement of the individual search for truth through fascinating new discoveries. The endeavour was one which offered the hope of fame, through work which was innocent, enjoyable, and ultimately beneficial to mankind. In the present age, there is no lack of such propaganda; but its message does not come through in such a clear and simple fashion. First, the dangers inherent in science have been common knowledge ever since the menace of nuclear weapons became a matter of public concern. Also, working in science is now recognized as a career, one among the many open to a person of the requisite ability, rather than a vocation for dedicated individuals. For the great expansion of university places in science is not seen as resulting from an enlightened desire for more pure knowledge, but as a response to a need for more manpower of a particular sort. Thus, to put the matter very simply, the effective image of scientific work may have changed from being pioneering explorations in the philosophy of nature, to the service of technology, either commercial and vulgar, or military and sinister. And what sort of ambition is it, to be a unit of scientific manpower?

This is not a fair representation of the state of science today; but here we are not discussing a complex and subtle social reality, but rather the simplified and distorted picture of it which reaches a lay public. And it may be significant evidence that the field which has suffered most from a fall in recruitment is the one whose public image has undergone the most drastic transformation.[35] Through the later nineteenth century and well into the twentieth, physics was the queen of the natural sciences. Its achievements were the most impressive, it stayed the closest to the philosophy of nature in name and in spirit, and it was accepted, by philosophers and laymen alike, as the paradigm of science, setting not only the standard of excellence but even the popular conception of what every science should try to be like. It was a tragedy for many eminent physicists, and also for their field, that this most aristocratic and philosophical of studies should have been the one which created the runaway military technology of nuclear weapons. By this work, physics was tarnished beyond repair; and through its status as the paradigm field, its moral fall affected the rest of science as well.

In its present technical and social situation, the world of science is particularly vulnerable to a decline in recruitment. For much of the research done at universities depends for its accomplishment on a supply of the poorly-paid skilled manpower of postgraduates. There is a rapid turnover of this labour, and as it declines in quantity and quality, each scientist will see his research suffering a similar decline, and with it his career prospects.[36] As we have seen, neither the

[35] The 'swing from physics' is not confined to Britain; the same phenomenon has been observed in America, and the imaginative new secondary-school syllabuses designed by leading research physicists have certainly not had a positive effect on the problem. See D. S. Greenberg, *The Politics of American Science*, p. 67.

Statistics for the decline of physics enrolments in American high schools are provided by F. G. Watson, 'Why do we need more physics courses?', *The Physics Teacher* (May 1967), pp. 212–14. He gives percentages of twelfth grade pupils taking physics, as follows: 1948–9, 26 per cent; 1958–9, 25 per cent; 1964–5, 20 per cent. This drop of about 20 per cent in eight years may be partly the result of a higher proportion of the age-group being enrolled in the twelfth grade, either by a policy of automatic advancement in grade regardless of performance, or by a raising of the school-leaving age in the backward States.

[36] Individual impressions of a decline in quality of students are very suspect, but there are examples where objective criteria can be cited. Thus, Bart J. Bok (letter to *Science*,

technical conditions of research, nor the social experience of scientists over the last two decades, make it easy for them to adjust to the situation by simple belt-tightening. Without the prospect of attracting interesting and useful research students, even university teaching in science loses its attraction and much of its present function. Hence, whatever the causes of the fall in recruitment, its effects are plain; and as the decline in morale within science becomes known outside, it may well contribute to a cycle starting with a further disappointment in recruitment. Where this process can end, no one can say; and it is possible that a change in the economic and social situation of potential recruits will bring them back again in satisfactory numbers; but whether even this would provide the leaven of gifted scientists in the mass of technological manpower, only time can tell.

It is in physics again that we find another symptom of a loss of morale: the sense that there are no more challenging problems which can be solved.[37] In the period before the Second World War, the most exciting part of physics was the study of the atom; and as a result of the technological success with atomic and nuclear weapons, physicists were in a position to apply ever larger energies, at ever larger cost, to the penetration of matter in the small. But the increase could not continue indefinitely, and recently a halt was called, first because of the enormously greater expense of each new accelerator, and also because of growing doubts about whether even such monsters would be adequate to their intended functions. Nuclear physics now finds itself at a dinosaur stage; unable to evolve further, it awaits extinction unless some happy accident rescues it. Of course, there were prophecies of doom in the leading sciences at the end of both the eighteenth and the nineteenth centuries, which turned out to be false. But in each case, the new line of progress came from an obscure corner of the field, rendering much of the earlier work, and the eminent men who did it, obsolete.

The sudden discovery of an upper limit to financial support has not been restricted to physics. All over the world, governments are having another look at the budgets for research, and at the very least,

154 (1966), 590–2) describes aspects of a general relaxation of the pace of graduate study in astronomy, and the lessening of the demands made on students (in rate of progress, and contribution to discussions and teaching).

[37] H. S. Lipson, F.R.S., in 'Where is Physics Going', *Advance*, No. 4 (1968), 17–21 (U.M.I.S.T.), speculates on whether physics as a school subject may share the fate of Latin.

decelerating the expansion of support for pure science. It matters not that the total budget for genuine science is only a small proportion of the sums allocated for technology of various sorts; when cuts have to be made, they fall most heavily on those with the least attractive case to present. If the resulting contraction were to be properly handled, it could have beneficial effects; for science would be healthier without the mass of dull and shoddy work which is supported when money is easy. But the danger is that those who maintain their contacts with the investing agencies will manage to survive, while those on the outside, however worthwhile their work, will go to the wall. The problems of decision on the allocation of resources are particularly painful and difficult when there is not nearly enough to go around; for a fair distribution, exceptional qualities of leadership are required. Only then could science survive this contraction without a very serious loss of morale.

These problems of morale pose a threat to the continued existence of scientific inquiry as we know it, which is as serious as those arising directly from technical and social conditions of industrialization. For when morale is low, the work which is done is at the minimum level of acceptance; and decisions and evaluations which are reached through consensus will tend to be those which involve the least risk and the least work, all around. Since the leadership of science exists and operates only by consensus, the quality of that leadership, to which the quality of scientific work is very sensitive, is at risk. The situation is further aggravated by the fact that for most of these problems, the inherited working ethic of science is irrelevant, just as its ideological basis is obsolete. Hence in many respects, individual scientists, and even more the leaders of science, encounter problems directly involving morality.

As we have discussed, science has traditionally been portrayed as an activity whose morality could be nothing but the best. The search for truth is innocent and ennobling; and the eventual benefits to mankind through the advance of knowledge and power, further secure the moral status of science. The very idea of a scientist being a thief, a swindler, or a man who offers his opinions for sale, is near to being a contradiction in terms.

The contrast between the working ethics of science, and that of politics, or business, is so strong as to induce a belief that scientists must be born, or at least made, as superior beings. This picture of the perfect morality of science had its strong basis in reality, both in the practice and in the self-consciousness of science in the academic period. For then, the favourable social conditions, and the effective arrangements for the protection of property, were sufficient to ensure that the working ethics of science would serve to resolve all the social problems encountered in its experience.

In the longer perspective of history, the moral innocence of academic science appears as a temporary feature, a happy accident of circumstances. For the problem of responsibility for powerful knowledge had been recognized by the practitioners of fields ancestral to science, in all the previous centuries. Such knowledge, including magic, alchemy, and parts of astrology, was to be restricted to those who could use it wisely; and hence it was transmitted in an oral or cryptic tradition to initiates in a brotherhood. Of the pioneers of the scientific revolution, Bacon and Descartes, although disbelieving in magical powers, retained this moral sense. Even in his Utopian 'New Atlantis', Bacon had the sages of 'Solomon's House' deciding which secrets they would reveal to the State, and which not; and Descartes stated a 'scientist's oath' of classic simplicity: I would not engage on projects which can be useful to some only by being harmful to others.[38] But Galileo, whose general style of work was so much more like that of a 'scientist' than any other natural philosopher of his time, was totally lacking in such a sensitivity. He considered himself as having the right to proclaim philosophical truth as he saw it, and was utterly unconcerned with the possible social effects of his unsettling doctrines. It is probable that he was sure that God's truths, which he was announcing, could not be harmful; but in practice he was demanding the influence over men's minds resulting from his pronouncements, while denying responsibility for the consequences of his actions. His moral position was made even more complex by the fact that while fighting heroically for the truth

[38] For Bacon: 'And this we do also: we have consultations, which of the inventions and experiences which we have discovered shall be published, and which not: and all take an oath of secrecy, for the concealing of those which we think fit to keep secret: though some of those we do reveal to the state, and some not.' *New Atlantis*, in Bacon's *Works*, eds. Spedding, Ellis, and Heath, 14 vols. (London, 1857–74), Vol. 3, p. 165. Unless otherwise noted, all references to Bacon will be taken from this edition. The American version of this edition unfortunately has a different pagination.

For Descartes: '. . . et que mon inclination m'eloigne si fort di toute sort d'autres desseins, principalement de ceux qui ne sauraient être utiles aux uns qu'en nuisant aux autres, si quelques occasions me contraignaient de m'y employer, je ne crois point que je fusse capable d'y reussir.' *Discours de la Methode*, last page of text, in the edition *Discours de la Methode*, texte et commentaire par E. Gilson (Vrin, Paris, 1925, 1926) p. 78. Unless otherwise noted, references to the *Discours* will be taken from this edition.

of natural philosophy as he saw it, he was not at all averse to making money privately through the sale of his inventions to the highest bidder.[39]

With the rise to dominance of the 'mechanical philosophy', or more correctly with the dehumanization and disenchantment of Nature for the educated common sense of European civilization, the expected powers of scientific knowledge were reduced, and hence also the responsibilities of men of science. The association of natural science with 'progress', first with that of the intellect in the eighteenth century, and then with industry in the nineteenth, removed any fears concerning the applications of science from the traditions of self-awareness within science. Hence it was as moral innocents that a group of distinguished physicists urged the American government to produce an atomic bomb. Although such a device clearly had its dangers, ordinary human morality dictated that there should be no risk of Hitler having sole possession of such a bomb. But, some six years later, what had been conceived as a deterrent weapon in the struggle against Fascism, was used on the civilian population of a nation near to surrender; and not once, but twice. Were the scientists who initiated the Manhattan Project morally responsible for the victims of Hiroshima and of Nagasaki?[40] This is not an easy

[39] Of the many accounts of Galileo's life and struggles, the one which best indicates the moral problems of this sort is J. Brodrick, *Robert Bellarmine, Saint and Scholar* (Burns & Oates, London, 1961), ch. 12.

Galileo twice tried to sell his method of determining longitudes at sea by observation of the eclipses of the moons of Jupiter. The first attempt started in 1616 with Spain, and Galileo even offered to travel there to train practitioners. The second attempt was with the Dutch States-General in 1636, a few years after protracted negotiations with Spain had collapsed. See J. J. Fahie, *Galileo, his Life and Work* (John Murray, London, 1903; reprinted Wm. C. Brown, Dubuque, Iowa, 1967), pp. 173–7 and 372–5. In each case, the prize offered by the foreign State for a successful method overcame any scruples that Galileo might have had in dealing with them: the one arch-reactionary Catholic, the other heretic; and all this during the Thirty Years' War. I am indebted to Mr. Peter Buck, then a research student at Harvard University, for the discussions in which this interpretation of Galileo's moral problems became clear.

[40] To my knowledge, the full history of the decision to bomb Nagasaki as well as Hiroshima has never been told. One theory is offered by Gar Alperowitz: ' ... I think the only way you can understand why Nagasaki was tripped off, automatically, bing-bing, just like that, with no consideration, is this tremendous rush to end the war—*not* just to end the war before an invasion, but *immediately!* ... What was the rush? Well, P. M. S. Blackett, another Nobel prize winner, saw in 1945 that the only way you could explain that immediate, fast one-two punch, was the fact that the Russians were in fact scheduled to enter the war on August 9. And it's in that context, to end the war, not just before an invasion, but bam, like that, that you explain Nagasaki on August 9.' See J. Allen (ed.), *March 4, Scientists, Students, and Society* (M.I.T. Press, 1970), p. 174.

question to answer; but whatever the judgement, it is clearly not one which can be reached by an argument within the traditional working ethics of science.[41]

Thus, first with the Bomb, and more recently through involvement in runaway technology, the world of science has been faced with genuinely moral problems. They concern the responsibility of an individual for the immediate and remote effects of his actions. Such problems are difficult, if not insoluble. The inherited working ethics of science offer no guidance; and the traditional claim of benefit for humanity has, in its realization, produced this darker side. Whatever else will happen to the ideology of science, it can never again claim innocence.

A few decades is a moment in the life of a civilization, and does not even cover the working life of a scientist. The sudden transformation of the social activity of science, rendering its traditional ethics obsolete and its ideology hollow, has caught many a scientist in a personal tragedy. That of Einstein is the most famous; the incidental relation $E = mc^2$, arising from his profound studies in the philosophy of nature, became the magic formula for the sorcery of the atomic bomb, for whose creation he was at least partly responsible. There are doubtless many other such cases: the President of the University of Pennsylvania, formerly a distinguished physicist, found himself in an unpleasant situation when, after repeated denials, he was forced to admit that germ-warfare research was being conducted secretly on his campus. Calling for his resignation, the student newspaper explained that, as an elderly man, he found himself involved in problems whose existence he could not have conceived when, as an old-fashioned scientist, he took office some years previously.[42]

---

[41] According to his account, J. Bronowski conceived his book *Science and Human Values* (Hutchinson, London, 1961) in response to the grotesque experience of the ruins of Nagasaki seen to the accompaniment of American popular music coming from a ship tied up at the harbour. He achieved an eloquent, and indeed classic, statement of the creativity and nobility of the best scientific endeavour. Unfortunately, he was unable to solve the moral problem in this framework; concerning the evil effects of atomic physics, he could say only, 'Science has nothing to be ashamed of even in the ruins of Nasagaki. The shame is theirs who appeal to other values than the human imaginative values which science has evolved' (p. 83). On the question of why these values gave so little guidance to those who stayed with the Manhattan project through to its culmination after the defeat of the Nazis, and who later proceeded to the construction of hydrogen bombs, he is silent. In fairness, he did try to convince official agencies to have Nagasaki preserved in its pristine ruinous state, as a site for international conferences on peace; but the idea was clearly too sensible to be adopted.

[42] See editorial, 'Harnwell—an Old Man', *The Daily Pennsylvanian* (28 April 1967),

For many generations, up to but not including the present, the study of nature has been among the most serene of occupations. To be sure, the work is arduous and even hazardous; but many an eminent scientist turned with relief from the turbulence and faction of politics, commerce, and even institutional religion, to the innocent contemplation of the unchanging and impersonal laws of nature.[43] In less than a generation's time, that haven has been lost. Science is in flux. Many who entered it as a refuge from the intellectual and moral squalor of ordinary society find, in their advancing years, that they are involved in administering just another bureaucratic establishment. They are enmeshed in the demands of society and the State; they must accomplish the administrative and social tasks of getting high-quality craftsmen's work out of a set of manpower-unit employees; and in participating in the leadership of their field, they must cope with the insoluble practical and moral problems which emerge when corruption sets in.

For the industrialization of science has produced another set of moral problems, internal to science but still incapable of solution through a working ethic that was conceived in terms of the search for truth, and organized around the protection of well-defined intellectual property. As the ideal of truth has become obsolete, and the location of intellectual property has shifted, this ethic has lost its relevance. The very naturalness of such conditions as shoddy science and entrepreneurial science ensures that they are connected to good work by a multiplicity of continuous gradations. A scientific entrepreneur may produce a piece of genuinely good work, but then

p. 4. I am indebted to Mrs. E. Brown of Philadelphia for her help in locating this reference.

[43] The testimony of Einstein in this regard has a particular pathos: 'It is quite clear to me that the religious paradise of youth which was thus lost was a first attempt to free myself from the chains of the "merely-personal", from an existence which is dominated by wishes, hopes and primitive feelings. Out yonder there was this huge world, which exists independently of us human beings and which stands before us like a great eternal riddle, at least partially accessible to our inspection and thinking. The contemplation of this world beckoned like a liberation, and *I soon noticed that many a man whom I had learned to esteem and admire had found inner freedom and security in the devoted occupation with it.* The mental grasp of this extra-personal world within the frame of the given possibilities swam as highest aim half consciously and half unconsciously before my mind's eye. Similarly motivated men of the present and of the past, as well as the insights which they had achieved, were the friends which could not be lost. The road to this paradise was not as comfortable and alluring as the road to the religious paradise; but it has proved itself as trustworthy, and I have never regretted having chosen it.' A. Einstein, 'Autobiographical Notes' in *Albert Einstein, Philospher-Scientist*, ed. P. Schilpp (Tudor, New York, 1949, 1951), p. 5 (italics added).

use it for further inflating the stock of his establishment; has he violated the ethics of science? A piece of work may be condemned as shoddy, but may then be defended as the best work a particular man could do in a difficult field; the assessment of quality of scientific results is subtle enough, and who can be sure of the intentions of another? And if a man does not wish to belong to a community of colleagues of the traditional sort, and does not need it for the building of his personal career in science, how can the informal penalties of disapproval be applied against him? Thus there is no escape from moral problems in science, even in the purest of pure science, except perhaps in those enclaves which have not yet been affected by industrialization and its social consequences.

Conclusion

It might seem far-fetched and alarmist to claim that science is in danger of decline and dissolution, through its inability to make a healthy response to its new conditions. The size of the whole enterprise, and the ever increasing number of worthwhile and exciting results which appear, seem to be a patent refutation of any such fears. But the history of natural science in Europe shows that its steady growth over the past centuries has been an aggregate of cycles of growth and decline in different fields and places.[44] Indeed, it would be astonishing if it were not so, for then natural science would be the only sort of creative work exempt from such rhythm.[45] Only

[44] See M. Yuasa, 'Center of Scientific Activity: its Shifts from the 16th to the 20th Century', *Japanese Studies in the History of Science*, 1 (1962), 57–75. The author used several sorts of quantitative indices of 'activity', including the mean age of scientists at any time. They agreed in producing cycles of about eighty years in length for each nation; that for France was at its peak at the time of the French Revolution, that for Germany in the 1870s; that for America seemed near to its peak after the Second World War, and that for Russia was still rising. Detailed historical studies of France and Germany confirm this analysis. For France, see R. Fox, 'The Rejection of Laplacian Physics: a Turning-Point in the History of the Physical Sciences in France', to be published in *Archive for History of the Exact Sciences*; and for Germany, see J. Ben-David and A. Zloczower, 'Universities and Academic Systems in Modern Societies', *European Journal of Sociology*, 3 (1962), 45–85, reprinted in N. Kaplan, *Science and Society*, pp. 62–85.

[45] A striking example of the appearance of a creative generation is given by G. S. Shackle for economics: 'The twelve years from 1840 to 1851 produced Menger (1840–1921), Marshall (1842–1924), Edgeworth (1845–1926), Pareto (1848–1923), Wicksell (1851–1926), Wiesser (1851–1926) and Bohm-Bawerk (1851–1914), seven of the greatest figures of our discipline, all born in virtually one decade, all but one dying in the six years 1921–6. By that last year, which we have taken as the first of our Years of High Theory, the great Victorian cohort had at last withdrawn into antiquity. A fresh start could be made without these giants peering over men's shoulders. Thus need and freedom beckoned.' See *The Years of High Theory* (Cambridge University Press, 1967), p. 296.

if one shares the nineteenth-century faith that the advancement of knowledge is automatically progressive and cumulative, can one believe scientific inquiry to possess the unique combination of features of being highly creative and also perfectly safe. But once the existence of such cycles is recognized as a natural occurrence, the close succession of periods of excellence in different fields and places becomes an entirely contingent process: there is no guarantee that in the later twentieth and twenty-first centuries there will always be some national centre of excellence in science.[46] Only in retrospect can historians begin to explore the subtle combination of factors which led to the inception and completion of a cycle of creativity in any given situation in the past; for the present and the near future, when conditions are so very different from those experienced previously, we really do not know.

Thus in spite of its great achievements, and indeed because of some of them, the world of natural science faces serious problems of an entirely new order. For these social problems, neither the inherited craft wisdom of working scientists, nor the fledgling social sciences, are in a position to provide solutions. Indeed, the social sciences themselves, and sophisticated technology as well, have analogous problems. The solutions, if there be any, cannot be imposed from outside by the fiat of politicians or administrators; scientific inquiry is too complex and delicate to be treated like the production of material commodities. What is required is first understanding, and then, above all else, the intellectual and moral qualities of a new leadership, capable of adapting the best of the heritage of science to the tasks of the present and future. Whether the present conditions are propitious for the emergence of such a leadership will be known only by those who look back from the future.

[46] A very emphatic graphic display of an earlier gap in scientific creativity is given on the back inside cover of A. E. E. McKenzie, *The Major Achievements of Science*, vol. 1 (Cambridge University Press, 1960). He has a bar-chart of life-spans of great scientists; there is a cluster of eight between Pascal (born 1623) and Newton (born 1642); then after Stahl (born 1660) there is nothing until Buffon (1707). Although such selections of names are bound to be arbitrary, the impression created is a correct one; and the decline in recruitment and enthusiasm for experimental philosophy was recognized at the end of the seventeenth century.

Part II

THE ACHIEVEMENT OF
SCIENTIFIC KNOWLEDGE

INTRODUCTION

AT the present time, the deepest problems in the understanding of science are social rather than epistemological. The older problems of the attainment of truth or its substitute, have given way to the concern for the maintenance of the health of science, and for the control of its applications. Until very recently, the two aspects of science, the activity itself and the resulting knowledge, have been studied separately. Both sorts of inquiry have been impoverished by this separation. Analyses of the social behaviour of scientists, and of the external influences on scientific research, have assumed the products of that research to be absolute, and unconditioned by the particular circumstances of their achievement. On the other hand philosophical analyses of the nature of scientific knowledge have either been completely abstracted from the processes of its achievement, or have invoked a model of a working scientist isolated from his environment and traditions. But a proper analysis of the social activity of science must be based on understanding of the very special goals of the scientists' tasks; and an analysis of achieved scientific knowledge must comprehend its character as a social possession, the product of an historical process. An analysis of science which unites these two aspects will be able to resolve the apparent paradoxes in its nature: that out of a personal endeavour which is fallible, subjective, and strictly limited by its context, there emerges knowledge which is certain, objective, and universal. Also, such an analysis should help to explain the failures of scientific inquiry as well as its successes, and could provide guidance for the solution of the social problems of the science of the present and near future.

This part of the book will be devoted to the problem of scientific knowledge. Its argument will be organized around four theses:

1. Scientific inquiry is a craft.
2. The objects of this work are not natural things, but are intellectual constructs, studied through the investigation of problems.

3. The work is guided and controlled by methods which are mainly informal and tacit, rather than public and explicit.
4. The special character of achieved scientific knowledge is explained by the complex social processes of selection and transformation of the results of research.

On the basis of these four theses, it will be possible, in the next part, to discuss some problems of science which arise from its being such a delicate and vulnerable social activity.[1]

In the discussion of these four theses, the central concept will be that of 'problem'; I shall consider science as a special sort of problem-solving activity. This concept cannot be left as a common-sense idea needing no close analysis; the work it does in this argument is too important. As my argument proceeds, I will develop a general schema of the distinct phases in the setting and solving of scientific problems. Indeed, this part could be organized around a formal doctrine of 'scientific problem', but the gain in coherence would be more than offset by the difficulties presented by a battery of unfamiliar terms. For this reason I shall defer the formal definition of 'problem' until it can be meaningfully discussed. For the present, we may think of a scientific problem as analogous to a textbook 'exercise', with the following crucial differences: a major part of the work is the formulation of the question itself; the question changes as the work progresses; there is no simple rule for distinguishing a 'correct' answer from 'incorrect' ones; and there is no guarantee that the question, as originally set or later developed, can be answered at all.

There is one important similarity between a scientific problem and a textbook exercise: the things discussed are not the objects and processes perceived through ordinary experience, but intellectual constructs. This is easily seen in the case of the more mathematical and theoretical sciences; the schoolboy solves problems about 'mass-points' which are considered to have no size, while the theoretical physicist discusses things whose properties can be known only by the most indirect and sophisticated experiments, and whose very existence is sometimes a matter of controversy. I shall argue that

---

[1] A condensed and simplified version of the ideas developed here will be found in the latter part of Course Unit 1 of the Foundation Course in Science of the Open University. I am grateful to the regular members of the Course Team for their criticisms and improvements of the draft there, and only regret that I could not enlist their help in the improvement of this text.

all disciplined inquiry is necessarily concerned with objects of this sort; even in the most 'descriptive' fields, such as ordinary history, the study of particular events is organized in terms of concepts as 'nation', 'class' and 'progress'. I describe these 'objects of inquiry' as 'classes of intellectually constructed things and events', rather than as 'concepts', in order to stress what, for me, is their most important feature. It might be objected that science sometimes discusses unique things apprehended by ordinary experience, such as the moon. But if one reads scientific discussion of the moon, one sees that all the properties discussed are just the sort of intellectual constructs that physics deals with. Indeed, the 'common sense' perception of the moon was transformed by a brilliant exercise in scientific research, using inspired observation and careful, if informal demonstration. Before the appearance of Galileo's *Sidereus Nuncius* of 1610, everyone knew that the moon is a perfect sphere, with some unexplained markings. With his telescope, Galileo observed that the moving boundary between the light and dark parts is not regular, but broken; and there were changing patterns of light and dark spots on the dark and light parts respectively. He interpreted these creatively (and dangerously) as shadows, and produced a geometrical argument to show that they are cast by mountains not impossibly high.

My discussion of the successive theses will run roughly parallel to the description of the different phases of the investigation of scientific problems. For the craft character of the work is seen most clearly in the earlier phases, where there is an interaction with the external world; while the artificiality of the objects of inquiry is most obvious when we consider what is involved in 'solving' a scientific problem; the social character of scientific inquiry is exhibited in the methods that guide and control scientific work; while its dependence on social processes operating over time, is shown in the transformation of 'research reports' of solved problems into accepted 'facts' and ultimately to genuine scientific knowledge.[2]

[2] In the philosophical literature conceiving science as inquiry rather than as validation of accomplished knowledge, a pioneering effort which deserves to be remembered is F. C. S. Schiller, 'Scientific Discovery and Logical Proof', in C. Singer, *Studies in the History and Method of Science* (Oxford, 1917). For the recent studies in this direction, the work of T. Kuhn, *The Structure of Scientific Revolutions* (University of Chicago Press, 1962) has, as it were, created the new paradigm which we all follow. Complementary to Kuhn is J. Ziman, *Public Knowledge* (Cambridge University Press, 1968), in which the social aspects of scientific work are given greater emphasis. The concept of

In the early parts of this discussion, I will show in great detail how the pursuit of scientific knowledge is, in the short run, very subjective and fallible. It may even appear that I am arguing towards a completely sceptical position, with the practical conclusion that scientists should only attempt to 'solve problems' and not concern themselves with the goal of achieving knowledge. This is quite contrary to my intention; I start with the historical fact that genuine scientific knowledge can be achieved, and does exist. The problem is then to see how this occurs, given the common experience of the uncertainty of scientific research. It is necessary for me to elaborate on the sources of uncertainty, both for the correction of the bias implicit in the philosophy of science dominant hitherto, and for the provision of materials for the solution of the main problem. This solution emerges at the end of the section; it is that the complex social processes of the testing and transformation of the results of research, working over an extended time, create both genuine scientific knowledge and also the circumstances in which it can be recognized. The paradoxical properties of this knowledge (that it appears in a variety of different versions, and contains hidden obscurities) can then be appreciated and explained.

'problem' is studied at length in J. S. Sharikow, 'Das wissenschaftliche Problem', in *Logik der wissenschaftlichen Forschung*, eds. P. W. Kopnin and M. W. Popowitsch (Akademie-Verlag, Berlin, 1969; Russian original, Moscow, 1965), ch. 1. His analysis and mine are remarkably similar on fundamentals, although the materials of my chs. 3, 4 and 5 diverge from his approach. Mario Bunge, *Scientific Research I : The Search for System* (Springer-Verlag, 1967), has a discussion of 'problem' in ch. 4 and offers many special insights.

3

SCIENCE AS CRAFTSMAN'S WORK

To anyone with experience of the 'art' of scientific inquiry, this thesis may seem so obvious as to be banal.[1] Yet this feature of science has generally been ignored in philosophical discussion, even in those that try to take into account the work by which scientific knowledge is achieved.[2] Yet without an appreciation of the craft character of scientific work there is no possibility of resolving the paradox of the radical difference between the subjective, intensely personal activity of creative science, and the objective, impersonal knowledge which results from it. When we think of material objects produced by handicraft rather than by mass-production, we easily appreciate the distinctive features of this sort of work. The craftsman works with particular objects; he must know their properties in all their particularity; and his knowledge of them cannot be specified in a formal account. Indeed, no explicit description of a craftsman's techniques, and of the objects on which he works, can be more than the simplest elements of the subject. They can be useful for the beginner, but he must develop a personal, tacit knowledge of his objects and what he can do with them, if he is to produce good work. Indeed, much of his technique may not even have the character of

[1] On the craftsmanship of science, see W. I. B. Beveridge, *The Art of Scientific Investigation* (Heinemann, London, 1950), and Paul Freedman, *The Principles of Scientific Research* (Pergamon Press, 1960). An account which provides a wealth of practical example and aphorisms is M. Hamilton, *Lectures on the Methodology of Clinical Research* (E. & S. Livingstone, Edinburgh and London, 1961).

[2] I am indebted to Michael Polanyi for his systematic development of the insight that science is craft work; see *Personal Knowledge* (Routledge & Kegan Paul, London, 1958). Much of this present work derives from an attempt to solve the problems which were raised by his analysis; in particular, how objective scientific knowledge can result from the intensely personal and fallible endeavour of creative scientific inquiry. Also, reflection on D. S. L. Cardwell's distinction between 'craft' and 'applied science' (exemplified in pilotage and navigation, respectively) helped me crystallize my own ideas on the craft aspects of scientific work.

conscious knowledge; by experience, his hands and eyes have taught themselves. It is this subtle interaction of the craftsman with his material, producing slightly different copies of the same general model, which gives handicraft productions their special charm.

At first sight, nothing could seem further from such productions than achieved scientific knowledge. Those results which reach a wide audience bear few marks of the individual hand and mind which brought them into being. Indeed, those which have become 'classic', have an appearance of such simple necessity, that it is frequently hard to imagine that their achievement required the exertions of a great talent. Working from such materials, it is very difficult for the historian to argue convincingly that their production was conditioned by such contingent factors as the social or intellectual environment of their creator, or by peculiarities of his personality. This apparent paradox will be fully resolved only through an analysis of the transformation of scientific results as they progress towards the state of being permanent knowledge. But the interplay of personal and social aspects of the work begins with the birth of the problem itself and is continuous through the inquiry. I will defer the discussion of the origins of problems for the next chapter; for the present I will show how the materials from which the solution is eventually constructed are produced and transformed by operations which require a craftsman's knowledge of the work.

Data

As a first example, let us consider an experiment in which quantitative readings are taken from a piece of experimental apparatus. Now, one of the things that every schoolboy knows about science is a general property of experimental equipment, which has been given the name of the 'fourth law of thermodynamics': no experiment goes properly the first time. The schoolboy knows that the equipment is not functioning properly, from its failure to produce readings of the required sort. If by some miracle it performs to expectations on the very first run, he can happily record the results in his laboratory notebook, and go home. For him, the situation is quite straightforward: everyone knows what result should be achieved; until it comes, the apparatus is not working properly; and when it does, the apparatus is working properly.

This simple and secure rule of assessment is not available when new work is being done on a piece of experimental apparatus. How, indeed, does a scientist decide that his apparatus is working 'properly'? Of course, hardly any experiment is completely new, and so the scientist will always have an idea of what to expect. If nothing else, he can expect a set of readings which will produce some sort of regular pattern; and if they gyrate wildly, then he does not need much craftsman's knowledge to realize that the equipment is not yet functioning properly. But when to say that it is functioning 'well enough' is a more subtle matter. It is clear that the more stable and consistent the readings, the more likely they are to be sound. But anomalous readings always do crop up; and if one waited for them to vanish entirely, or tried to 'explain' each and every one of them, one would never get beyond this first stage of the work. In short, the scientist must be a craftsman with respect to his apparatus; and his judgement of when it is working 'well enough' must be based on his experience of that particular piece of equipment, in all its particularity.[3]

In this work with pieces of physical equipment, the scientist is a very special sort of craftsman, for the objects he is dealing with are highly artificial. The relation of the readings taken off the apparatus to the objects of his inquiry is not at all immediate: the establishment of their relevance requires another set of operations. The experimental apparatus itself is frequently a complex affair, designed for the production of very special effects. A purely 'practical' mastery of the technique of standard manipulations will be sufficient only for the most routine work. Without a deeper knowledge of the operation of the apparatus, the scientist may well fall into the first pitfall of experimental research, that of too easily accepting readings which are stable for reports which are sound. For the readings taken off the machine are only a measure of the response of the output mechanism to certain aspects of the process going on inside. At every stage of what may be a lengthy sequence, the apparatus itself will modify the signals coming through, and so introduce a 'systematic error' into the readings. The scientist must know by experience what size of error is 'negligible'; he must be able to reduce the recognized errors to such a size; and he must be sensitive to clues announcing the presence of stable

[3] There is a lengthy discussion of 'Bringing an Apparatus under Control' in E. Bright Wilson, Jr., *An Introduction to Scientific Research* (McGraw-Hill, 1952), pp. 137–40; but there is no mention of the criteria for judging that the apparatus *is* under control.

non-negligible errors from other sources. For this sort of work, the scientist must be a master of the operations of the apparatus. Such a mastery comes partly from craft knowledge of the traditional sort, where the experience of others is transmitted by precept and imitation; but it also involves some explicit and formal scientific knowledge in which the 'theory' of the apparatus is set out. Because of this formal component of his knowledge of his apparatus, the scientist must be a 'technician' in this respect rather than a 'craftsman' of the traditional sort.

It is clear that the set of quantitative readings taken off the experimental apparatus cannot be considered independently of the interpretation put on them; the scientist cannot even deal with the systematic errors of the apparatus unless he has some idea of what the readings represent, as reports of properties of particular things and events inside the apparatus. It is such reports, rather than the bare set of numbers, that are the objects of this part of the work; we may call them the 'data'. We shall soon see that there must be a further series of operations on these data to yield materials suitable for inclusion in the argument of the solved problem, and the production of the data must be accomplished with these subsequent operations in mind. But the data are a record of the point of contact with the external world, and as such constitute the foundation in experience of scientific knowledge.

A different set of craft skills are involved in the production of data in fields which are 'descriptive' rather than 'experimental'. In these, we may say that the data are 'found' rather than 'manufactured', for the things and events whose properties are reported are not studied in artificially pure, simple, and stable situations, but are instead a sample taken from an existing population. Data in such fields cannot enjoy the precision of that which comes from experiments, for the interaction with the external world cannot be so closely controlled. Accordingly, such data cannot serve as the foundation for evidence in a highly abstract and subtle argument, and so the properties which are reported will tend to be more similar to those accessible to common observation. This does not obviate the need for craft skill in making reports, for there are pitfalls in these data as in the other sort. For the process of sampling is governed by the circumstances of the scientist and the tools at his disposal, and can quite easily yield specimens which are unrepresentative of their population in some important aspect.[4] Also, any scientific description of a property is necessarily abbreviated and stylized; and the existing categories of description will never fit perfectly the objects of the inquiry at hand.[5] Hence here, as in experimental fields it is necessary for the scientist to have a craftsman's mastery of his materials and tools, and to be able to judge when a report is 'sufficiently sound' for the data-collecting to be concluded, and the refinement of the data into information to commence.

It might appear at first that in theoretical or purely mathematical fields, where there is no interaction with the external world at this early stage, the category of 'data' could not apply. But mathematicians will also speak of 'experiments' and 'inductions' in the early stages of working up a proof.[6] In this case the objects are

[4] In fields where the data are found, the problem of coping with 'outlying' data is particularly severe, for one can only rarely repeat the process under observation. One might try to use statistical tests to estimate the significance of such data, deciding whether they are indications of a regular cause, or merely freaks; but all such tests depend on hypotheses about the universe from which the data are drawn, and so are less likely to be genuinely applicable in this extreme case. The extent to which such outlying data occur in experimental and observational work is likely to be underestimated by those unfamiliar with such work, since it is frequently suppressed. A most striking example of such suppression (in the teaching literature) is the classic study on deaths by horse-kicks in German army corps, which is the paradigm case for the Poisson distribution, describing randomly occurring events. The original study of horse-kick deaths by von Bortkewitsch extended over fourteen army corps, but the data from four of these was rejected as being anomalously high, and the data from the remaining ten was shown to fit very well the theoretical numbers. A note of this suppression remains in some of the literature; see J. S. Coleman, *Introduction to Mathematical Sociology* (Collier-MacMillan: The Free Press, Glencoe and London, 1964), p. 291. But this is the exception; there is no hint that the horse-kick deaths data is other than 'raw' in R. A. Fisher, *Statistical Methods for Research Workers*, 13th ed. (Oliver & Boyd, Edinburgh and London, 1955), p. 55.

[5] For an example of the subtleties of 'descriptive' sciences, one may consider the sorts of problems encountered in constructing a bibliography. The first decision is on the 'field'; and even here the existing conventions must always be modified by the particular circumstances or even because of the passage of time. The taxonomy of classes is an obvious source of trouble. Then, should it be 'comprehensive' or 'critical'? In pure versions of the former, it will be choked with obsolete and inferior material and is likely to be an unmanageably large project; in the latter it will be 'subjective', incomplete and possibly misleading unless the principles are clear. When works are cited, should *all* editions up to the present be mentioned; and, if not, what principle of selection should be used? Even on the apparently trivial question of the alphabet, there are decisions to be made on such special cases as 'Mc' and the German umlauts. There is one large university library where the German 'ü' is treated as 'ue'; and the French name 'Haüy' is filed as if it were 'Hauey'! In practice, unless there is some prior decision on the *function* of the information in a bibliography, these problems are resolved (if at all) only as the bibliographer becomes aware of them singly, and the result is likely to be a mass of data which gives satisfaction in no aspects whatever.

[6] See G. Polya, *Mathematics and Plausible Reasoning*, vol. i (Oxford University Press and Princeton University Press, 1954): *Induction and Analogy in Mathematics*.

purely mental constructs; but they will be more simple or more familiar representatives of the general classes with which a mathematical proof is usually concerned. An exploration of the properties of these samples, in relation to the partly-formed conjecture with which the mathematical investigation begins, gives the mathematician some idea of the content of the result he should try to establish, and also hints towards the structure of its proof. Indeed, in many cases the proof emerges from a repeatedly refined and generalized description of the properties of the sample objects.

A craftsman's intuitive knowledge of these abstract objects, derived from experience in manipulating with them in a variety of contexts, is essential for such 'data' to be a sound starting-point for the detailed investigation. Without a sense of the generality implicit in his sample objects, and without a sense of the possibility of refining his descriptions into rigorously argued components of a complete proof, the intending mathematician will wallow in a morass of ill-formed speculations and partly solved conjectures, unable to develop towards a tight, coherent, and general argument. Here too there are pitfalls in the work, which can be avoided only by the craftsman's intuitive knowledge of his objects. For the work of constructing a mathematical argument cannot proceed other than by the testing and development of a rough conjecture, based on this special sort of experience. At a certain point the mathematician must decide when this 'data' is sufficiently sound to justify the investment of time and intensive effort for its refinement. There can be no 'proof' that the sample objects are truly representative, and that the descriptions of their properties reflect the structure of the general classes. Moreover, every sketch argument involves the use of ancillary and special results, which seem either trivially true or easy to prove. Hence it can easily occur that plausible objects, with plausible properties, and with obvious 'lemmas' supporting the argument, turn out to be unrelated to the objects with which the intended proof is concerned, or indeed to any significant class of mathematical objects whatever; but this can be discovered only by disheartening experience, when the process of constructing a proof fails to 'converge', and the mathematician is left with a débris of insignificant special examples scattered among broken-backed arguments.[7]

---

[7] Examples of this phenomenon are not easy to find in the worthwhile published literature; but I can cite my own experience as a beginning research student, and that of friends.

Thus in any sort of scientific inquiry we may use the idea of 'data' for the results of the first working-up of the materials in the investigation of a problem. I have stressed the importance of craft skills in the production of data which is sound of itself, and useful for the later stages of the work on the problem. Without such skills, the scientist will blunder into pitfalls and produce reports without soundness or significance. The craft character of the production of data is of some philosophical significance, for it is in this phase that the scientist makes new contact with the external world, or achieves a new organization of his conceptual objects. The directions and fortunes of the later stages of the work will depend on the results of this work. Yet we have seen that no set of data can be 'perfect' as a report of properties of the objects of investigation, nor can it be independent of the plans and expectations for the later stages of the work. If such reports are truly the foundations of scientific knowledge in experience of the external world, then those foundations would seem to be peculiarly insecure and complicated. And so it is; the wonder is not that our scientific knowledge is an imperfect and fallible picture of the external world, but rather that it exists at all. In the light of the history of human inquiry into the natural world, we see how difficult it is for such knowledge to come to be. I shall later show that when such knowledge does exist, it is achieved by a complex social endeavour, and derives from the work of many craftsmen in their very special interaction with the world of nature.

We have already had an intimation of the social character of the activity of science, in the judgement made by the scientist in deciding whether a batch of data is 'sound'. For the standard he applies will derive partly from his own experience, and partly from that of his teachers and colleagues. Since perfection of data is impossible, the standard will be based on a common judgement of what is good enough for the functions which the data performs in problems of that sort. This judgement depends in turn on the criteria of adequacy imposed on the solution of such problems. The solution cannot yield necessary truths or indisputable facts, but only results which in their turn are judged acceptable (or not) by the community in accordance with criteria developed through its social experience. To be sure, some of these results may, either themselves or their descendants, come to be classed as 'facts' or even 'knowledge'. But Nature is not so obliging as ever to give marks of True or False for scientific work, and so in the last resort a scientific community must set its standards for itself. The sets of such standards over the different fields are very diverse in their contents and rigour. The self-discipline of scientists which they require, and the quality of the work they ensure, depend on the strength, health and integrity of the community involved, and are most sensitive to the leadership it receives.[8] Thus the simple judgement of 'soundness' of data is a microcosm of the complex of accumulated social experience and judgements which go into scientific endeavour. This may seem a fanciful extrapolation from a simple routine test; but it

[8] A classic statement of the exacting skills of experimental research is provided by Helmholtz: 'At one time, we have to study the errors of our instruments, with a view to their diminution, or, where they cannot be removed, to compass their detrimental influence; while at other times we have to watch for the moment when an organism presents itself under circumstances most favourable for research. Again, in the course of our investigation we learn for the first time of possible errors which vitiate the result, or perhaps merely raise a suspicion that it may be vitiated, and we find ourselves compelled to begin the work anew, till every shadow of doubt is removed. And it is only when the observer takes such a grip of the subject, so fixes all his thoughts and all his interest upon it that he cannot separate himself from it for weeks, for months, even for years, cannot force himself away from it, in short, till he has mastered every detail, and feels assured of all those results which must come in time, that a perfect and valuable piece of work is done. You are all aware that in every good research, the preparation, the secondary operations, the control of possible errors, and especially in the separation of the results attainable in the time from those that cannot be attained, consume far more time than is really required to make actual observations or experiments. How much more ingenuity and thought are expended in bringing a refractory piece of brass or glass into subjection, than in sketching out the plan of the whole investigation! Each of you will have experienced such impatience and over-excitement during work where all the thoughts are directed on a narrow range of questions, the import of which to an outsider appears trifling and contemptible because he does not see the end to which the preparatory work tends. I believe I am correct in thus describing the work and mental condition that precedes all those great results which hastened so much the development of science after its long inaction, and gave it so powerful an influence over every phase of human life. ... In addition, however, to the knowledge which the student of science acquires from lectures and books, he requires intelligence, which only an ample and diligent perception can give him; he needs skill, which come only by repeated experiment and long practice. His senses must be sharpened for certain kind of observation, to detect minute differences of form, colour, solidity, smell, etc., in the object under examination; his hand must be equally trained to the work of the blacksmith, the locksmith, and the carpenter, or the draughtsman and the violin-player, and, when operating with the microscope, must surpass the lace-maker in delicacy of handling the needle. Moreover, when he encounters superior destructive forces, or performs bloody operations upon man or beast, he must possess the courage and coolness of the soldier'. See H. von Helmholtz, 'The Aim and Progress of Physical Science' (1869), in *Popular Lectures on Scientific Subjects*, 1st series (London and New York, 1893), pp. 319–48; 320–1. It might be thought that Helmholtz overdramatizes somewhat; and indeed, for routine research in safe fields, the work is not so demanding. But as we will show in the course of this study, his list of tasks and skills is not exaggerated for any genuinely original inquiry.

reflects the daily practice of science, where one man will cheerfully use data which another, working in the same field but in a different school, will reject as nearly worthless.

Information

It is impossible to separate off the different phases of the investigation of a problem into discrete and independent units. Even in the production of data, the later stages of work on the problem are present in an embryonic form, as expectations on the character of the data and its refined products, and tentative plans for the later operations. Throughout the course of the investigation of a problem, there is a continuous re-cycling, so that the problem itself evolves, or perhaps is destroyed, through the interaction of the materials which it brings into being. But there are natural points of transition, where the satisfactory accomplishment of the tasks at one stage makes possible the investment of resources in the concentrated, systematic work of the next stage.

The natural successor to the production of data is its refinement into a more reliable and useful form. For the reports on the properties of particular things and events, however sound, are still far from being 'facts'. Their transformation involves a new set of craft skills, with the application of new tools, and the making of a new set of judgements. We can illustrate this next phase of the work, the production of 'information', with our earlier example of readings taken off a piece of experimental apparatus. What we do may in practice be quite simple, but in principle it is a momentous step. Most commonly, we plot the readings on a graph, and fit a curve to them. This step is crucial for later work, for if we are satisfied with the fit, we thenceforth ignore the particular readings (except perhaps to cite them as special evidence in the published argument), and consider the curve as the report of the properties which concern us. Thus the data has been transformed into a new sort of material, with the aid of certain mathematical tools. This transformation involves two separate judgements, which can be assisted by special tools, but which both involve the risk of pitfalls. The first is that the points fit sufficiently well to a pattern, and the second is that this particular pattern, rather than any other, is the significant one. Even the schoolboy knows of the first requirement; if the points representing his readings are spattered all over the graph, then he cannot simply draw the required curve in the likeliest place and call it a day. How well is 'well enough' is a matter of judgement, frequently assisted by statistical tools, but in the last resort depending on an assessment of the risks and costs of errors in the particular work at hand.[9] The judgement of whether a curve is of 'the right sort' is more crucial and more difficult. The curve which is fitted to the points on the graph amounts to a statement of the functional relationship between the variables of which particular values are recorded in the experiment. Very sophisticated methods have been developed for assisting in this judgement, but in the last resort it depends on the craft knowledge of the scientist, to decide which sort of functional relation is represented by the discrete set of points obtained from his readings.[10]

If the fitted curve is of the right sort, and the points cluster around it well enough, we may say that the information is 'reliable'. But this is not the only test it must pass. For the statements carried

[9] A good example of the possibility of encountering pitfalls in even the simplest aggregation of data is given in A. H. Robinson and R. D. Sale, *Elements of Cartography*, 3rd ed. (John Wiley & Sons, 1969), p. 162. The exercise is to draw 'isopleths' (curves connecting domains of equal value) on a square of thirty-six unit squares; each row has values increasing by one from the initial value, and the first column reads downwards: 123321. The isopleths are v-shaped, and retain this shape if the unit squares are aggregated by threes, horizontally. But if they are aggregated by threes vertically, the isopleths become straight vertical lines!

[10] The use of numerical calculations for curve-fitting and estimation of 'significance' can produce numbers nearly automatically, but the need for exercising judgement is not thereby obviated. For each such technique is in fact a mathematical argument, leading to certain conclusions on the basis of certain assumptions on the behaviour of the data. The appropriateness of the particular assumptions to the data at hand cannot be assumed as given; and it has even been argued that the 'Gaussian law of errors' is totally inappropriate to experimental data, since it is a continuous distribution and one which gives a positive probability to errors of arbitrarily high magnitude. See N. R. Campbell, *An Account of the Principles of Measurement and Calculation* (Longmans, Green, 1928), ch. X, pp. 149–84. He quotes an explanation for the widespread acceptance of the Gaussian law: mathematicians have accepted the law because they thought it established by experiments, experimenters because they thought it established by mathematics (p. 182). These considerations are not philosophical worries remote from practice; for the numerical results obtained from a calculation of only moderate complexity can depend strongly on the variant of 'least squares' technique adopted. See K. Holden, 'The Effect of Revisions to Data on Two Econometric Studies', *The Manchester School* (March 1969), pp. 23–37. The author used two estimation procedures (ordinary least squares, and 'limited information maximum likelihood two-stage least squares'), on two sets of data for a five-equation model. It turned out that the parameters computed by the more sophisticated method were more sensitive to changes in the data; and also 'The variances due to the estimation method are greater than those due to the data revisions, which indicates that the choice of estimating procedure has more effect on the parameter estimates than the choice of data' (p. 28). (I am indebted to Dr. A. Coddington for this reference.)

forward in the work are not merely about the functional relation implicit in the readings of the output mechanism of the equipment. Rather, they are concerned with the properties of the classes of things and events, of which some samples were the objects of the experiments. These classes are inaccessible to direct view and testing; it is only through the theory in whose terms the experiment is conducted and interpreted, that these properties can be inferred. It is quite possible for the experimental apparatus to produce sound data, from which the systematic errors of the equipment have been eliminated, and to which a good curve can be well fitted, and yet for the interpretation of the results to be quite erroneous. For if at some crucial point the actual workings of the experiment are different from what is described in the theory, the properties deduced for the classes of things and events will be quite different from those which actually hold. Since it is impossible to be certain that the theory is perfectly relevant to what is going on, there can be no certainty of the 'relevance' of the information to the work in hand. To be satisfied that one's descriptions are really those of the inaccessible classes of things and events which are being approached through the experiment, is yet another judgement. This is even less capable of being reduced to a routine exercise that the judgement of reliability, and so requires an even more refined craftsman's skill.[11]

Thus the data which is first produced by an experiment must be transformed, and tested, in order to produce genuine information; and all these are craft operations. The situation is no different in a descriptive science, where the data is found rather than manufactured. There the transition from data to information is even more hazardous. For the data does not arise from a process which is designed for the production of a particular effect, and which is at least partially under control and reproducible at will. Rather, it

[11] There seems to be a widespread belief that in a matured, quantitative, experimental science such as physics, the problems of achieving reliable information are so much less severe than elsewhere that they can be considered as negligible. But the difference is only one of degree, not of kind. Perhaps the most significant single experience in my undergraduate education in physics and in the philosophy of science was an accidental discovery of a table of values of fundamental constants, in J. D. Stranathan, *The 'Particles' of Modern Physics* (Blakeston, Philadelphia, 1942), where I saw each particular quantity given several values, which differed among themselves by far more than the 'error' assigned to any of them. I was fortunate in having been prepared for this surprise by a previous initiation into the pitfalls of physical experiments in my laboratory courses; but even so the ground swayed slightly under my feet.

reports the properties of a sample of things and events which have been produced 'naturally', with all their variability and irrelevant features, and where the processes of sampling and observing may have induced further variations and distortions.[12] If the pattern to be detected is not a simple quantitative one, then the tests for reliability cannot use standard tools, but must rely more heavily on personal judgement. Similarly, the judgements of relevance are even more hazardous than in the case of experimental fields; for the lack of control over the materials described in the data, and the inferior precision and elaboration of the arguments, inevitably reduce the rigour of the testing of the relevance of any piece of proffered information. The distinction between reliability and relevance in descriptive fields can be seen from the example of the testing of individuals for psychological and educational purposes. One set of skills and controls are involved in constructing tests that are 'reliable', in that a subject will score nearly the same aggregate marks every time. But what is called the 'validity' of the test, the relevance of the numerical score to the attribute which is the object of assessment (be it 'intelligence', aptitude, or even mastery of a particular academic subject) is quite another matter.

The category of 'information', as the refined product of the data, can also be useful in the analysis of purely theoretical or mathematical sciences. For this corresponds to the set of arguments which are sufficiently elaborated to be recognizable as components of a completed proof. They must still be checked for their internal

[12] The hazards of the transition from data to information in descriptive sciences is shown particularly well by what might be called the Case of the Bulgarian Pigs, described in O. Morgenstern, *On the Accuracy of Economic Observations*, 1st ed. (Oxford University Press and Princeton University Press, 1950), p. 19. The data are the number of pigs in Bulgaria on 1 January 1910 and 1920; these were 527,311 and 1,089,699 respectively. The information derived from this pair of figures was that the number had more than doubled during the decade. Unfortunately, the inference was fallacious; what had happened in the interim was that the calendar had been changed from the Julian to the Gregorian, so that the peasants celebrated Christmas in January (new style) just as the Soviet Union celebrates the 'October Revolution' in November. Just half the pig population is slaughtered immediately before Christmas; and so the extra pigs on 1 January 1920 were merely awaiting their Yuletide extinction.

Similar pitfalls beset the more sophisticated transition from data to information in history. In 'The Elizabethan Aristocracy, an Anatomy Anatomized', *Economic History Review*, 2nd series, 3 (1951), 279–98, H. R. Trevor-Roper discusses the use of the sums of money quoted in documents known as 'bonds', as information about the size of a loan. He argues that such documents were far more likely to be a record of the *security* for the loan, which would be at least twice as great as the loan itself. The conclusions about the 'impoverishment' of the Elizabethan aristocracy are modified accordingly.

coherence, lest there be hidden flaws in them; this corresponds to the reliability of the information in sciences based on experience. Also, since each argument establishes a property of a particular class of conceptual objects, then the mutual relevance of the different arguments must be checked, lest the overlap between the classes discussed in the various arguments be too small to provide a significant result. The set of such arguments does not yet constitute a proof, for the different components must still be welded together into a coherent argument, and in the course of this work their internal reliability and mutual relevance must be tested again.

Even in fields based on experience, the craft work of achieving scientific knowledge is not completed with the production of information. For the properties of the objects of scientific inquiry are not fully established by a mere list of reports. Rather, an argument of some degree of elaboration is necessary for conclusions to be drawn about these objects. In this argument the information is put to use as evidence; and for this function it is subjected to further refinement and testing. The elaboration of the argument is of course a matter of degree; in some descriptive fields, the necessary argument may be quite rudimentary, and a collection of information can constitute a worthwhile piece of work. I shall discuss these later phases of the investigation of scientific problems in the next chapter, whose theme will be the artificiality of the objects of scientific inquiry. For the remainder of this chapter, I shall analyse some other aspects of scientific work where its craft character is clearly exhibited, and is of importance for an understanding of the social aspects of the work.

Only a part of the information used in a problem will derive from the data produced in its investigation. The rest is taken over from other work, either from individual research reports or from collections of standard materials.[13] Although a particular bit of this material may be generally accepted as solid fact, it must still be considered as no more than possible information; for its reliability

[13] The proper handling of the stock of 'existing information' relevant to a problem is a craft skill whose most elementary part is 'information retrieval' techniques. On this task, the scientist has no substitute for his judgement and experience, since the establishment of those classification categories which will produce 'relevant' materials, the decisions on where to search and where not to search, and the quick assessment of the quality of the materials examined, are all very subtle tasks. A common pitfall is to require 'all' the information before engaging on the work, to avoid errors and duplication; this is precisely analogous to designing an aeroplane that cannot crash—one that cannot fly.

and relevance to the work in hand can never be taken for granted. If this 'existing information' relates to things which are remote from the problem in which it is used, then it must generally be accompanied by an explanation. This will not be designed to impart a deep understanding of the material, but only to enable the user to handle it competently and to avoid the most common pitfalls. Of the stock of established facts in a field, only a small part of that which survives in general accounts and teaching are results interesting in themselves. Generally, a fact is preserved from oblivion only when it is useful as existing information, performing a function in new work analogous to that of a tool. We shall later see that the patterns of evolution of established facts are conditioned by this function; and indeed give rise to the special properties of permanent scientific knowledge.

Tools

In the production of experimental data, the apparatus is easily recognized as a special sort of tool, similar in its functions to the tools of a handicraftsman. They are not the objects of the work, but are the means by which those objects are created and shaped. The tools of the scientist are not all physical equipment; in the refinement of data into information, other tools are brought into use, most noticeably statistics and other mathematical representations. The physical and intellectual tools of scientific work vary in their complexity and sophistication; some require no special skill or training for their use, while for others the scientist must have a craftsman's knowledge of their powers, limits and possible pitfalls. Every field has its own special tools, of which some require an extensive technical knowledge, supplemented by craft experience, for their use. It is for this reason that a lengthy apprenticeship is necessary before anyone can embark on independent work in a developed scientific discipline. We shall see how tools condition the work which is done in any field of science, and also enter into the social relations between different disciplines.

We can make a rough classification of the different sorts of materials which can be considered as tools, starting with the physical apparatus by which data is produced. Following on this are the tools which actually transform data into the shape appropriate for information and test it. Such data-processing tools can be statistical techniques, or machinery which is programmed to exhibit patterns.

There is a third class of tools, which do not actually produce or transform data, but which make possible the interpretation of the data or of the information. One example of this class is a corpus of standard information about the objects of the inquiry, assembled in handbook form. In sciences dealing with large collections of different objects, noticeably the descriptive sciences, the need for such tools is particularly strong, and there will develop entire sub-disciplines devoted to their production. From this corpus of standard information, materials may also be taken for inclusion as evidence (after the appropriate tests) in the argument of the problem. In this same class are 'tool-subjects', other fields of science which must be mastered to some extent in order that competent work can be done in the given field. The relation of being a tool-subject to another field, which links many pairs of fields, is in general asymmetric; those fields which deal with the more abstract and general properties of matter function as tool-subjects to those less so. This asymmetry is of great significance for the development of science over the last few centuries, and will be examined more closely.

Finally, there is another important class of tools, of a significantly different character. These are required when the language of the argument of problems in a field is not merely the vernacular enriched by technical terms, but when it is a formal, artificial language of its own. The most common example of this is the use of the objects of some fields of mathematics (as the calculus) in theoretical sciences. Such formal languages are not merely a translation of the vernacular. They provide power to the arguments, making possible inferences which could never be made in the vernacular; and also impose a precision (not necessarily accuracy) on the objects of the arguments. Also, the potentialities of the formal language for the construction of arguments are revealed by continuing research in the formal system as an object in its own right by specialists; and the avoidance of pitfalls in its use is learned partly by craft experience but also by deeper knowledge of the mathematical system. Hence the scientist who uses some branch of mathematics as a language-tool must render himself competent in its practice and theory.

Although tools are auxiliary to the advancement of scientific knowledge, their influence on the directions of work done is important and frequently decisive. New tools make possible the production of entirely new sorts of data and information. As the tools of a field develop, projects which previously required outstanding talent or great perseverance, now become routine and nearly trivial. This effect can be seen even by students, who may first study physics without the calculus, and later discover how the laborious and roundabout methods of describing and deriving various effects are rendered unnecessary by this powerful tool of argument. Also, in fields where the data comes from experience, it must be produced in a certain general form, so that the appropriate tools can be applied to it. This is most obvious in the case of statistical tools; many are the cases where a consulting statistician throws back a set of data to a surprised and somewhat hurt scientist, telling him to discard that lot and bring back another set of a better statistical design. In this requirement on the data we see the first of the reasons why one cannot simply go out and get data as an independent foundation for later work. Finally, each set of tools has its characteristic pitfalls, and if blunders and disasters are to be avoided, the user must develop a craftsman's knowledge of their properties. The abuses of statistics are again the most notorious example of this;[14] but even in descriptive sciences, the uncritical use of handbook information, which is always incomplete, obsolescent and not quite fitted to the needs of the work at hand, can lead to the most astonishing blunders.

A consideration of tools also leads us directly into some important social aspects of scientific activity. Because so many of the essential tools for any field of science are so highly sophisticated, to achieve complete mastery in the use of some of them involves becoming a specialist in the tool rather than in the field to which it is being applied. There is thus a natural division of labour between tool-experts and their clients; and the tool-experts are not merely individuals serving as auxiliaries to the clients in the work, but themselves can form a self-contained speciality, a tool-providing field. When two craftsmen with different skills are involved in the same project, they will inevitably see the work from different points of view. The different approaches will be complementary, and can correct and enrich each other; but they can also be the occasion of conflict. For each of the partners may be wanting to get something slightly but significantly different from the project. The client wants data or information, reliable by the accepted standards

[14] The classic primer in the abuse of statistics is Darrell Huff, *How to Lie with Statistics* (Norton, New York, 1954; Gollancz, 1958).

of adequacy of his field, with a minimum of expense and delay; while the tool-expert, unless he is completely subservient, will be looking for opportunities for developing particular tools in which he is interested The two parties even perceive the situation differently, for their different interests correspond to different bodies of craft knowledge; and unless both parties enter the relation with considerable mutual comprehension and respect, only their respective incompetences will be communicated, and conflict will ensue. How such conflicts work out depend on the social relations, and relative strengths, of the clients and the tool-experts, individually and as recognized collectives. It must not be thought that the tool-experts' function as auxiliaries keeps them in an inferior position; they are not merely essential for the work, but they may frequently have command of technicalities which are incomprehensible to the clients, and which also bear prestige in themselves. This is most noticeable where the tools are mathematical in character.[15]

Not all tools are so specialized as to require a distinct corps of experts for their use. In such cases we may speak of tool-users, not

[15] The development of electronic computers could be the subject of a useful case-study in the history of modern technology, if attention is paid to the different purposes of the users, the experts, and also the manufacturers. Although the development of the operating characteristics of these machines can be seen as a steady improvement with occasional forward leaps, their history from the users' point of view can be seen as a series of partially resolved crises. On this, see C. Strachey, 'Systems Analysis and Programming', *Scientific American*, 215, No. 3 (September 1966), 112–26. The divergence between the purposes of users and experts can be seen in the criterion of 'efficiency' of a computer program; whether it is assessed by the proportion of 'computing' time in the total length of a program, or by the convenience of the user. Inexpert users do not constitute a coherent group, and so their convenience does not rank high among the specification of the functions of the computer. Only occasionally can they even publish their criticisms, since to do so requires a level of expertise which resolves the individual difficulty. One such publication is M. Hamilton, 'Computers for the Medical Man: a Solution', *British Medical Journal*, 2 (30 October 1965), 1048–50. The author had found that the so-called 'standard' programs for simple data processing are nothing of the sort; and succeeded in devising some which could be used and interchanged without extensive re-writing every time. The deficiencies of computers from the users' point of view will be discussed quite frankly by manufacturers on some occasions, but these will tend to be in the context of advertising an improvement which obviates the previous difficulties. Thus Hewlett-Packard describes the prevailing situation as follows: 'Given the chance, the computer can live up to its promise. But in all too many laboratories, the computer doesn't even stand the chance of a trial because it creates new problems that some scientists consider to be worse than the old. Chief among these is the complexity of putting the computer to work in the laboratory—programming it, mastering the instrument-computer and the man-machine interfaces—which, to the scientist, is often a greater drudgery than the manual data gathering and calculation that the computer eliminates.' See *Science*, 167 (1970), 1196–7. They offer small, instrument-oriented digital computers, and also shared time with packaged programs.

merely clients, and of tool-using fields in relation to these products of the tool-providing fields. There is a general asymmetry between tool-providing and tool-using fields, which, when combined with the prestige-ranking among sciences that we have inherited from the past, produces a systematic tendency for tool-providing fields to assimilate those which they nominally serve. The relation is a simple one: those fields which describe the more abstract and general properties of matter serve as tool-providers for those which are concerned with the more concrete, and more particular or more highly organized. The sequence runs from mathematics to physics, chemistry and biology; attempts to tack the human sciences neatly on to the end of the scale have not as yet proved successful. The reasons for this asymmetry are twofold. First, the tools which are designed for the production of data or information in the more particular and organized properties of matter, must necessarily be designed to cope with complexity and variability which is encountered in such situations, rather than to attain the precision which is possible in simpler and more stable contexts. Hence tools designed for biological work will find no use in physics. On the other hand, some of the more abstract and general properties of matter may well be relevant to a problem in a field such as biology, especially in the production and testing of data and information; we have seen this in the case of statistics. This asymmetry is reflected in the teaching of the various sciences; the physicist's tool-subjects are mathematics and certain techniques of instruments, while the biologist must learn something of all those sciences further up the scale.

This asymmetry of provision and use does not imply that tools are created within the confines of the more abstract disciplines, and then hired out to those lower down on the scale. For it will frequently happen that a tool is brought into being to perform a function in a problem in a 'complex' field, is then seen to be capable of extension in all directions, and may even greatly influence the development of the tool-providing field itself. The history of statistics, many of whose most powerful techniques were developed in just this way, is a case in point; and many (although not all) branches of mathematics have evolved directly out of tools designed for special functions in the natural sciences.[16]

[16] The advances in statistical technique associated with such names as Galton and R. A. Fisher were achieved in the creation of tools for analysis of numerical data in

We have already seen that in a general way the tools which are available define the range of problems that can be studied. But the influence of tools on a field can be more subtle than a mere creation of possibilities. The extensive use of a tool involves shaping the work around its distinctive strengths and limitations; one can rarely apply a new tool to an existing stream of research without modifying the stream strongly. Hence as new tools come into being, and are judged appropriate and valuable by people in the field, they alter the direction of work in the field and the conception of the field itself. The men of an older generation who cannot master the new tools may grumble that the field has been distorted or taken over by outsiders. Whether they are right is a matter that only a later history can judge. But any such judgement must be made in terms of an implicit philosophy of science; and in the one which has been dominant for several generations, a field becomes more genuinely 'scientific' as it more closely resembles theoretical physics. Hence the natural tendency of the available tools to modify a field in that general direction are reinforced by the prestige of the limiting point of the process. It is important to realize that this confluence of the technical aspects of tools, and the prevailing metaphysic, is an historical accident rather than an essential feature of natural science. If the 'paradigm' natural science were to become a discipline like ecology, which uses the whole range of tool-providing sciences but whose objects cannot be reduced to those of any of them, then the social relations of tool-providing and tool-using fields would be drastically altered.

The combination of these two asymmetries, one inherent in the work, and the other socially imposed, has produced a false picture of the relations of the different scientific disciplines. It is commonly supposed that physics is not only stronger than biology as a field, but also more fundamental in that its problems and objects are independent of the latter. When two such fields are connected, as in biophysics, the penetration is all in one direction. Yet this is not necessarily so, for in the past it has been otherwise, on most significant occasions. For example, few physicists will be aware that the conception of the conservation of energy, which transformed the

the biological and human sciences; they soon generated lively fields of purely theoretical research, cast in a form comprehensible only to mathematicians. In mathematics, the field of differential equations was originally a tool for rational mechanics; and Fourier's series was invented and developed in the course of investigations into the theory of the flow of heat in solid bodies.

physical sciences in the middle of the nineteenth century, arose partly out of biological problems which were investigated in a consciously philosophical style.[17] The full history of the discovery of the First Law of Thermodynamics shows that this surprising origin was no 'accident'; but to explain it lies beyond the competence of the traditional philosophy of science, and the physicist's common sense of science on which it is based.

Finally, this discussion of tools shows how subtle must be the craft knowledge of a scientist who is a leader in his field. He must not only be able to develop a craftsman's competence in the use of particular tools, and have a sense of their powers and limitations; but, if he is to do anything but follow fashions set by others, he must assess the sorts of problems into which the use of particular tools would lead him and his colleagues and decide whether they are appropriate for the best progress of work in his field.

Pitfalls

The craft character of scientific work is exhibited most systematically through the concept of 'pitfall', which I have already mentioned several times.[18] The importance of this concept can be seen from Francis Bacon's aphorism, 'What in observation is loose and vague, is in information deceptive and treacherous.' Leaving aside his distinction of observation and information, we may ask why there should be deception and treachery in the inference? This is very strong language indeed. What does it mean? Whenever we extend from the known to the unknown, we do so on the basis of expectations of what we will find; these are necessary to give direction to our moves, and to supply interpretations of what we encounter. These expectations are always incorrect in some measure, and so we learn through the discovery of our errors. However, not all our errors are so considerate as to announce themselves as soon as they are made. It can happen that we follow an erroneous path of in-

[17] For the importance of *Naturphilosophie* in the intellectual background of many of the 'discoverers' of conservation of energy, see T. Kuhn, 'Energy Conservation as an Example of Simultaneous Discovery', in *Critical Problems in the History of Science*, ed. M. Clagett (University of Wisconsin Press, 1959), pp. 321–56, especially pp. 338–9. The case of Helmholtz has been studied in great depth by Y. Elkana, 'The Emergence of the Energy Concept' (Ph.D. Thesis, Brandeis University, 1967).

[18] I learned the concept of 'pitfall' as part of the craft wisdom of mathematics from my teacher Professor A. S. Besicovitch, F.R.S. He had the aphorism, 'In analysis, the pitfalls are everywhere dense'; its full flavour can be appreciated only by those who know the technical meaning of the latter term.

vestigation for some time, investing great resources into its pursuit, and only much later discover that we are mistaken. Then we realize that our time and labour had been wasted, and perhaps our confidence and reputation shaken as well. Bacon knew that we cannot but try to confirm our expectations as we advance into the unknown; and this is much easier if the things we find, as we go along, can easily be interpreted in our favour. In this fashion, with 'loose and vague' materials, we can deceive and betray ourselves, to our eventual undoing.

Thus the path of discovery is beset by concealed traps for the unwary, which we can call 'pitfalls'. In some ways these are more dangerous than the physical hazards from which they take their name, for one learns only in retrospect that one has stumbled into a pitfall at some earlier point of the work. One may produce data on equipment which has a concealed systematic error; or the theory of the equipment may have a hidden error which vitiates the inferences from the readings; or the first sketch of an argument may have an ambiguity or false deduction which undermines all the reasonings established on its basis. At every stage of our exploration of the unknown, we are at risk of being mistaken, and of remaining in ignorance of our mistakes until irretrievable damage has been done.[19]

The encountering of pitfalls through the making of judgements influenced by expectations is shown most clearly in the hazards of

[19] Pitfalls in observation are most likely to occur when the 'data' are subjectively-assessed extreme or nul points of a visual phenomenon. This was responsible for the short-lived inquiry into 'N-Rays' early in this century; see D. J. Price, *Science since Babylon* (Yale University Press, 1957), ch. 4. The identical pitfall vitiated the 'Allison effect', whereby the elements 'Alabamine' and 'Virginium' were 'discovered'. See F. G. Slack, 'The Magneto-Optical Method of Chemical Analysis', *Journal of the Franklin Institute*, 218 (1934), 445–62. (I am indebted to Mrs. R. Countryman for this reference.) It is interesting that the 'findings' in this second case were longer-lived; even after the war American students of chemistry learned of these two elements by their patriotic names, rather than as Francium and Astatine.

Quantitative experiments can also be subject to disastrous pitfalls through the scientist's ignorance of distorting effects on his equipment. The experiments of D. C. Miller with an improved interferometer, by which he detected an absolute motion of the earth through the aether from 1925 onwards, were probably vitiated by temperature effects, in spite of the warning on that point by Helmholtz to Michelson and Morley in the 1880s. See L. S. Swenson, Jr., 'The Michelson–Morley–Miller Experiments before and after 1905', *Journal of the History of Astronomy*, 1 (1970), 68. Temperature effects also produced a pitfall in an experiment of otherwise classic simplicity, testing whether photons lose or gain energy when moving against or with a gravitational field. In this case, the pitfall was exposed by a Cambridge (Trinity College) undergraduate, who has since gone on to a distinguished career in physics. See B. D. Josephson, 'Temperature-

using research assistants, working essentially as technicians, for the production of data. For the assistant will generally know what his supervisor expects to find; indeed, he must have such explicit expectations if he is to make the first judgements on the soundness of the data. But when unexpected and contrary results appear, he must make a judgement on their significance, balancing his own limited technical competence against the superior understanding of his master, and perhaps being influenced by political considerations as well. The natural course of action is to present information from which the anomalous data have been expunged. If the supervisor is concerned to have evidence derived from genuinely sound data, he must go to the trouble of training his research assistants in genuinely critical and independent thinking as well as in the craft of the research. Sometimes this may be more trouble, and still not so reliable, as doing the work oneself. This phenomenon indicates that there may be an upper limit to the degree of division of labour, in the dilution of the sophisticated craft skills, in worthwhile scientific research.[20]

Dependent Shift of Rays Emitted by a Solid', *Physical Review Letters*, 4 (1960), 341–2. (I am indebted to Professor F. A. E. Pirani for this reference.)

The possibility of pitfalls being discovered in well established experimental fields is shown by a report, 'A Cautionary Tale for Chemists', *New Scientist*, 45 (1970), 543. Two chemists, in the course of a straightforward kinetics study, discovered that an important component of one of their reactions was the glass wall of the flask containing it; the glass was not inert against dilute sodium hydroxide. The article concludes: 'The studies indicate the need to interpret more carefully than has hitherto been the case all chemical data measured in this type of system. Chemistry, it seems, is still capable of springing a few surprises.'

Very few scientific fields have standard literature on the pitfalls of their characteristic patterns of argument. In quantitative social science, the classic is W. S. Robinson, 'Ecological Correlations and the Behaviour of Individuals', *American Sociological Review*, 15 (1950), 351–7. This is discussed and further developed in E. Allardt, 'Aggregate Analysis: the Problem of its Informative Value', in '*Quantitative Ecological Analysis in the Social Sciences*, eds. M. Dogan and S. Rokkan (M.I.T. Press, 1969).

20 I recall a colleague telling me of his admiration for a particular foreign research student who, although not brilliant, did overcome his natural humility sufficiently to inform the supervisor that the data were not as expected. The student had been so conditioned to believe everything said by a Professor in his own country, that it required considerable courage for him to make the criticism. At this very fundamental level, there is clearly some correlation between independence of thought and worthwhile scientific work. A similar tale with a somewhat different ending is told by Lincoln Steffens in his *Autobiography* (Harcourt, Brace, New York, 1931). While studying under Wundt in Leipzig he and his colleagues came across the laboratory records of another American who had been there before. From them they could reconstruct the history of his research, where he produced data 'which would have given aid and comfort to the enemy and confounded one of Wundt's most axiomatic premises'. This would also

From the nature of the work, it is impossible to eliminate pitfalls from scientific inquiry. The experience of matured scientific disciplines is that they can largely be avoided. This is done in two ways: by the charting of standard paths which skirt them, and by each investigator becoming sensitive to the clues which indicate the presence of the special sorts of pitfalls he is likely to encounter in his own work. The first of these requires a tradition of successful work in the subject where there is a body of standard techniques which can successfully be applied as a routine. When these are taught to students only some of the pitfalls which they have been designed to avoid are pointed out. But these techniques are an embodiment of successful craft practice, built up over generations. Then, when an individual scientist explores beyond the range of the well-established techniques, his craft knowledge must necessarily be more subtle and personal, for the pitfalls he is likely to encounter are peculiar to his particular materials and tools. The clues to the presence of pitfalls are all he has, since he is beyond the range of the charted paths. Thus the accumulated social experience of his field must be supplemented by his personal craft experience of his portion of it.

This discussion of pitfalls is of some philosophical significance, for it indicates one of the reasons why it is vain to seek for certainty, or even for a guarantee of the existence of certain knowledge, in science. There can be no absolutely certain foundations in experience, nor any absolutely certain inferences from that experience, for the achievement of knowledge of the natural world. Worse still, from this discussion of pitfalls we see that we even lack a guarantee that our errors will be revealed to us by a direct and straightforward process. It is possible to argue from the impossibility of complete certainty in science, to a 'sceptical' position, asserting that all our supposed scientific knowledge is merely hypothetical, or probable, or illusory. But such abstract and general arguments do not indicate why some of our knowledge is more certain, and more solid, while some is less so. The concept of 'pitfall' furnishes one criterion whereby we can make such a distinction. In an established discipline, a trained man can work reliably in the realm of the known,

have been damaging to a career full of promise; so the student solved the problem by 'changing the figures item by item, experiment by experiment, so as to make the curve of his averages come out for instead of against our school'. Full of admiration for this mathematical feat, Steffens none the less concluded that experimental psychology did not provide a foundation for a science of ethics, and turned elsewhere (pp. 150–1).

and successfully feel his way forward into the unknown. But this happy state of affairs does not always hold; and indeed every one of our firmly established sciences has a long prehistory of its immature state. Then, even the accepted knowledge was liable to be revealed as illusory, and in the vanguard of the science, each man groped forward blindly, hardly ever succeeding in avoiding the pitfalls in his way. We can see that the establishment of a successful discipline requires a man of exceptional talent and courage, for he must not only succeed in his researches where all others have failed, but he must create the craft knowledge of the pitfalls of the inquiry, so that the lesser men who follow him can safely proceed to the development of his own achievements.[21]

Thus we can distinguish between more and less solidly established knowledge, by the degree to which the ways around its common pitfalls are well charted, and those encountered in the extensions of that knowledge can be sensed in advance. This criterion is not sufficient in itself, for the progress of science renders most of its earlier achievements irrelevant and obsolete. However, the concealed imperfections which are present in scientific knowledge at any point of its development, are not the same as pitfalls. A pitfall is the sort of error that destroys the solution of a problem, and nullifies its conclusions about the objects of the inquiry. We do not say that Newton's assertion of the existence of absolute space and time, or Boyle's ignorance of van der Waals' equation, were 'pitfalls'. Although our knowledge is more refined, we do not contest the solidity of their achievements. Of course, these are extreme examples, and we cannot draw a perfectly sharp demarcation between perfectible knowledge and results vitiated by pitfalls. But as a matter of historical experience in a matured discipline the line

---

[21] This progression can be discerned in the decades of work that were required for Galileo's mastery of the principles of bodies in non-uniform motion. His first formulation of the law of acceleration of falling bodies was the more common one of his time, that speed increases in proportion to distance traversed. He actually succeeded in 'deriving' from this false relation the result already known to him, that distance is proportional to the square of time. He eventually discovered the falsity of this rule: I think it likely that he found it impossible to relate it to other known results of his (such as, equal times of descent along all chords of a circle), or to base a theory of projectile motion on it. When he finally came to disprove it, one of his two arguments was itself fallacious! See Galileo, *Discorsi . . .*, translated as *Dialogues Concerning two new Sciences*, tr. H. Crew and H. de Salvio (Macmillan, New York, 1914; reprint Dover, New York, no date), pp. 167–8 (Edizione Nazionale, viii, 203–4). For a discussion of the false rule, see A. R. Hall, 'Galileo's Fallacy', *Isis*, 49 (1958), 342–4, followed by discussion by I. B. Cohen and S. Drake.

is fairly sharp; and this is another expression of the fact that solid knowledge derives from a tradition of successful craft work in its achievement.[22]

At this point of my discussion, as at others to come, I must consider the question of why this important feature of science has hitherto escaped notice. It is easy to see why all the earlier traditions in the philosophy of science mainly concerned with epistemological problems should have ignored the craft character of scientific work, and with it the concept of pitfall. In several of the human sciences, the craft character of the work of advancing knowledge has been recognized by leading men, and through their writings is familiar to all those who are concerned to understand their own work.[23] Hence unless we believe that the study of the natural world is utterly different, in its work as well as in the character of its conclusions, my thesis will become obvious as soon as it is realized to be significant through the changing common sense of science.

I must also reckon with the historical fact that the formal training of scientists has generally been carried on without any recognition of the craft character of scientific work. There are several reasons for this, of which one is the implicit philosophy of science which has prevailed up to now. Also, most of the basic knowledge which a scientist will need to have available as tools can be organized into a systematic form in which the pitfalls are bypassed without any immediate need for their being identified. For the pitfalls are encountered only when this knowledge is put to use, and they depend so much on the particular application, and are so various, that a comprehensive discussion of them would be quite impossible. In laboratory courses, students are given a gentle introduction to the pitfalls likely to be encountered in the use of physical tools for the production of data, and they learn the craft techniques for manipulating these tools reliably. However, this essential aspect of labora-

[22] I am grateful to Mr. Keith Boughey, then an undergraduate at the University of Leeds, for raising the problem solved in this paragraph during a discussion in 1967.

[23] Among practising historians who have written on 'historical method', the craft character of historical research is either recognized explicitly or is implicit in the whole argument. An illuminating recent exposition is G. R. Elton, *The Practice of History* (Collins (The Fontana Libary), Sydney University Press; London and Glasgow, 1967). In C. Wright Mills, *The Sociological Imagination* (Oxford University Press, 1959; The Grove Press, New York, 1961), the idea of craftsmanship runs through the book, and an appendix 'On Intellectual Craftsmanship' offers an account of the author's own style.

tory teaching is frequently ignored, and students usually believe that they are merely 'verifying' for themselves that certain standard effects can be reproduced.

A recognition and systematic use of the phenomenon of pitfalls might be very effective in the teaching of those simple but essential craft skills which are involved in scientific, scholarly or administrative work. A person trained for conceptual thinking generally finds the books on 'method' in such fields to be of little use; there seems to be little fruitful middle ground between boring recipes and meaningless aphorisms. But an exposition of standard techniques in terms of the pitfalls they are designed to circumvent, with examples, could go far to make them meaningful and obviously worth mastering.[24]

It is only in the training of research students in science that the craft character of scientific work is now explicitly recognized, and with it the importance of learning how to sense the presence of pitfalls. Everyone who supervises research students knows of the dangers of giving too much, or too little, help and guidance. If the student is simply told of a problem-situation and then left to his own devices, he may spend so many months floundering among failed attempts even to set a problem that he becomes completely discouraged and gives up before he has produced anything solid. But if he is given constant advice, he proceeds all too smoothly through his project, being little more than a reliable pair of hands for executing the supervisor's ideas; and he never learns to grapple with his materials or to sense the pitfalls occurring in his field. How much help should be given depends on the project, the student, the policy of the school, and the attitude of the supervisor. Research supervision is itself a craft, the most subtle and demanding sort of teaching. Some research schools let their students stumble into a large proportion of the pitfalls in their way; this is expensive, time-consuming, wasteful of talent and hard on the students. But those who survive such a course are skilled craftsmen, and of tested strength of character in this respect as well. In others, research students are given plenty of guidance, so they are sure to get through their work in a reasonable time, without too much suffering. Wastage and agony are thereby reduced, and 'success' is more likely, especially if the research student is really functioning as

[24] This has been done for architects, in H. B. Cresswell, *The Honeywood File* (London, 1929; reprinted, Faber & Faber, London, 1943).

low-paid skilled manpower working on the supervisor's problems.[25] But those who emerge from such an experience with a Ph.D. as a formal certificate of competence, are as yet unskilled and untested in original scientific research. The one extreme policy results in many failures, and in a few scientists of promise; the other produces competent 'scientific manpower' on a production-line basis. The character of the scientific work done by the graduates of the different sorts of research schools will inevitably reflect their training; and here we see one way in which the craft work of science is influenced by the institutional and social context in which it is conducted.[26]

Craft Methods : Techniques

We have already seen several ways in which the work of scientific inquiry requires knowledge which is learned only through precept and experience in a multitude of particular cases, and which therefore is not 'scientific' in character. The assessment of data and of information, and the manipulation of tools, are all subject to pitfalls; and it is only the craft knowledge of the investigator which enables him to avoid some and sense the presence of those which remain. Such knowledge forms part of the corpus of methods, those principles and precepts which guide and control the work being done. For the present, we can further illuminate the craft nature of scientific inquiry by discussing those methods which are most closely related to the particular tasks undertaken, and which therefore consist more of completely tacit knowledge and particular precepts. Those methods which condition the assessments made of problems, and are more concerned with their function as social possessions, will be discussed systematically later on.

[25] Thus E. Rudd, 'Rate of Economic Growth, Technology and the Ph.D.', *Minerva*, 6 (1968), 366–87; 'When we interviewed current research students in chemistry departments and asked them if their research topic was closely related to that of their supervisor, sometimes the reply was: "What do you mean, closely related to that of my supervisor? It is my supervisor's research."' (p. 382).

[26] Because research supervision involves the transmission of the most subtle skills, it may give a clue to one of the causal factors involved in the 'direct square law of association' known in the folklore of science. This is, that men of the first rank will associate with men of the first rank; those of the second, with those of the fourth, etc. For a scientist of moderate ability and commitment will generally not be actively concerned about the craftsmanship involved in his work, and even less able to impart the skills and the spirit of craftsmanship to the level that he possesses. Hence the craft skills of research in a field can become attenuated very rapidly in transmission through a sequence of lesser men.

Of all the 'methods' of scientific inquiry, the most straightforward are the techniques of routine manipulations of tools. This is usually the content of courses in 'methods' where materials from a tool-providing field are made available to potential tool-users. The tools designed for such standardized uses are usually made as close to 'foolproof' as possible; so that by following the simple rules the user should be able to avoid pitfalls in any ordinary applications. However, perfection in this as in anything else is not easily achieved, and so the simple rules of manipulation may be supplemented by somewhat more subtle and particular instructions and precepts for avoiding pitfalls which might produce obvious damage (as to the tool itself and its user) or hidden errors of interpretation of the products of the operation. These too can be classed as methods; they are the first crude, explicit elements of craft skill in using the tool, as distinct from merely mechanical manipulation.

In general, the contents of such elementary points of technique will be precepts rather than principles. For usually the tool-user cannot be in sufficient command of the theory of the tool to appreciate the explanations for its operations, or of its malfunctioning, even when such explanations exist. As an elementary example of how a 'method' is invoked for avoiding such pitfalls, the beginning student in chemistry has no basis for knowing that concentrated sulphuric acid evolves heat on mixture with water, could not understand the explanation if it were given to him, and should not have to learn the phenomenon by 'trial and error'. Hence he will simply receive a precept to drop the acid gradually into the water, rather than to mix all the masses or to drop the water into the acid; and the 'explanation' of the precept is restricted to an assertion of the dangers of the other courses.

For a full mastery of the use of tools, explicit precepts are insufficient. Any extension of the uses of a tool involves new hazards of pitfalls; and the unknown cannot be described by formulas. Hence the full craft knowledge of any particular tool, for a particular range of functions, will involve a large measure of personal experience. To the extent that the personal knowledge of a tool is deep and subtle, any set of explicit precepts will fall short of conveying it. In any real situation there are too many subtle cues, and too many partly relevant precedents, for the knowledge of how to cope with novelty to be reduced to tables of experiences and inferences. This aspect of tool-using involves the solution of 'diagnostic' problems, as those faced by a physician.[27] The matured craftsman of scientific inquiry will be working on technical problems which are peculiar to himself and some colleagues, and which change fairly rapidly with the development of his field; and so there is no substitute for his personal, largely tacit knowledge of the tools which have become nearly continuous extensions of the sensory, motor, and intellectual equipment within his body. The transmission of such knowledge will then be largely through a close personal association of master and pupil, and the explicit precepts of the refined methods of tool-using have meaning only in the context of the solution of sophisticated technical problems.

The informal and largely tacit precepts of method are not restricted to the use of tools in the solution of the technical subsidiary problems of a scientific inquiry. The scientist's craft also includes the formulation of problems, the adoption of correct strategies for the different stages of the evolution of a problem, and the interpretation of general criteria of adequacy and value in particular situations. These other tasks, which distinguish original scientific work from the routine production of bits of information, have no standardized, elementary versions to which simple, explicit precepts can apply. Hence most of the body of methods governing this work is completely tacit, learned entirely by imitation and experience, perhaps without any awareness that something is being learned rather than 'common sense' being applied. Since this sort of knowledge is so different in character from that embodied in the published results, and is transmitted through a different channel, it is not capable of the same universality of diffusion, nor of the same closeness of control of quality. It is therefore subject to particular weaknesses from which public scientific knowledge is protected. Yet it is on this informal knowledge of the higher elements of the craft of scientific inquiry that the success of the whole social endeavour depends. We will later discuss the relations between these two sorts of knowledge; for the present it is sufficient to establish the point that in every one of its aspects, scientific inquiry is a craft activity, depending on a body of knowledge which is informal and partly tacit.

[27] For an analysis of 'diagnostic' problems, and a demonstration that their structure is similar to that of general scientific problems, see M. Hamilton, *Clinicians and Decisions* (Inaugural Lecture, Leeds University Press, 1966).

Style

One of the reasons for the lack of awareness of the craft character of scientific work is the form in which the results of that work reach students and the lay public. There they are presented out of the context of their creation and in a simplified or vulgarized version. The fine structure of the argument by which the result was established, and its context in earlier and alternative approaches, cannot be conveyed except to a technically competent audience. It is just in these fine points of detail that the style of a scientist is revealed.[28] The investigation of a scientific problem is creative work, in which personal choices as well as personal judgements are involved at every stage up to the last. The scientist comes to the problem with a unique set of interests, skills, and preferences. No two scientists can set and solve exactly the same problem, except one which is routine and straightforward. The materials of the problem are malleable, and the conclusion which is achieved can vary in its emphasis and form. Even when the problem is solved, its materials are further developed by later work, in the context of new investigations. The apparently rigid and necessary form of achieved scientific knowledge is arrived at only when the material comes into a standardized version, for use as general purpose tools or information.

Thus, even though the scientist is concerned with properties of the external world, the work he produces will be characterized by a certain style unique to himself. This style of scientific work is analogous to that of artistic creation; historians of art have been able to analyse works of art in terms of 'formal elements' of a composition, independent of the themes portayed. This sort of style has been used as a foundation of the critical history of art, and through this concept, links of influence (both within art and from outside) have been established.[29] For the scientist, as well as for the artist,

[28] For this section, I am embarrassed by a paucity of citations from the literature of the history of science; and colleagues who are equally certain that style is an important aspect of scientific inquiry have been equally unsuccessful in finding substantiating evidence. I am convinced that this silence is more a reflection of the traditions of scholarship in this field than of an absence of style from creative scientific work.

[29] For a discussion on the use of the concept of style in the history of political thought in relation to its origins in the history of art, see K. Mannheim, 'Conservative Thought', in *Essays on Sociology and Social Psychology*, ch. 2, 74–8. His concept of 'style' is one characteristic of social groups rather than of individuals; but there is no contradiction between his usage and mine. Closer to my own sense is the concept of style in teaching, developed by M. Oakeshott, in 'Learning and Teaching', in R. S. Peters, *The Concept*

the personal style will be realized through choices within the range of possibilities defined by the whole body of methods for his problem. There is no conflict between a highly individual style in the investigation of problems, and the production of results which meet the socially imposed criteria of adequacy for the field. Through personal acquaintance with a scientist's style, a colleague can recognize his work; and a later historian can describe the style, and even use it for the explanation of a unique achievement.[30] Of course, the development of a personal style is a matter of degree; those scientific workers who produce bits of information on order will not generally need, or be permitted, a strongly marked personal style for their tasks.

Some important features of the social aspects of scientific activity can be illuminated by the concept of style. Since the personal style of a matured scientist is influenced by his earlier experience, we can speak of the transmission of style from a master to his pupils. In this way one can construct intellectual genealogies; and to understand a man's work it may be relevant to know who was his teacher's

---

of Education (Routledge, London, 1967). '. . . in every "ability" there is an ingredient of knowledge which cannot be resolved into information, and in some skills this may be the greater part of the knowledge required for their practice. Moreover, "abilities" do not exist in the abstract but in individual examples:. . . Not to detect a man's style is to have missed three-quarters of the meaning of his actions; and not to have acquired a style is to have shut oneself off from the ability to convey any but the crudest meanings.' (Quoted from T. H. B. Hollins, *Another Look at Teacher Training* (Inaugural Lecture, Leeds University Press, 1969), pp. 19-20.) In the reporting of scientific results, the style of investigation may not be allowed to dominate as in personal teaching; but for the achievement of those results, without style there can be little originality.

[30] A vivid account of personal style was given by Boltzmann, in connection with Maxwell: 'A mathematician will recognise Cauchy, Gauss, Jacobi, Helmholtz, after reading a few pages, just as musicians recognise, after the first few bars, Mozart, Beethoven or Schubert. Perfect elegance of expression belongs to the French, though it is occasionally combined with some weakness in the construction of the conclusions; the greatest dramatic vigour to the English, and above all to Maxwell. Who does not know his *Dynamical Theory of Gases*? At first the Variations of the Velocities are developed majestically, then from one side enter the Equations of State, from the other the Equations of Motion in a Central Field; ever higher sweeps the chaos of Formulae; suddenly are heard the four words: "put $n = 5$". The evil spirit V (the relative velocity of two molecules) vanishes and the dominating figure in the base is suddenly silent; that which had seemed insuperable being overcome as by a magic stroke. There is no time to say why this or why that substitution was made; who cannot sense this should lay the book aside, for Maxwell is no writer of programme music who is obliged to set the explanation over the score. Result after result is given by the pliant formulae till, as unexpected climax, comes the Heat Equilibrium of a heavy gas; the curtain then drops.' (Quoted from R. A. Millikan, D. Roller, and E. C. Watson, *Mechanics, Molecular Physics, Heat and Sound* (Ginn & Co., 1937), p. 219a.

teacher.[31] This is clearly an important element in the creation of scientific 'schools'; and through the transmission of his personal style a great scientist can have a direct influence on scientific work even after he has departed and his own problems have become obsolete. But this interpersonal action is also conditioned by a style: that by which the scientist relates his personal scientific endeavour to the communities of which he is a member. Whereas the personal style of the scientific work will be largely conditioned by the internal history of the field (perhaps as enriched by more or less explicit philosophical concerns), this style of interpersonal and social relations will be very directly influenced by the institutional and social structure of the scientific community, and of the larger society of which it is a part.

The particular character of a scientific school will depend on these two aspects of the style of its master. Since style is so largely tacit, we do best to refrain from attempting an exhaustive list of its components. But certain significant features of scientific schools can be identified and explained on the basis of an informal understanding of style. First, it is impossible for a personal style of scientific work to be transmitted perfectly; for no two people, especially two of different generations, can have experiences so similar as to provide bases for nearly identical styles. The difference between the styles of master and pupil will be all the greater when the master has done deeply original work, creating the objects and methods of a field through his endeavours to attack some very fundamental problem. For the pupil learns the developed forms of these materials, and applies them to more straightforward problems; and in spite of a complete personal allegiance, and an attempted copying of the master's style, he cannot re-live his experience, and so cannot be a replica of him. If the master has a certain style in his social relations, he may fall into the pitfall of considering the brilliant pupil as a true image of himself, fail to detect the lack of depth and subtlety in his results, and so contribute unwittingly to the vulgarization of the products of his life's work. Other sorts of style in social relations can lead to conflict, as when a master expects both brilliance and obedience from his pupils in the execution of his grand programme.

[31] H. T. Pledge, in *Science since 1500* (H.M.S.O., London, 1939), gives three charts indicating master-pupil relations in experimental science from the sixteenth to the nineteenth century. K. E. Rothschuh, *Geschichte der Physiologie* (Springer-Verlag, 1953), gives several genealogical charts for this discipline.

He will then find some disappointing his hopes, and others betraying his trust; and the continuation of his work through a school is endangered.

Such situations of delusion, disappointment, and conflict do occur in science, and of course they represent a less than perfect state of affairs. But it would be a serious error to suppose that such errors in perception and expectation by masters prevent the accomplishment of great work, or even the formation of influential schools. For the different strategies required at the different stages of the development of a field have different sorts of style appropriate to them. The special circumstances of pioneering work give rise to a style which helps to explain a phenomenon frequently found in nascent disciplines. This is the presence of several schools, each loyal to the exclusive message of its master, and all doing battle with each other for several academic generations. Eventually, the passage of time confutes all the warring sects, and later scientists freely use information and insights from all sides. Then there is a general wonder that people weren't sensible enough to do so at the beginning. But being sensible about the variety of possible opinions and approaches is an aspect of the style which is appropriate to a period of consolidation of a field. The task of creating a discipline out of nothing is so difficult and dangerous that only men with a special sort of personal style will attempt it; the degree of dedication and courage which are necessary for the work, frequently carry with them a touch of messianism. When such a man undertakes the heroic task of making an objective science out of his deep insights into the truth, he cannot afford to be broadminded about the claims of rival theories or about criticisms of his own. It then requires only a strong personality and a favourable social and institutional situation for each such pioneer to collect talented disciples, all of whose insights are gleaned second-hand from the master, but whose personal dedication to his approach is, if anything, more rigid than his own. Each such group then proceeds to erect a fortress of orthodoxy, and to conduct battles until the field outside their enclaves develops to the point that their controversies are merely boring. The direct descendants within each original school, left with little but a rigid doctrine and a scholastic style, then gradually wither away.[32] Indeed, there

[32] The mathematical sciences in the eighteenth century suffered from such 'scholasticism', as the British followers of Newton patriotically adhered to his clumsy notation for the calculus, his one-sided views on the foundations of mechanics, and his theory of the corpuscular nature of light.

occasionally occurs a paradoxical and dismaying situation where the old master himself eventually loses his enthusiasm for the struggle with the ancient enemy, and his disciples are forced to denounce him as a renegade from the 'ism' bearing his name. Out of these early conflicts within a nascent field there do appear results which pass into the permanent record, appropriately standardized and denatured; and it is left to historians to rediscover the drama and tragedy which were an essential part of their creation.

4

SCIENTIFIC INQUIRY: PROBLEM-SOLVING ON ARTIFICIAL OBJECTS

Introduction

HAVING argued that scientific work is necessarily a craft activity, depending on a personal knowledge of particular things, and on subtle judgements of their properties, I must now make an important qualification. There are essential differences between the craft work of scientific research and other sorts of human activities. For the objects of scientific inquiry are of a very special sort: classes of intellectually constructed things and events. Their difference from the objects of handicraft production, or even of ordinary discourse and action, gives scientific knowledge its special power, and makes scientific inquiry a particularly complex and delicate social activity.

With this thesis I run counter to an influential tradition in the philosophy of science, whose roots are in England rather than on the Continent. It derives ultimately from Bacon, who believed that men should, and could, turn to 'things themselves', and away from the false theories used for describing them and explaining their actions. A variant on this theme became popular in the nineteenth century, through the writings of Huxley and Clifford: that science is nothing other than 'organized common sense'.[1] With respect to the activity of scientific inquiry, this belief in 'naturalness' is correct; in my discussion of the different tasks and judgements involved in the craft work producing data and information and using tools I did not find it necessary to invoke any elaborate or formalized principles of 'method'. But in the extension of 'common-sense' from the work to the objects of the inquiry, the belief is seriously misleading. I shall later discuss the very real and difficult problems of controlling the applications of scientific knowledge; and these all revolve around the

[1] For example the classic work by W. K. Clifford, *The Common Sense of the Exact Sciences* (1885; reprint, Dover, New York, 1955).

fact that this knowledge is esoteric, available only to those whose long training in a field has made them expert.

Perhaps less crucial, but equally significant, this belief makes a genuine history of science impossible. Until very recently at least, the implicit assumption in the writings of the history of science (especially at the level of schoolbooks and popularizations) was that our sort of science is the obvious and natural way to study the world of nature, merely the application of unfettered common sense. Periods and cultures which failed to produce effective natural science of our sort were condemned for an almost wilful neglect. By the same assumption, great discoveries in science have been presented as the perception of the obvious, and the failure of a great discoverer to get the thing quite right is the occasion for an apology on his behalf. In this way, the history of scientific achievements comes out as a succession of one-eyed men in the kingdom of the blind. The question of why it required a great man to get even one eye open, and still the question of why he happened to come along at just that time, are rendered incapable of being asked, let alone answered. Such parodies of history are not merely of interest to professional historians of science; since the published history of science, especially at the popular level, is evidence for the self-consciousness of science, it serves as a reminder of the necessity of present discussion.

Intellectually Constructed Things and Events

I have already mentioned the classes of intellectually constructed things and events which are the objects of scientific knowledge, in my discussion of 'information'. Even in the production of data the manipulations are with samples which are designed to serve as true representatives of such classes, rather than with some 'real' objects which can be known independently of an elaborated structure of theory. Of course, these classes of things and events are intended to relate, as closely as is possible, to the inaccessible reality of the external world. But they are different in character from that reality: and from this difference derives the specialized nature of the craft work of science, and the never-ending perfectibility of scientific knowledge.[2]

[2] It is insufficiently realized just how far removed are the conceptual objects, and the problems, of scientific inquiry, from the world of ordinary experience. This abstraction can pass unnoticed because the teaching of science in the schools concentrates on the accomplished knowledge concerning the natural world, with instances of where that

An elementary example of such an artificial object is the concept of chemical 'substance'. The definition of any particular substance is highly complex, using properties derived from a variety of fields, and always subject to gradual or rapid change.[3] Also, the 'substance' is not a formalised description of a unique collection of material; rather, it is a class of things, the members of the class being defined by their possession of certain properties. Thus a chemical substance, serving here as an example of an object of scientific knowledge, is a class defined 'intensionally' by certain properties of its members.[4]

knowledge has enabled effective control to be achieved. Also, since so much of the environment perceived by people in advanced societies is man-made, those parts beyond our control or understanding are easily neglected. This abstraction was by no means an easy thing to achieve, and was not a feature of the 'science' of the early Greek philosophers. See H. Gomperz, 'Problems and Methods of Early Greek Science', *Journal of the History of Ideas*, 4 (1943), 161–76 (but not to be found in Contents or Index).

The great advances in physical science in the seventeenth century required an extreme abstraction, and a putting to one side of outstanding problems, for its success. The classic case of this is Galileo's treatment of 'motion' purely in terms of a 'mobile' defined by its changing position; the contrast with Francis Bacon's more traditional approach to the study of motion, which proceeded through a taxonomy of motions, is striking. See the *Novum Organum*, Book II, Aphorism 46; translation in *Works*, iv, 214–32.

In the science of optics, the act of abstraction was equally profound in its philosophical implications (which were quickly developed by Descartes), and in its consequences in a one-sided development of the science over the following centuries. See V. Ronchi, *Optics, the Science of Vision*, tr. E. Rosen (New York University Press, 1957). He observes that the adoption of Kepler's hypothesis of a telemetric triangle (the assumption that the eye perceives the divergence of rays from a point source) enabled the physiologico-psychological aspects of vision to be ignored thenceforth (p. 50). In the book he shows 'how productive of error it was to have forgotten that optics is the science of vision and to have developed it blindly (for that is the appropriate word) as the optics or images or, I reiterate, as the optics of the centres of the waves' (p. 206). For an extended historical study of optics before and after the onset of this blindness, see V. Ronchi, *Histoire de la Lumière* (S.E.V.P.E.N., Paris, 1956). A summary of his thesis on the history of optics is 'Complexities, Advances and Misconceptions in the Development of the Science of Vision: What is being Discovered?', in *Scientific Change*, ed. A. C. Crombie (Heinemann, London, 1963), pp. 542–61.

[3] The element 'Oxygen' has undergone such changes since its first discovery, or, better, invention. For an account of the changing conceptions of this substance and its properties, see E. Farber, *Oxygen and Oxidation—Theories and Techniques in the 19th Century and the First Part of the 20th* (Washington Academy of Sciences, Washington, D.C., 1967); reviewed in *Isis*, 59 (1968), 454–5. The concept of 'acid' with which 'oxide' was identified by Lavoisier has also evolved since the identification was shown to be false. For a brief sketch, see J. W. Mellor, *A Comprehensive Treatise on Inorganic and Theoretical Chemistry*, (Longmans, Green, London and New York, 1922), 385–6.

[4] For a penetrating philosophical analysis of the concept of element, see F. A. Paneth, 'The Epistemological Status of the Chemical Concept of Element', *British Journal for the Philosophy of Science*, 13 (1962–3), 1–14 and 144–60; especially sections 5 and 6, pp. 149–60. (This is a translation from the German; the original was published in 1931.) A similar theme, with a wealth of historical examples, is discussed by R. Hooykaas, 'The Concepts of "Individual" and "Species" in Chemistry', *Centaurus*, 5 (1958), 307–22.

To determine the 'extension' of the class, seeing whether a particular object, given independently, belongs to it, we must decide whether this object is a 'sample'. For this, we test whether this real collection of material satisfies the defining properties of a member of the class to an acceptable degree, and which moreover contains specified 'impurities' (which would render it a 'non-sample') to no more than a tolerable degree. We notice that although the 'sample' cannot be a perfect realization of a member of the class which constitutes the defined 'substance', it is still in itself a highly artificial creation; few 'chemicals' can be found in the world of nature, outside the laboratory.

The same example serves to show how scientific knowledge is concerned with such classes of things and events, and not with particular things and events that happen to be observed. Demonstrations in school science courses are explicitly designed to show the properties of the general classes, through the behaviour of selected representatives. The merriment that ensues when a demonstration 'goes wrong' proves this; the teacher has failed to show the pupils what 'really' happens to the samples representing the general class, and has instead produced an effect involving extraneous and unwanted other things and events. It might be thought that the more descriptive sciences are closer to the real things of nature, and are not so completely imprisoned in the world of their own concepts. But the history of the centuries-long search for a 'natural' system of classification (and hence description) of living organisms shows that here too the objects of scientific knowledge are intellectually constructed classes rather than 'things themselves'[5]

Chemical 'substances' and biological 'species' are less artificial than many objects of scientific knowledge, for their samples are things which have many properties accessible to fairly direct inspection. Many conceptual classes have samples whose properties

[5] An example of the artificiality of a basic concept in a social science is that of 'price', used freely by economists as a simple quantitative variable in many contexts. O. Morgenstern, *On the Accuracy of Economic Observations* (2nd ed., Princeton University Press, 1963) shows how radical an abstraction this is from the complex realities of the terms of trade (pp. 181–7). Hard-headed businessmen fare no better; 'profit' is revealed on analysis to be a most elusive concept, whose measure depends more on book-keeping and accounting conventions than on any particular aspect of the actual operations of the firm (pp. 70–87). The more elaborate measures, compounded out of aggregates and estimates of these basic ones, such as 'national income' and 'gross national product' and 'growth' have a contact with reality which is far less firm than the precision of their estimates would suggest.

or even whose character as a 'thing' or 'event', are incapable of easy translation into ordinary experience; 'field' is a good example of this. In such cases, as in that of such basic entities as 'mass' and 'energy', it is beside my present purpose to determine whether the scientific concept is best considered as a class of 'things' or of 'events', where these terms are taken over from ordinary speech. For to the extent that their ontological status becomes obscure their character as intellectual constructs becomes apparent; and in the same measure, any particular manual experience involving them becomes sophisticated and theory-laden.

It might be objected that I have not yet shown any distinctive feature of scientific knowledge, for even our ordinary speech and thinking are done with names, which are identifying tags for general classes of things and events. Moreover, these names, and their classes, are far from natural; not merely abstract concepts, but the names and classification of everyday things are the result of cultural processes, varying widely from one milieu to another, and constantly changing over time. In this sense it is true that all our knowledge is 'artificial', and that we cannot conceive things for which our own culture provides no language. But the objects of scientific knowledge are even more artificial than this; to indicate the difference I have used the term 'intellectually constructed'. We can see the difference when we consider the border areas where terms from the vernacular are taken over into disciplined discourse. This occurs when the looseness of ordinary speech, and common sense thinking, is inadequate to the needs of the situation; public administration and the law are familiar examples.

In these fields, the intricacy of the relations into which people get, must be grappled with; and procedures are developed so that complex practical problems can be solved in consistent fashion. Much of the work of developing administrative or legal systems can be seen as defining classes of things and events, and setting the decisions to be taken in situations involving members of those classes. The routine work of decision-making then consists largely in determining the classes to which the real people, and their particular tangles, should be assigned. These intellectually constructed classes derive ultimately from common sense experience; but in their detailed elaboration, the need for systematic coherence weighs at least as heavily as the retention of the original link. In any sane system of this sort, there is a place for common sense ideas to be injected directly into the decision-making process; but this is rightly kept as a last resort, or else the whole system would collapse, and chaos would supervene.

The necessary artificiality of developed legal or administrative system is analogous to that of a developed science. The world of nature is far too complex to be comprehended in terms of the concepts we build up in our ordinary experience. Nor can we interact directly with the natural world, without the mediation of our organizing and simplifying conceptual structures. Of course, the screen between us and the natural world, which is necessary if we are to achieve any effective knowledge of it, can also serve to shield us from an awareness of our ignorance. Just as a legal or administrative system can drift out of all contact with the social realities it is intended to service, a scientific discipline can become totally self-contained and sterile. More seriously, its practitioners can come to take their system of artificial objects as the basic reality, ignoring its differences from that external world which it was originally intended to represent.[6] A fuller discussion of these social problems must wait until we have seen in more detail how the artificial nature of its objects influences the craft work of scientific inquiry, and gives it its peculiar strength and delicacy.

The Complexity of 'Discovery'

When we accept the view of science as craft work operating on intellectually constructed objects, some traditional views on the process of the achievement of scientific knowledge are immediately excluded. Because of their importance, I shall discuss them briefly

[6] The tendency of established scientific disciplines to create a world of their own and then live totally within it, is now becoming recognized as an obstacle to the solution of practical problems involving science and technology. Thus, 'I must also stress the incompetence of the established disciplines to tackle society's real problems. What we mean by a discipline is an agreed, tested body of method—usually analytical—that we bring to bear on problems *of our own choosing*. The essence of our thinking is that we cannot tackle problems that don't fit the competence of our own discipline. It's true that we constantly try to enlarge that competence. Confronted with a new problem, we spare no effort to improve our methods. But if we don't succeed, we don't tackle the problem, and we tend to condemn colleagues who try.' J. Kenneth Hare, quoted in John S. Steinhart and Stacie Cherniak, *The Universities and Environmental Quality—Commitment to Problem Focused Education—A Report to the President's Environmental Quality Council* (United States Government Printing Office, Washington, 1969), p. 6. For a review of this significant document, in which the philosophy of science developed here is implicitly presupposed, see L. J. Carter, 'Environmental Studies: OST Report Urges Better Effort', *Science*, 166 (1969), 851.

before going on to an alternative approach, in which the special character of the craft of scientific inquiry is described.

The tradition of the 'naturalness' of science involves, as we have seen, the assumption of the triviality of scientific discovery. In principle, all one needs to do is to go out with an open mind and clear eyes, and one will find facts and discover true laws of nature. Such faith doubtless had its social and ideological functions at some times in the past, but it is a travesty of the work of achieving scientific knowledge. It is possible for a master-craftsman (especially in descriptive science) to make observations and inferences with a speed, and apparent ease, that astonish his less gifted and less experienced colleagues. But such discoveries come from the opposite of an open mind and clear eyes; rather, they come from a mind which is constantly, if quietly, at work on a multitude of problems; and from eyes which are trained to perceive phenomena which, when interpreted in terms of the artificial objects of his discipline, are significant for the solution of those problems. Also, such a master, if he does work worthy of his talents, will be engaged on really big problems, and the deeper results he achieves will be reliably established only by years of arduous labour.

Any serious study of the history of science destroys the assumption of the simplicity of a scientific discovery. What we find instead is a brief phase when the objects of inquiry are in rapid change, and are correspondingly complex, ill-defined, and plastic; until one person achieves a solution of a problem which seems to determine the objects decisively, but which leaves further work to be done until the 'discovery' is complete. Perhaps the best example of this process is the classic 'discovery' of oxygen by Lavoisier. At around the same time, Scheele and Priestley isolated a substance which they recognized to be of importance; they named it 'fire air' and 'eminently respirable air' respectively. It was Lavoisier who realized that this substance was the master-key to a new system of chemical theory and nomenclature; he named it *oxygène*, and the glory of the discovery is his. Unfortunately, the intellectually constructed class so named by Lavoisier is not the simple ancestor of our present element Oxygen. For in his system of chemistry, it was a 'principle' responsible for two classes of phenomena which he hoped thereby to coalesce: the burning of substances in air, and the formation of acids. And it was as 'acid-maker' rather than 'fire-maker' that he named his principle (as one sees clearly in the German translation of Sauerstoff); and so

Lavoisier's error in the creation of this object is preserved in the name by which his 'discovery' is enshrined.[7] Historians of chemistry have not quite come to grips with the problems of judgement indicated by this classic error; those who opposed the new French nomenclature on the grounds of its being theory-laden are still dismissed as 'reactionaries' in authoritative histories.[8]

Indeed, one could argue that no discovery of any sort can be made of 'things', but must always involve an incompletely specified intellectually constructed object. The name involved in the classic question 'Who discovered America?' indicates that this may be a paradigm example for such a thesis.

Scientific Inquiry as a Task

We can now attempt to combine the materials of the two themes we have developed so far, and discuss scientific inquiry as a special sort of craft work, which operates on intellectually constructed objects. Out of this will come an anatomy of a 'scientific problem', and the possibility of giving a useful definition of this concept to serve for our subsequent discussion.

If we are to proceed by analogy with craft work in general, we need an anatomy of the tasks which are accomplished in that sort of activity. This is available in the writings of Aristotle. His scheme is most popularly known in terms of four 'causes': final, formal, efficient, and material. But it is clear from his own examples that these are the constituents of the task, which includes the work and its objects. The 'material' cause is the physical substance which is worked on; the 'efficient' cause is either the agent or his activity of shaping it; the 'formal' cause is the shape which is realized in the work; and the 'final' cause is either the purpose of the activity (the creation of a specified object) or the functions of the object itself. For

[7] See T. Kuhn, 'The Historical Structure of Scientific Discovery', *Science*, 134 (1 June 1962), 760–4, for an illuminating discussion of this example. Historians of chemistry have generally avoided emphasizing this error of Lavoisier, and have concentrated on his studies of combustion. But if Lavoisier had considered combustion rather than acidification as the essential property of this element, he would doubtless have called it *pyrogène* rather than *oxygène*.

[8] Thus, M. P. Crosland, *Historical Studies in the Language of Chemistry* (Heinemann, London, 1962) mentions the 'reactionary views of the Editor of the *Observations*' (p.189); this was de la Metherie, who objected to the new nomenclature on the (correct) grounds of its being misleading (not all acids being oxides) and involving 'hard and barbarous terms'; and he published letters showing (correctly) that the new nomenclature was theory-laden (in rejecting the phlogiston theory of combustion).

Aristotle, there was an intellectual and social hierarchy in these constituents, downwards from the 'final' cause; for this is the one whose setting is a policy decision at the highest level, and it then determines the less independent and less intellectual work of making plans and then shaping bits of passive matter.[9] Aristotle used the principle 'art imitates nature' to adapt this scheme to 'natural' productions; that is, processes which come to completion by their own internal plan, without the need for human intervention. With this, he could make his fundamental criticism of the earlier 'atomist' philosophers: that in concentrating on the efficient causes of the production of natural things, they had never inquired into what is a 'thing'.[10] On the other hand, Aristotle made a complete separation between the production of things and the achievement of knowledge. For him, knowledge was concerned with 'what could not be otherwise', and hence was eternal and unchanging; it had nothing in common with the production of things, which is in the realm of the variable and the contingent.[11] He may have been influenced in this by his training under Plato, and also by his brief historical experience of scientific knowledge. It is only in the twentieth century that we have become familiar with the historical evolution of scientific knowledge, in which great scientific achievements which seemed in their own time to give true and necessary results are inevitably modified and superseded.

We can adapt Aristotle's scheme to our own purposes by identifying the differences between the objects of the two sorts of craft work: making material things, and scientific inquiry. In our case, the task is not accomplished by the production of an object which has a certain function, and an appropriate shape imposed in certain sorts of matter. Instead, it is a statement of a particular sort. The apparent form of the statement can vary widely: it can appear as a report of an experiment or observation, a description and analysis of a complex situation, or as a statement of a fact of or a law, or as an hypothesis, theory or model. This list could be extended; but for our present purposes, the differences between these types of statements are less important than what they have in common. First, their 'materials', the objects to which they refer, are always the

---

[9] For a discussion of the basic role of the 'craft analogy' in Aristotle's thought, and its roots in earlier Greek philosophy, see F. Solmsen, 'Nature as Craftsman in Greek Thought', *Journal of the History of Ideas*, 24 (1963), 473–96, especially pp. 485ff.

[10] See the *De Partibus Animalium*, Book I, ch. 1, 642ᵃ 25.

[11] See the *Ethica Nicomachea*, Book VI, ch. 2, 1139ᵇ 18–36.

intellectually constructed classes of things and events. Particular instances are reported only as representative samples of those classes. Second, the 'form' of a completed piece of work is not a bald assertion that such-and-such is the case; rather, it is an argument, in which evidence is cited and from which a conclusion is drawn. In the discussion of the evidence, there will be some description of the relevant part, 'efficient cause', by which the evidence was produced. Under this will be included mainly the tools and techniques used; there may (depending on the conventional literary style for research reports in the field) also be mention of the difficulties and pitfalls encountered and overcome. Other aspects of the efficient cause which are usually omitted in research reports are a history of the investigator's work, and the more subtle methodical judgements which guided and controlled his work. And last, the statement which is the goal of the work as a whole, the establishment of new properties of the objects of the investigation, emerges as a conclusion from the argument.

Thus our first sketch of an analogy with the Aristotelian scheme yields as the 'material' constituent the classes of intellectually constructed things and events which are the objects of investigation; as the 'efficient' constituent the agent, with his work, tools, and methods; as the 'formal' constituent the argument; and as the 'final cause' the conclusion of the argument, with its new statements of the properties of the objects of inquiry. The data and information which provide the sole contact with the external world are not the conclusion and are quite distinct from it. They are nothing more than the foundation for the evidence, which concerns the objects of inquiry, and which is embedded in the argument about them. These different constituents of the completed work are interrelated, and they condition each other. The character of the 'material' for a given problem delimits the sorts of operations and tools which can be applied to it for the eventual production of evidence, and correspondingly delimits the sorts of arguments which can be constructed around that evidence. Conversely, the setting of a particular problem within a field will determine (within the above limits) the sort of data which will be brought into being, and the particular sorts of tools and arguments which will be brought to bear.

The artificiality of the objects of scientific inquiry is the key to the deep differences between this sort of craft work and that which Aristotle took as his paradigm; hence we can best start our closer analysis of this scheme with an examination of the 'material' component of a scientific problem. It is the establishment of new properties of these intellectually constructed classes that is the goal of the task. In this, the objects themselves are altered, for they exist only as classes defined by their properties. We should recall that not all their properties are exhaustively defined by the formal statements presented in the public record; each scientist must have a craftsman's intuitive, personal and partly tacit knowledge of his intellectual objects and of their physical samples, if he is to work creatively or even competently with them. But there must be a large common core of practical knowledge of the objects (again, not all of it in the explicit record) if there is to be an effective social endeavour of their study.

The study of these objects, which are intended to correspond to an inaccessible external reality, cannot progress very far without some interaction with the ultimate sources of our knowledge in that reality. But the fashion in which this interaction yields new properties of these objects is highly indirect and in principle inconclusive. We have already seen that the production of data, and its refinement into information, are craft operations which are governed by the problem at hand, and which produces less than conclusive inferences from the behaviour of particular samples in unnatural conditions. The connection between such a foundation in particular experiences, and the establishment of the properties of intellectually constructed classes of things and events, is in principle tenuous. And in practice, it has been forged only after long periods of development, extending over generations or centuries for the different fields of scientific inquiry.

The Argument

This link between the 'material' component and the 'final cause' of a problem is formed by the argument. Through it the experience is made relevant to the properties of the objects of inquiry. The argument will be a lattice-like structure of assertions about the objects of the work, connected by inferences which are accepted for the particular linkage functions they perform. Each assertion must be based, either directly or indirectly through these inference-links, on a statement of experience or on an explicit postulate. We shall see later (in connection with the 'adequacy' of solutions of problems) that the argument must have a structure that is both complex and subtle; for there is no formally valid pattern of argument that can establish properties of general classes from reports of particular experiences. The argument may be partly mathematical and deductive, but (outside purely theoretical fields) it must also include inductive, confirmatory, probabilistic, or analogical inferences, which are never capable of carrying certainty from premiss to conclusion.[12] Because of this lack of demonstrative certainty, the argument will include subsidiary arguments on the strength of particular inferences, especially for those that have a crucial position in the structure.

First, they cannot yield true and certain conclusions about the objects of inquiry, and still less about the external reality which they are intended to represent. This limitation is revealed by the subsequent history of any solved problem, even those which seemed (in their popularized versions at least) to establish irrefutable properties of natural things; I shall discuss this at greater length in connection with the evolution of scientific knowledge. Second, the necessary complexity and subtlety of a scientific argument makes impossible any trivial testing of its adequacy, as is possible through the examination of schematic structure in the case of simple syllogisms. Rather, the testing is done by an application of the criteria of adequacy accepted for the field. This itself requires judgements; and I shall discuss it more thoroughly in connection with 'methods'.

At first glance, an argument presented in a research report may appear to have a pyramidal structure, with a number of reports of experience, citations of known information, and perhaps postulates, all leading to a conclusion which consists of one or a few assertions. But this reflects the conventional style for such reports, rather than a real completeness of the argument. At a variety of points, the structure of such an argument could be developed to yield further conclusions; and one of the paths to the investigation of descendant problems involves doing just this.

---

[12] The complete set of such inferences is present, if implicitly, in any scientific paper where experimental results are obtained and also explained. Inductive inferences are involved in the taking of the sample studied as a true representative of its class; probabilistic inferences are involved in any assessment of the 'goodness of fit' of data; deductive inferences are the basic links in a verbal or mathematical argument; analogical inferences are used for relating the concepts understood by the mathematical symbols to the objects of inquiry involved in the experiment; and confirmatory inferences are made when the evidence is cited in support of the conclusions of a theoretical argument.

Evidence

Only in the most purely 'descriptive' of fields can the information have the appearance of being the direct and sole foundation of those statements about the objects of inquiry which comprise the conclusion. Generally, the information appears as embedded in the argument, providing a part of the basis for the conclusion. Hence this information in the argument is not a statement of the properties of the objects described in the conclusion; rather it is evidence brought into the argument which as a whole establishes these properties. The information which is selected from the available stock for use as evidence must be subjected to further testing, notwithstanding the fact that it was originally produced with this function in mind. For each use of information as evidence is special and demanding. Since no piece of evidence based on reports of particular things and events can entail a positive assertion about a general class, the weaker inferences which are made must be scrutinized in the light of the work they do.

The evidence must be of sufficient strength to support the load of argument placed on it in its particular location. Moreover, the evidence must be shown to fit the statement being supported, in the correspondence of the objects of its reports with those of the statement. A failure to make a proper assessment of the strength and the fit of evidence before including it in the argument can lead to pitfalls in the drawing of the conclusion. For these assessments, the scientist again makes use of the criteria of adequacy accepted for the field; and the interpretation and application of these criteria to each particular case involves making judgements, using the craft skill of the scientist.

The special category of 'evidence' is most easily recognized in fields where problems involve both complex arguments and large masses of information; and where the information itself does not bear obvious credentials of its reliability and relevance. An extreme example of this situation is in the law, where there is a highly developed 'law of evidence' for the presentation and testing of information offered as evidence in court cases. In the disciplines dealing with human history, the pitfalls which beset the inferences made from information in its use as evidence are a recognized hazard, and so any crucial piece of evidence must be carefully scrutinized for its strength and its fit. In most work in the natural sciences, one usually has either a large mass of information with a relatively simple argument, or a complex argument needing evidence at only a few points. Hence neither 'descriptive' nor 'theoretical' natural sciences generally require highly developed skills in testing evidence beyond the tests for reliability and relevance already involved in producing information. As in any other sort of work, when there are fewer pitfalls encountered and fewer special skills required the achievement of reliable results becomes easier; and hence more effective work can be done by the same level of talent. Thus one reason for the greater power of the sciences of nature, compared to most of the sciences of man, lies in the simpler character of routine work, and the correspondingly greater competence of mediocrity.[13]

But even in such well-established fields pitfalls can be encountered in the interpretation and use of evidence. For example, the contemporary fashion for using mathematical materials at every possible point of an argument induces a tendency to accept statistical information as facts, rather than as evidence, in a wide variety of fields. This can lead to negligence even in testing the reliability of such information in relation to the populations whose properties it describes. The pitfalls in the way of even the humble tasks of statistical data-collection are well-known to those who make their living by the quality of the inferences based on such materials (as commercial market surveys) but not always so well-known to others. Moreover, the relevance of the classes established for the collection and processing of data to those of the argument of the problem is by no means automatically assured. Even when the problem is a purely 'empirical' one of determining a statistical property of a population,

[13] The classic paper on the determination of the values of physical constants, R. T. Birge, 'Probable Values of the General Physical Constants', *Reviews of Modern Physics*, 1 (1929), 1–73, shows how subtle is the evidence, and how complex are the arguments, in really fundamental physical research. In his introduction (pp. 1–7) he lists the common pitfalls in the conversion of experimental data to what I call 'information': confusion in the sense of 'experimental error'; the fitting of curves or lines to graphed points by estimation rather than by calculations; and the use of incorrect values of the 'auxiliary constants'. The argument will involve a discussion of the known sources of inaccuracy in the data and information, as well as complex calculations for deriving the final results. He stresses that the experimenter (as well as the reviewer) must exercise his judgement at every stage. Finally, he observes that some painstaking experimental work, which had produced accepted results, was in fact vitiated by one or another error: and, in some cases, it was even impossible to 'save' the experimental data because the loss of the records of its computation (including special techniques and values of auxiliary constants) prevented a re-working of it. In D. Lerner, *Evidence and Inference* (The Free Press, Glencoe, Ill., 1959), there is an essay on 'Evidence and Inference in Nuclear Research' by Martin Deutsch (pp. 96–106) which shows how indirect and conditioned both by theory and by expectation is the data of high-energy physics.

there must be an argument establishing the quality of the information; and if any inferences whatever are drawn, there must be supplementary arguments for establishing the strength and fit of the statistical evidence for its function in the main argument.[14]

The pitfalls to be encountered in connection with evidence are most noticeable when existing information is taken over for use in a problem. Whether this material is immediately capable of an adequate strength and fit for its function in the argument will depend on the mode of its original production. For when it is cited, the argument by which it was derived is implicitly included in the argument of the new problem, as a subsidiary argument for its adequacy. Also its objects are implicitly assumed to be those of the new problem. If the original objects were significantly different from the new ones, or if the original argument was of a pattern which is not adequate or is inappropriate for the new problem, then the information itself is inadequate for its new functions as evidence and the new argument fails. In a general way, these considerations lead to a precept of method, of using only that existing information which is at least as solidly based on experience as the new information of the problem at hand. As an extreme case, one does not use a hypothetical conjecture as crucial evidence in a largely inductive argument. But the pitfalls do not lie exclusively in shifts in this direction; for the objects of inquiry in problems closer to experience will frequently be different from those of the same names in more theoretical investigations; and so even 'empirical evidence' may be irrelevant and misleading to the inquiry at hand.

The neglect of the category of evidence clearly derives from the traditional concerns of the philosophy of science, and its paradigm examples of scientific achievement naturally obscure this aspect of scientific work. The classic inclined-plane experiment of Galileo has served for generations to exhibit science as the straightforward and simple production of data which immediately becomes fact. In recent years, historical scholarship has shown how deeply theory-laden was this particular experiment, in its concentration on very

[14] On the abuse of national statistics for economic policy-making, see A. Coddington, 'Are Statistics Vital?', *The Listener*, 82 (1969), 822–3. He deals not only with the uncertainty in the sampled data and scaling factors whereby the gross statistics are calculated, but also with the more subtle questions of the effect of the systems of conventions on the production of the data itself; in this last respect his analysis goes beyond that of O. Morgenstern.

artificial and novel objects.[15] Moreover, there could not have been chosen a more misleading example of Galileo's approach to the use of experience. An attentive reading of Galileo's major works will show that he was fully aware of the problems of evidence, and of the qualities of strength and of fit of evidence deriving from various sorts of experience for various sorts of argument. Quantitative experiments actually performed were very much the exception for him; his evidence derived from a continuum of sorts of experience, ranging from surprising phenomena and craft experience, through to abstract thought experiments.[16] His conclusions show a corresponding variety, from the initial framing of problems to be solved, through the tentative or speculative solution of problems, to rigid mathematical demonstrations on explicitly artificial objects.

When we leave the world of routine scientific work by competent craftsmen, the problems of evidence change radically. In any genuinely novel work, or even in work of crucial importance for a big problem, the accepted criteria of adequacy may not extend to cover all the inferences made from the evidence in a particular argument. For the novelty of the problem will generally entail a corresponding novelty, and uniqueness, in evidence and its relation to experience and to the argument. Also, since such work is frequently done at the

[15] The pioneering work on this interpretation of Galileo was done by A. Koyré, *Études Galiléennes* (Hermann, Paris, 1939). Koyré even argued that Galileo could not have achieved the precision he claimed. But the doubts of Galileo's French contemporaries which Koyré cites (ii, 73) are not on the distance-time law, but on other assertions of Galileo, published in the *Dialogue* of 1632; see R. Lenoble *Mersenne, ou La Naissance du Mécanisme* (Vrin, Paris, 1943), pp. 465–6. For a brief study of this as part of Galileo's work, see The Mathematical Association, *A Second Report on the Teaching of Mechanics in the Schools* (Bell, London, 1965), ch. 6, 'Problems and Methods in Mechanics—an Historical Sketch'.

[16] In the *Two New Sciences* we find surprising phenomena in industrial practice mentioned at the very beginning as a fertile source of problems (*Edizione Nazionale*, p.49). Among these is the 'size effect', the rule that a large ship under construction would fall apart under its own weight unless supported, while a small one of the same design is stable. Craft experience is the basis for the observation that a rope can be made to grip on the smooth drum of a windlass (p. 58), that a rope breaks by the tearing, not slipping, of its strands (p. 56), and that ropes can lift heavy weights in being shortened through wetting by mist (p. 67). Galileo's experiments include the classic quantitative ones; qualitative experiments such as the adhesion of two polished plane surfaces (p. 59); and experiments that could not succeed, as that of a piston loaded with a weight until it descended against the 'force of the vacuum' (p.62). Thought-experiments abound in the *Third Day*; one shows that an imperceptible height of fall produces an imperceptible impact (p. 200). Finally, Galileo solves some problems by 'pure reason alone'; thus he argues against an instantaneous assumption of a determinate speed by a body beginning to fall, on a principle of 'sufficient reason': no speed is more likely than any other (p. 200).

limit (if not beyond) of the capabilities of the existing tools for producing reliable information, the assessment of the strength and fit of the evidence can become very subtle indeed.[17] Then there can arise disputes about the adequacy of the solution to the problem, which cannot be resolved either by a scrutiny of the data and information, nor by an appeal to accepted criteria of adequacy. Thus at such points, this aspect of the 'objectivity' of scientific knowledge, which is really a result of a successful social tradition of producing and testing the materials of that knowledge, breaks down. In the long run, to be sure, further work will decide the issue; but the decision on whether to engage in such further work, which partly depends on the assessment of the adequacy of the controversial piece, must be taken now. Thus, at such infrequent but critical junctures in the advancement of science, the assessment of the evidence adduced in an argument becomes a crucial judgement, in which the individuals are thrown back on their own personal resources. They are forced to put themselves at risk in making the judgement, and they lack the safe channels of an accepted tradition to steer them towards the correct answer.

The Conclusion

At the end of an argument, comes the conclusion; and this is the completion of the cycle of a research project, which is the first step towards the achievement of scientific knowledge. The artificiality of this product of craft work should by now be obvious. The conclusion is not concerned with 'things themselves', but with those intellectually constructed classes which can serve as the objects of an argument. The contact with the external world is always unnatural and indirect; and the reports of that contact do no more than serve as the basis for evidence which is embedded in an argument whose pattern cannot be formally valid. Although the statements in the conclusion refer to explicit objects, and are capable of being understood by any competent practitioner, and although every stage in the argument, including the reports of experience, can

[17] Pasteur's great discovery of left and right-handed crystalline forms of tartrates was exceptionally fortunate in its circumstances; Pasteur could convince the influential Biot of its validity by an experiment performed in his presence. Otherwise, he might have had difficulties, for it is only under extremely carefully controlled conditions that the chemical reactions needed for producing the effect can go through properly; and crystals of the sort observed by Pasteur are very rare. See A. Ihde, *The Development of Modern Chemistry* (Harper & Row, 1964), pp. 322–3.

be reproduced or tested, the work as a whole has been conditioned by personal judgements, depending ultimately on a private, craft knowledge.

Whether such an account of the completion of the investigation of a problem seems a reasonable one will depend on the experience of the reader. For those who imagine science as the accumulation of hard facts and indubitable truths, a 'conclusion' as described here will appear to be a miserably weak result, a parody of the achievement of scientific research. Indeed, it is hard to imagine how the magnificent edifice of our established scientific knowledge could be composed of such feeble elements. We will later see that in fact it is not so constituted; for the conclusion of the argument, along with the other components of a research report, enters a new cycle of development, in which it is further tested, and also transformed, so that its weaknesses are exposed and if possible corrected.

But for those with either an experience of scientific work in any lively field, or some knowledge of the history of science, this description should seem a fair one. Indeed, through an understanding of the inherent limitations of the conclusion of a problem, we can appreciate the inevitability of error in scientific work, at least in the productions of all but the men of the greatest genius. For a conclusion can be no better than the evidence on which it is based, and the objects in whose terms it is framed. The objects of enquiry delimit the sorts of problems that can be set; and to the extent that these are capable of enrichment and deepening, then will later problems be correspondingly improved, and their solutions more powerful. Similarly, the evidence depends on these, and on the tools available at any time; and these too improve. When there is a debate over a scientific result, neither side can draw a blank cheque on what might be produced in its favour were better tools available; the issue must be decided on the evidence that is there for scrutiny. A new set of tools, providing new and better evidence, may well tip the balance, so that while it was reasonable and correct to support one solution under the old conditions, the new ones require a change of position. Such changes are not easily made by a scientist in mid-career, of course: but that is another problem, of a practical sort.

From this argument, we can derive the paradoxical conclusion, that on occasion it has been 'correct' for scientists to adhere to an 'incorrect' theory, even when offered what we now know to be 'correct'. For the philosophy of science of the nineteenth century, such a conclusion would have been more than paradoxical; it would have been anathema. So strong then was the commitment to a faith in a cumulative, infallible progress of scientific knowledge, that erroneous opinions (either old ones, or established beliefs discovered to be so) needed to be explained away by specially constructed historical myths. Such was the fate of 'caloric', of 'phlogiston', and of 'vitalism'; in each case it was shown that only a refusal of certain scientists to open their eyes to the obvious kept the bad old theory alive. What such a commitment did to the study of the history of science needs no lengthy description here; historians must still cope with the remnants of a crop of fantastic 'anticipations' of scientific truths produced in that earlier period.[18]

We can take this conclusion even further, and state the thesis that the forward progress of science must necessarily be accomplished largely by way of detours. For, again, it will only be the man of genius who can sense, beyond the limitations of the evidence at his command, what avenues of advance will be the most fruitful ones

[18] The most influential of these were the supposed 'precursors' of Galileo's dynamics and kinematics in fourteenth-century Paris, discovered by P. Duhem. See E. Moody, 'Galileo, and his Precursors' in *Galileo Reappraised*, ed. C. Golino (University of California Press, 1966), pp. 23–43. Perhaps the most famous case of a theory whose correctness was established only gradually, indirectly and incompletely, is the heliocentric system of Copernicus. For a century from its first announcement, it suffered from its violation of the principles of the only coherent and accepted system of physics, that of Aristotle. The two sorts of observational evidence that might support it, stellar parallax and effects on falling bodies of the earth's rotation, could not be produced until about three hundred years after its announcement. For working astronomers, almost all its constructions could be translated into models that did not involve the motions of the earth. The one exception, in the later sixteenth century, was the theory explaining the slow changes in the elements of the orbits of the celestial bodies. Although this did win adherents to the Copernican system, for it promised a solution to the pressing problems of calendrical reform, the problems were at the limits of the competence of astronomers, and plagued by spurious data. Galileo's own polemics were mainly against Aristotelian natural philosophy; his one sustained argument for the motions of the earth was his fallacious and generally incomprehensible theory of the tides. Thus we can say that, up to the early seventeenth century, a judicious astronomer who had no metaphysical bias in his assessment would return the opinion 'not proven 'on the Copernican system, and treat it as an hypothesis.

Through the seventeenth century, both natural philosophy and astronomy provided an increasingly powerful set of indications that the Copernican theory was true in essence if not in detail. Some of Kepler's technical achievements (as in the case of the problem of planetary latitudes) depended on a heliostatic system; the orbits of comets could be interpreted only on its basis; and the Aristotelian system fell out of favour. Finally, Newton's great achievements in the prediction and explanation of the planetary motions, which could not be comprehended except on the assumption of a rotating and orbiting earth, provided 'moral certainty' to the heliostatic hypothesis. By this time, all the technicalities of Copernicus's astronomical system were utterly obsolete.

over the years after his own work. The others, working with the imperfect materials available to them, proceed in the most natural direction in their choice of problems and in their assessment of controversial conclusions. That the path does get straightened out eventually, or rather that a successful path can be recognized in retrospect, is an historical fact; for we have the body of genuine scientific knowledge as witness to the possibility of such a corrective process. How the process works is a matter which we will discuss later at the appropriate point.

The artificial and fallible nature of the conclusions of solved scientific problems has an importance beyond the improved historical understanding of natural science. For the same holds true in any field of thought and action where there is an explicit body of theory in whose terms arguments are cast and conclusions are drawn. Since so much of the direction of the technical and social aspects of our existence is now done by specialists trained in a formal 'science' of their craft, an appreciation of the limitations of the conclusions to scientific problems is as important for politics as much as for epistemology.

Working on a Problem

Having sketched the 'anatomy' of a solved scientific problem, I may now indicate some aspects of the 'physiology' of scientific inquiry. The distinction of the different phases of work, relating to the production of data, information, evidence, argument and conclusion, does not imply that these are successive, discrete and independent stages of every investigation of a problem. Although the work on a problem tends to be dominated by different tasks at different stages of its progress, there is an alternation among them, and they also condition each other's plan and performance.

For example, we have already seen that the choice of the data which is to be produced is influenced both by the sort of information which is desired, and also by the special tools with which the data and the information will be worked. In its turn, the information is intended to function as evidence; and the evidence itself depends on the particular form of the argument which will be developed, and hence on the conclusion which one hopes to derive. Thus in this work, as in Aristotle's analysis of handicraft, the 'final' cause is the first, in the sense that it conditions all the others. In the investigation of scientific problems, however, the conclusion cannot be completely known in advance. In such a problem (as distinct from other sorts), if one starts with a precise question admitting a small range of well-defined answers, then it is almost certain that the result will, as an enrichment of knowledge, be nearly trivial.[19] For in such cases, nothing new is gained by the interaction with the external world; and the solved problem yields little more than what was put into it, in the form of the first question that could be formulated. In scientific work of high quality, one must start with a general idea of the sort of conclusion that might be achieved, and of the sort of evidence that might be produced to support it, and then gradually develop a programme of detailed work.[20] This is always tentative and 'experimental' at first, for unexpected results at any point will modify the possibilities of the use of the materials at later stages, and also require the provision of new materials at earlier stages of development. Pitfalls can be encountered at any point, and when they are discovered there must be surveys of the damage, and a reorganization of the work for the necessary repairs to the incomplete structure of the argument.

When the work on a problem is well under way, there is a hier-archically ordered set of subsidiary problems under investigation, each one concerned with the production or perfection of a component of the intended argument; perhaps new data or information, or some particular chain of reasoning. These subsidiary problems are of a

[19] In this point I am *not* arguing that the answering of a precisely framed question is necessarily a trivial undertaking. The testing of a hypothesis by a properly designed experiment may be necessary for crucial evidence for a deep problem. The issue here is how much work has gone into the development of the question, between the first con-ception of the problem and its refinement into its penultimate form. If the scientist has learned nothing about the problem, or if there has been nothing to learn about it, once he had the first partly clear idea of it, then it is hardly likely that there will be much in the result from which anyone else can learn. The distinguished American physicist Karl Compton put it succinctly: 'When I was directing the research work of students in my days at Princeton University, . . . I always used to tell them that if the results of a thesis problem could be foreseen at its beginning, it was not worth working at.' Quoted in D. S. Greenberg, *Politics of American Science*, p. 155.

[20] For historical scholarship, an illuminating account is given by E. H. Carr, *What is History?* (MacMillan, London, 1961), ch. 1, 'The Historian and his Facts' (pp. 1–28). Describing the interaction of the two activities of reading and reflection (either in the mind or in writing drafts), which are mutually enriching and correcting, he says: 'If you try to separate them, or give one priority over the other, you fall into one of two heresies. Either you write scissors-and-paste history without meaning or significance, or you write propaganda or historical fiction, and merely use the facts of the past to embroider a kind of writing which has nothing to do with history' (p. 23). In this fashion Carr disposes of some classical pseudo-problems of the philosophy of history, which arise from a con-ception of historical research that is a caricature of its actual craft practice.

different character from the principal problem under investigation. They belong to a class which I shall later discuss as 'technical problems', where the function of the required information or device is known in advance; in this case, from the requirements of the main problem. Such technical problems do not have the same freedom to grow and evolve; and where they are subsidiary to a larger investigation, they can be left to less experienced or less imaginative workers. Of course, when a subsidiary problem turns out to be incapable of solution to the standards of adequacy required for the function its result will perform in the completed argument, that part of the argument must be re-examined to see what other sorts of results will be adequate there, and new subsidiary problems set and investigated. It is in this complex fashion that a problem grows through interaction of the scientist with his materials. The inspired guesses and great illuminations which bring a problem into being are only the beginning of the work; and the impersonal, demonstrative narrative of the published research report are only a stylized record of its completion. What lies between is demanding and subtle craftsman's work on the very special objects of scientific knowledge. But it is not an entirely undifferentiated set of intuitive procedures. Using the categories of data, information, evidence, and argument, one could construct a chart describing the state of the work on the main and subsidiary problems of a scientific inquiry at any point of its development. The organization of complex administrative and engineering projects had been considerably helped by the use of charts for flow-diagrams and 'critical path analysis'. A chart along the lines suggested here would, at the very least, show by its complexity the degree to which a scientific project was other than straightforward routine; and a sequence of revisions of the chart would provide a record of the evolution of the problem itself.

The similarity of structure between these other sorts of projects and a scientific problem should not obscure one very deep difference. Although the objects of such practical or technical problems are also artificial to some extent, they do not change their nature in the course of the work. One of the things that makes scientific problem-solving so uniquely subtle is that the very objects of the work evolve as the work goes on, and in a fashion which is not predictable in advance. For the discovery of new and unexpected properties of the objects of the investigation entails a change in the objects themselves; the objects described in the conclusion of a problem with genuine novelty are not those which existed when work on the problem began. This is not a purely philosophical point; for in such problems, there is a need for a frequent review, and a re-setting of subsidiary problems, lest the work done at the earlier stages of the project be irrelevant to the needs of its conclusion.

Thus in a significant scientific problem, the objects of inquiry are themselves plastic. They can remain unchanged throughout the investigation only when the scientist is unusually prescient when he sets the problem, or when the problem itself is banal. It is for this reason that the investigation of real scientific problems cannot be reduced to 'asking a question of Nature' with the expectation of a simple answer, or 'testing an hypothesis' to see whether it is true or false. Even the description of 'normal science' as 'puzzle-solving' carries the connotation that the puzzle is there in advance, and also that it has a unique solution.[21] Much of the routine work done by scientific workers is such puzzle-solving, on essentially technical problems. But good scientific work, even when it is not revolutionary, is more demanding, and more interesting, than that.

The work on a particular problem is completed when an argument, meeting the accepted standards of adequacy, can be framed and conclusions drawn. But this does not occur suddenly, as with the dotting of the last *i*. Rather, the cyclic interaction of the various materials of the problem decreases in intensity, the argument is stabilized, the evidence becomes sufficiently strong and well-fitting, and fewer lots of new information and data are required. The unexpected results decrease in importance, pitfalls become negligible, and there is a sense that the whole process is 'converging' towards solution. But this does not always occur; and a warning sign of the imminent failure of a problem is when the difficulties begin to 'diverge'; when the subsidiary problems called into being by new difficulties become larger and more fundamental, and ever more extensive modifications of the argument become necessary, with increasing lots of fresh data being required for throwing into the breaches. An important part of the craft skill of a scientist is to sense whether a problem is beginning to converge as early as it should, and to detect signs of incipient divergence; and then to decide when to abandon a doomed venture. It is the lack of such a skill that takes

[21] These terms derive from T. Kuhn, *The Structure of Scientific Revolutions*, ch. 4, 'Normal Science as Puzzle-Solving'; the dichotomy between such 'normal' science and 'scientific revolutions' expressed in that work is, as Kuhn soon saw, somewhat too sharp.

beginning research students through to hopeless muddles, and on to the final despairing struggle to salvage something of value from the wreckage of a failed problem.

'Scientific Problem' Defined

Having identified the constituents of a completed scientific problem, by analogy with the four 'causes' of Aristotle, and also having sketched the cycle of operations by which a problem is investigated and solved, we can now attempt a formal definition of the term 'problem' itself. This term is becoming common, both in the description of scientific work and in its philosophical analysis. The concept requires some closer analysis, both for its philosophical use, and for the distinction between 'scientific problems', which I am discussing here, and some related, but different, sorts of problems which I shall discuss later.

When can we say that a problem 'exists'? We would like to be able to distinguish between a 'problem' which is ready for investigation, and less well-defined things, such as a sense that a certain sort of difficulty, theoretical or practical, needs to be resolved; or that a certain sort of work could or should be done. Any definition of the instant of birth of a 'problem' out of a 'problem-situation' is bound to be arbitrary, since all the components of a problem undergo continuous and possibly radical change during its investigation. Taking our clue from Aristotle, we can say that the 'final' cause is the first one; only when there is some specification of the new conclusion to be drawn can we say that a problem exists. With this as the basis, all the other constituents fit into place. The 'material' cause, the existing stock of relevant scientific results as it is to be modified by the solution of the problem, is presupposed in the statement of the problem. Although the 'formal' cause, the argument from which the conclusion will be derived, cannot be known in detail, the accepted patterns of argument for the field can usually be presupposed. Of course, in a deep investigation where this itself is modified, the statement of the problem is correspondingly more difficult. But something must be known of the 'efficient cause'; for a conjecture or hypothesis which is not accompanied by a plan for its establishment may be quite interesting but it is not something on which work can be done.

Herein lies another reason why it is insufficient to characterize a scientific problem simply as 'a question put to Nature', or as 'an hypothesis to be tested'. Judging such questions on purely internal features, as their surprise, improbability, or organizing and unifying power, can lead to utterly unrealistic accounts of the evaluation of scientific problems.[22] The question must contain, in addition to its implied answer, some plan (implicit or explicit) for the attainment of the answer. For the solution of genuine scientific problems is not merely having bright or even brilliant ideas; these are empty unless they are developed and enriched by the hard, complex and sophisticated craft work of scientific inquiry. Unless there is some idea of how the work will be done, there is no way of knowing whether the solution can even be achieved; and in general the form that the tentative solution takes will depend on the projected means of its accomplishment. This is of course a matter of degree; as a limiting case there is the handful of classic unsolved problems in mathematics. But these are accepted as genuine problems either through the naturalness of their objects, the empirical evidence for their truth, or through the authority of their proposers.

The use of the term 'problem' to define the programme of inquiry goes back to antiquity in mathematics; although even there it has the connotation of being able to do something rather than to know something. It has occasionally been used to give point to a precept of method in other fields, the most famous example being Lord Acton's dictum, 'study problems, not periods'. But it has tended to carry with it the implication that solving a problem is a form of inquiry which is both safer and less deep than something else which is being abjured. It would be most unfortunate if this present discussion were taken as evidence for the proposition that 'all the scientist does is to solve problems' as if this were equivalent to doing crossword

[22] Lest the very idea of 'hypothesis' be lost when the so-called 'hypothetico–deductive method' loses favour among philosophers of science, I should remark here that the pattern of argument in statistical inference necessarily involves the report of the testing of a prior hypothesis; and many pitfalls of interpretation can be encountered unless the work actually proceeds by first framing the hypothesis and then getting the relevant data. But a worthwhile hypothesis for statistical testing can only be set when the problem has gone through its earlier stages of development, and is ready to be put in a final, crystallized form. The deep and complex problems of the validity of such real 'hypothetico–deductive' arguments are little studied by the philosophers who have concerned themselves with this 'method' in the abstract. It is noteworthy that the *British Journal for the Philosophy of Science*, a leading journal in its field, has had just a single paper on the logic of statistical inference in the twenty years of its publication. The exceptional paper which proves my rule is I. Hacking, 'On the Foundation of Statistics', *British Journal for the Philosophy of Science*, 15 (1964), 1–27. See 'Index 1950–1969 Volumes 1–20', *British Journal for the Philosophy of Science*, 21 (1970), 21.

puzzles. For I am arguing that whatever a scientist does, it is best conceived as the investigation (including both the creation *and* the solution) of problems. We shall see that problems can vary in depth from the trivial to the profound, and that when genuine scientific knowledge comes to be, it is achieved through a complex social endeavour, where the materials embodied in the solution of one problem are tested and transformed through their use in the investigation of subsequent problems.

We may now put our definition formally, and say that a scientific problem is a statement (always partial and subject to evolution) of new properties of the objects of inquiry, to be established as a conclusion to an adequate argument, in accordance with a plan (specified to an appropriate degree) for its achievement. The naturalness of this definition, for the social activity of science in the later twentieth century, can be seen from its implicit use when applications are made for research grants and contracts. The two components, statement and plan, which we might call the 'final' and the 'efficient' causes of the problem, are each necessary, and are jointly sufficient, for specifying the project. On their basis, the judgements of value, feasibility and cost can be made; and it is in terms of such judgements that the decision is made for investment in the work. The evolution of the problem in the course of inquiry is a common occurrence; those who control investment in scientific work are aware of this tendency to change, and they can also tell when one problem has turned into another.

With this definition of 'scientific problem', my analysis of the cycle of the first, individual, phase of the achievement of scientific knowledge is nearly complete; in a subsequent chapter I will discuss the second, social, phase, in which the solved problem is accepted as a research report and is then subjected to further work before becoming a 'fact' and ultimately 'knowledge'. In this chapter we have seen how the problem is solved by a conclusion being drawn about the classes of intellectually constructed things and events which are the objects of the investigation. The conclusion is the outcome of an argument, and is related to the external world through the evidence embedded in the argument. This evidence derives ultimately from the data, which is the sole point of contact with the external world, through the information which is produced from it through the application of tools. Throughout, the work involves the use of craft skills, and the making of a variety of judgements; the character of both of these is conditioned by the artificiality of the objects of the inquiry.

The Origins of a Problem

This is an appropriate point to complete the description of the cycle of investigation of a problem, and to describe how a problem comes to be. The history of any inquiry begins with something that is less than a problem; we may call it 'problem-situation'. By its very nature, this is difficult to define. It may be considered as an awareness that there is a question to be asked, without anyone being able to frame the question successfully. Alternatively, it can be regarded as a recognition that certain functions might be performed by the solution of some problem as yet undefined. The variety of such functions corresponds to the complexity of the frontier of established results in a field. There may be phenomena calling out for explanation; there may be a conflict between two schemes of explanation; a given result might seem capable of being made more deep or general; an established conclusion might appear to require refutation or correction; the list can go on and on, and it would be pointless to try to give an exhaustive formal account to describe what is the particular craft practice of each field of inquiry.[23]

---

[23] Examples of each of these functions of a problem yet to be defined, are as follows. The high-energy particle accelerators of the 1950s and 1960s produced a multiplicity of phenomena; for the theoretical explanation of some of them, problems could be set and solved; for others only problem-situations remained. The production of work by a heat-engine could be explained, in the 1840s, either by a 'fall' of caloric in temperature analogous to that of water in height, or by the 'conversion' of sensible heat into work; the conflict between these two set the problem-situation for William Thompson (later Lord Kelvin). (I am indebted to D. S. L. Cardwell for this example). The problems of the electrodynamics of moving bodies, and their apparent incoherence in failing to be Galilean-invariant, set the problem-situation for the young Einstein. See G. Holton, 'Einstein, Michelson, and the "Crucial" Experiment', *Isis*, 60 (1969), 133–97. Fourier was put on the track of his successful theory of the conduction of heat in solids by a paper of Biot on the steady-state temperatures of a heated bar; Biot's physical model was superficial and inadequate of its analysis, and Fourier immediately saw that it could and must be deepened somehow. See I. Gratton-Guinness, 'Joseph Fourier and the Revolution in Mathematical Physics', *Journal of the Institute of Mathematics and its Applications*, 5 (1969), 230–53; a book on Fourier's work by Gratton-Guinness and myself is forthcoming. Finally, it is likely that Davy's antipathy to the 'French' system of chemical nomenclature posed a problem-situation of refutation; this crystallized into the problem of demonstrating that the 'oxy-muriatic base' of muriatic acid was, by Lavoisier's own principles, an element. The thick green gas was subjected to every possible process of decomposition, and resisted; hence Davy named it 'chlorine', and destroyed the foundation of the nomenclature based on 'oxygène' being the principle both of burning and of acidification (from which its name was constructed).

The problem-situation will persist as long as there is a recognition of the function to be performed, without a successful framing of the problem which will enable its performance. This phase of the 'potential' existence of a problem may be very brief, but for really deep problems, the bringing of the problem itself into existence can be a long, arduous, and hazardous operation.

Relating the problem-situation to the Aristotelian scheme of the task of investigating a problem, we notice that the task itself, defined by its goal of the establishment of certain new properties of the objects of inquiry, depends for its existence on a prior 'end': the function defines the problem-situation. In this there is a point of similarity between scientific problems and technical ones; but in the latter case the function remains dominant throughout the inquiry, while here it can easily be displaced if the investigation of the problem reveals new possibilities in the material.

Although the problem-situation is in a less specified state than the problem to which it may give rise, it is already a very artificial thing. The very existence of a problem-situation presupposes a matrix of technical materials: existing information (with the intellectual objects it describes), tools, and a body of methods including criteria of adequacy and value. For in the absence of such a matrix of technical materials, a genuine problem could never come into existence, and the decisions on investigating it, and on shaping it during the work, would have no foundation.

Because of the unspecified character of a problem-situation, its development does not require the precise thinking and concentration of attention that a fully developed problem demands. Hence the perception of what is needed to crystallize it into a well-defined problem need not come as the result of straightforward toil. There are well-attested cases of mathematicians suddenly having a (correct) insight into the solution of a difficult problem, usually during a period of rest or distraction from intensive and unsuccessful work.[24] In natural science, the classic cases of crystallization of a problem on perception of a commonplace event tend to be apocryphal, for the problem-situation itself is already so technical that the 'falling apple' will provide insights relating to it only indirectly. And in the course of an investigation leading to a great discovery, the problem-situation itself will tend to be fluid, and the identification in retro-

[24] See J. Hadamard, *The Psychology of Invention in the Mathematical Field* (Princeton University Press and London, Oxford University Press, 1949).

spect of that most crucial insight will tend to be conditioned both by later events and by fashion in theory of discovery. But without a problem-situation, a crucial phenomenon, whether commonplace or artificial, would go unremarked either as ordinary or as a stray anomaly. This is the basis for the maxim, 'chance favours the prepared mind'; and the more trivial the crucial phenomenon, the more certain it is that the observer had been preparing for it intensively. The mind must be well-prepared for yet another reason; for the phenomena, as they occur, will have none of the reliability and definition of the data which are later produced in the course of the investigation of the problem. They are as likely to lead to pitfalls as they are to show the correct path. The phenomenon of the limited working height of lift-pumps misled Galileo; for he related it to the problem of cohesion in solid bodies, on which he was building a sort of solid-state physics, and in doing so he was led away from the correct explanation of the hydrostatic phenomena previously explained by the principle of 'Nature's abhorrence of a vacuum'.[25]

Following on the perception of a crucial phenomenon comes the initial insight into the problem and its solution. At this point one may speak of the problem being born. This is the most exciting and also the most dangerous time for the scientist. Quite suddenly all, or nearly all, the pieces of the puzzle fall into place. The conclusion to be drawn takes shape, becomes real, and awaits only the performance of the routine detailed work to be acceptable as genuine. At this point in the work every scientist of real originality and commitment is filled with a 'sober drunkenness';[26] the experience is as close to the mystical as any other in our culture. The dangers arise because this initial insight will determine the form of the full specification of the problem, and will include error along with truth.

[25] In the *Two New Sciences*, the lift-pump phenomenon is introduced after the discussion of cohesion in solids with which the book begins (Edizione Nazionale, p. 64).
The importance of the awareness of the problem-situation for a recognition of phenomena is shown by C. Süsskind, 'Observations of Electromagnetic Wave Radiation before Hertz', *Isis*, 55 (1964), 32–42. There were many observations of the reception of electric signals at a distance, but these were unsystematic and easily dismissed by leading scientists.

[26] This term is borrowed from L. Edelstein 'Recent Trends in the Interpretation of Ancient Science', *Journal of the History of Ideas*, 13 (1952, 573–604; reprinted in *Roots of Scientific Thought*, ed. P. Wiener (Basic Books, New York, 1957), pp. 90–121. He quotes Ptolemy: 'I know that I am mortal, the creature of a day; but when I search into the multitudinous revolving spirals of the stars, my feet no longer rest on the earth, but, standing by Zeus himself, I take my fill of ambrosia, the food of the gods', p. 583 (p. 100 of the Wiener volume).

Days or years can later be wasted in an attempt to confirm what once seemed blindingly obvious. It is likely that the braziers which Galileo saw swinging regularly in a church had this hidden danger; for he never ceased to believe, and doubtless spent much time trying to prove, that the period of a simple pendulum is independent of its amplitude.[27] Worse yet, it takes superhuman coolness for a scientist later to recognize that his great insight was misleading, and in effect to amputate part of his life in rejecting it. The deeper and more powerful the initial insight, the greater the force of personal identification with the problem it defines. This can last for the rest of a man's life, long after the initial insight has been whittled down by destructive criticism, or bypassed by the progress of the field. From such situations are personal tragedies made. They also pose the most fascinating problems for historians of science, for the initial insight is frequently buried deep in the body of the mature work which finally results from it; yet to interpret that work historically it is necessary to reconstruct it from its beginnings.

The setting of the problem is not completed by the achievement of the initial insight. This must be refined in the light of what is available from the existing results in the field, including information, explanations and tools. Also, there must be a judgement of the degree of specification which is appropriate to the problem at this stage. Too complete a specification can result in the waste of effort on incorrect lines of attack; too little specification can lead to an aimless wandering around the field. Above all, the scientist must examine his nascent problem, and make the judgements of value, feasibility, and cost; and only then decide whether to gamble his time and (usually) someone else's money, on its investigation.

Thus in the early stages of the creation of a problem, routine craft operations are less significant than imagination and judgement; and so this phase of the work will be strongly influenced by the personal style of the scientist. Moreover, at this crucial early stage, the character of the work required, and the appropriate attitudes, change with great rapidity. At first one must be able to think in an analogical and associative fashion, to perceive the crucial phenomena for what they are; then to proceed boldly and speculatively to the initial insight; but soon after to be rigorously self-critical in the definition and assessment of the problem. It might be tempting to imagine these different functions split up among members of a team,

27 See *Two New Sciences*, Edizione Nazionale, pp. 139–41.

benefiting from each other's strengths and protecting against their weaknesses. But since deeply novel ideas cannot be communicated until they have received a fair degree of specification, the basic work of creation, with its special hazards and pitfalls, must rest with an individual. This is why the achievement of great discoveries in science is not a matter of perceiving the obvious, or still less of luck, but is reserved to those with greatness of intellect and of at least some aspect of character.

This dramatic picture of the investigation of problems may tend to convey the misleading impression that all scientific work is an heroic endeavour, as in a style of propaganda for science which was once dominant but which is now becoming rapidly obsolete. I shall later discuss the problems of mass-production in science; but for now it will suffice to indicate the diversity among problems, in respect of their degree of elaboration and their challenge and difficulty. It is easy to see that the evolution of a problem, and the change of the objects of inquiry during its investigation, will tend to be most rapid and dramatic in those fields where the argument is most highly elaborated, and where the production of data from experience is a relatively small (although perhaps crucial) part of the work. These are the 'theoretical' sciences; at the other extreme are the 'descriptive' ones, where the data is largely found rather than manufactured, and where a mass of data provides evidence in a simple argument. In the latter sort of field one might seem to be 'fact-collecting' were it not for the presence of pitfalls at every stage of the work in all but the most routine of problems. Even though great and dramatic advances can be made in descriptive sciences, the pace is slower, and the style of original work is different: breadth of experience and maturity of outlook weigh more heavily than in the abstract, argumentative fields. In such fields, one may consider a scientist's work as consisting of a large multitude of very small subsidiary problems, all of them guided and controlled by the gradual evolution of a leading, significant problem.

But the real division among fields in respect of the excitement and hazards of worthwhile work does not lie along this axis. Rather, it relates to the proportion of the necessary work in the field which consists of non-trivial 'technical' problems. In many fields which are by no means descriptive, there is a need for large quantities of reliable information, either produced to order for a particular problem, or made publicly available in handbooks and digests. The production of this information requires sophistication, diligence, and care; for its objects are artificial and each new project encounters new pitfalls. So the work demands something more than a technician's training; in the social classification, it is scientific. Yet when pursued by the unadventurous, it can become a routine, never completely repetitive and boring, but a straightforward craftsman's job all the same. The style of work which is appropriate for such research would be of little use in highly theoretical or speculative fields; but the great rank and file of scientific workers spend their careers in setting and solving such high-level technical problems. Whether one wishes to describe such work as 'science' is a matter of choice; but its presence, as an essential ingredient of the industrialized science of the present, cannot be ignored.

Craftsman, Technician, Scientist

My final application of the analysis of scientific work as problem-solving on a world of artificial objects will be to develop the distinction between the work of a scientist, a technician, and a craftsman. Through this we will come to an improved historical perspective on the picture of the natural world in which our contemporary natural science operates. I have already argued that every scientist must be a craftsman in some sense, and he must be a technician too. But the objects of the work of a traditional craftsman and of a technician, and the tasks they accomplish, are each distinct; and we can define the differences in terms of the special character of the intellectually constructed objects of scientific knowledge.

Craftsman's work is done with particular objects, which may be material or intellectual constructs, or a mixture of the two: and the operator must know them in all their particularity. Their properties and behaviour cannot be fully specified in a formal list; in fact, no explicit description can do more than give the first simple elements of their properties. Hence the operator's knowledge of them must be 'intuitive', or of the sort described by Polanyi as 'tacit'.[28] It cannot be learned from books, but from experience, derived from a teacher by precept and imitation, and supplemented by the personal experience of the operator himself. Such a craftsman's knowledge of his objects is necessary for any sort of scientific work; even in pure mathematics, where the objects of the work are purely intellectual

---

[28] See M. Polanyi, *Personal Knowledge*, pp. 87–95.

creations, the properties of the objects which are known from established results are not sufficiently particular and subtle to guide the work of constructing an argument for a new proof.

The craft character of the scientist's knowledge of his objects is even more marked in the case of the basic, elementary concepts which are learned at an early stage of training. For the teacher cannot exhibit the classes of things and events as parts of a formal and coherent structure; rather, he must try to make them plausible, and impart some knowledge of their properties through the student's successful manipulations with them. They are thus learned informally, by imitation and precept; and they become so deeply embedded in the mature scientist's picture of the world that they eventually seem entirely 'natural'. One cannot imagine an intelligent adult who cannot count; and so the scientist (or the teacher of science) finds it difficult to accept that some intelligent and educated adults cannot manipulate with such basic concepts as 'mass' and 'acceleration'. Embarrassment can arise when this craft knowledge of the artificial objects of science is assumed by its possessors to be the total possible knowledge; probing questions by awkward students then threaten to expose a vacuity of incomprehension beyond the façade of confident manipulative knowledge. Of course, at some stage of training, the concealment of the artificiality of the objects becomes impossible, and also unnecessary. Thenceforward, the student consciously operates explicitly in a world of intellectually constructed objects, but he is rarely reminded (and hence nearly as rarely learns) that the 'natural', 'elementary' objects he learned in his youth are of the same sort.

Even when the objects of a field are recognized as constructs, the craftsman's familiarity with them which scientists achieve in the course of their training has a conservative effect. For the imaginative work of perceiving new phenomena and setting new problems will tend to be done in terms of what is already intuitively known. It is for this reason that the research schools founded by great men tend to lose their originality; and it also explains why deep innovations may arise from the new perceptions of scientists outside a leading school, or outside the field altogether, or from young, self-trained men. In such cases, the innovator will need enormous courage and talent to overcome the pitfalls in the work, which would have been at least partly recognizable to the 'properly' trained men, had they escaped from their familiar world into the partly new one. Hence the chances for success in such cases are small; only a few fortunate and talented outsiders can crash into a field and make it partly their own. If they do succeed, then of course they are retrospectively admitted to respectability, and may in their turn contribute to a new orthodoxy of common sense objects.

Given all these similarities, traditional craft work differs from that of scientific inquiry both in its objects and in its tasks. Its tasks are the making of particular things and the control of particular events; and these particular things and events do not derive their existence from an elaborated, formal framework of knowledge of intellectually constructed classes of things and events. Most of traditional craft work does not require formal literacy, although good craft work does require intelligence: each new task presents new difficulties and new possibilities.

The work of a technician is also with particular things and events: but some of the properties which he must understand and control require the intellectually constructed classes of 'scientific' knowledge for their explanation or even for their description. Hence the training of a technician will be radically different from that of a craftsman. The apprentice craftsman can soon learn to perform simple manipulations on the objects of his mature work, either learning by imitation and precept from a master, or by making things himself which, though lacking finish and sophistication, are genuine objects of the craft. On the other hand, the technician in training must master a considerable body of knowledge of an abstract, scientific character before he can manipulate or even recognize his objects. This background scientific knowledge will, in its teaching, be designed quite explicitly to serve as tools, and so it will lack the elaboration and attempted rigour which is given to the same material when it is taught to intending scientists. Yet since the technician's work will generally be confined to the manipulation of a small set of objects, he must also be given a thorough craft apprenticeship for developing the necessary skills. The proper blending of 'craft' and 'scientific' elements in technical education is difficult even as a pedagogical problem, but it is further complicated by considerations of status. For rewards of 'brain' work are so significantly higher than those of 'hand' work that there is a constant tendency to elevate a particular body of technique towards the status of an applied science. This is frequently justified by the introduction of new and more sophisticated tools; but it can have the result of turning out people whose training leaves them unprepared, psychologically and technically, for the only jobs open to them.

Turning now to the scientist, we see the essential difference between his work and those of the other two. For although he must have a craftsman's knowledge of his objects and his tools, and do technician's work as well, his real work is quite different in character. For unless he can successfully set, investigate, and solve problems, drawing conclusions about classes of things and events and not merely manipulating particular samples, his title is inappropriate. Because scientific work generally requires a greater capacity for abstract thinking, and a more rigorous training, a career in such work tends to be restricted to those who have absorbed education from early years. This naturally gives it an enhanced social status over technician's work; but it is a mistake to think that it is essentially of a higher order of intellectual endeavour. Scientific work can be done well or badly; and it is quite possible for the most intelligent and competent person in a research laboratory to be the senior technician, who may or may not be aware of the deficiencies of those whom he serves.

This threefold classification is far from absolute, and two points where it breaks down are worth special notice. The most common case in the modern world is where the distinction between scientist and technician becomes blurred. Many scientists are employed on problems of such a routine character, where the work is so straight-forward and the classes of things and events are modified so little by the new knowledge gained, that the 'drawing of conclusions' is little more than the description of certain properties of quite reliable representatives of the general classes under investigation. In such cases the work is hardly different from that of a technician as I have defined it; and such work is done by men trained as 'scientists' rather than as 'technicians' only because of the customary difference between the two sorts of training in depth and rigour. The distinction is also blurred from the other end, in the many cases where the accomplishment of a particular technical task involves such complex and sophisticated procedures that the work is in effect the drawing of new conclusions about intellectually constructed classes of things and events. Both types of borderline case occur on a large scale in the industrial applications of science; the terms 'research worker' and 'technologist' give a fair indication of the quality of endeavour required in the two cases.

The Craftsman's Cosmos

A less familiar exceptional case arises in traditional crafts, which seem to involve no more than manual skills learned by imitation. One will find, on close inquiry, that no such activity can be entirely free of theory; the craft knowledge must always include an 'explanatory' component, if nothing else than to guide the work when new situations arise. It does not matter that such 'theories' may be (and, over the course of the centuries, usually were) incorrect as scientific explanations of the properties of the materials; without them, there could be no successful craft work at all. But the matter does not always end there. Over the millennia of human experience, the basic productive crafts, which are now being extinguished by modern science and industry, developed very highly articulated theoretical structures. The operators considered themselves as engaged in something far more complex and rich than mere manual activities. We could say that they were being technicians in our sense, perhaps even scientists, except that they did not so much draw conclusions as enter into a dialogue with their objects.

This mental and spiritual aspect of craft work has long been a puzzle to anthropologists and archaeologists, for the 'magical' component of an otherwise sound technique seemed so patently irrelevant and absurd.[29] Indeed, since the same material objects can be created by purely manual operations, one might simply dismiss all the magical theory as aberrant, retrograde superstition. But this methodological criticism derives from a particular view of reality. If one is quite sure that no personal being resides or participates in a particular artifact, then the copy made by the self-taught anthropologist is the same as the original; but otherwise, the difference between the two will be that between the dead and the living. And it is undeniable that the 'magical' craftsman had a spiritual experience as part of his work, even if it was illusory; while the modern scientific copyist does not.

Of course, it is obvious common sense to us that the craftsman's material objects are devoid of spirit. But this common sense is a very recent and specialized cultural product of European civilization;

[29] V. Gordon Childe wrestled with this problem, which has destructive consequences for the ideology of 'rationality' and 'progress', in the tenth Frazer Lecture, *Magic, Craftsmanship and Science* (Liverpool University Press, 1950). The same phenomenon has been made the basis for an analysis of the modern ideas of 'progress' as the successor to the magical-alchemical vision, by M. Eliade; see *The Forge and the Crucible* (Rider, London, 1962).

and its triumphs over the past few centuries in the creation of modern science and technology are now revealing their dangers for human society and for the very continuation of life on this planet. We will understand ourselves, and our contemporary problems, better if we appreciate the world of the traditional craftsman as utterly different from our own, perhaps quite wrong, but coherent in itself and with its own value for the proper interaction of our species with the rest of nature. The recent growth of movements rejecting runaway technology and the culture with which it is associated is a reminder that our 'scientific' world-view is not necessarily the whole answer.

5

METHODS

HAVING described the peculiar character of the activity of science, as a special sort of craft work operating on intellectually constructed objects, we can proceed to an analysis of the judgements by which that work is governed. From this discussion we will achieve a better understanding of the ways in which the social aspects of scientific inquiry condition the work of individuals. We will then be better able to analyse achieved scientific knowledge as the product of a social endeavour extending through time, and also to attain insights into the delicate questions of the health of scientific communities.

I have already mentioned some of the judgements which are necessarily involved in the investigation of scientific problems, beginning with the basic judgement of the soundness of a set of data. These individual acts of judgement do not derive solely from private intuitions of the scientist; rather they are based on a body of principles and precepts, social in their origin and transmission, without which no scientific work can be done. I shall use the term 'methods' for such principles and precepts, which (through their interpretation and application in particular situations) guide and control the work of scientific inquiry. Such a corpus of 'methods' has a special character, which may seem paradoxical and at variance with the objects and results of the activity it governs. For methods cannot be established 'scientifically', through arguments resting on controlled experience; this is partly because there is no simple test of the 'correctness' of a particular method, and even more because the principles and precepts are incapable of a fully explicit, public statement. The body of methods associated with any field is partly informal and even tacit; it is not transmitted by the same public channel of communication as scientific results, but through an informal, interpersonal channel. Hence the testing, criticism, and improvement of the methods for a field must proceed by means quite different from those applicable to its scientific results; and in this aspect of scientific inquiry, the character of the community engaged in the work is thus crucial for the nature and quality of its achievements.

There is no doubt that without an appropriate 'method', in some sense of the term, scientific work is impossible. A trained scientist can instantly identify the traces of the bungling amateur, or the crank, by the absence of 'method' revealed in a report of his work. Also, there is a rough distinction in practice between 'methods' in the plural, referring to more detailed techniques of work, 'method' in the singular, referring to more general principles, and finally 'methodology' which is, or should be, reflection on the former categories or their equivalents. Yet, while in some fields of science the practitioners are deeply concerned (sometimes even obsessed) with 'method' in one or all of these senses, in other fields of a similar character the whole subject is complacently ignored. There is even a division and confusion among 'methodologists' themselves, concerning the relations of their analyses and conclusions to the work actually done in the disciplines they are discussing. One unfortunate consequence of the apparently abstract and self-contained character of methodological debates is that working scientists with a philosophical concern will sometimes, after sampling such discussions, dismiss the whole subject as irrelevant nonsense.[1]

The root of the difficulty in any discussion of method is that it involves an attempt to render explicit that which is largely tacit. For the achievement of significant new scientific knowledge is a creative activity, involving intellectual work that is both bold and subtle. No machine that has been conceived, let alone constructed, is capable of such work.[2] Nor can there be a 'science of science' which

[1] Thus L. K. Nash, *The Nature of the Natural Sciences*, pp. 320ff. The section entitled 'Scientific Method, Shadow and Substance' begins: 'We never credit the possibility of a definable Literary Method, practice of which brings within reach of mediocrity the production of works of genius. Yet at least four reasons may encourage even master scientists, and others, to believe possible a communicable Scientific Method with some such magic power.' With Polanyi, Nash sees science as craft work; and he also knows of the manifold ways in which scientists can delude themselves. See also P. W. Bridgman, *Reflections of a Physicist* (Philosophical Library, New York, 1950): 'I like to say that there is no scientific method as such, but rather only the free and utmost use of intelligence. In certain fields of application, such as the so-called natural sciences, the free and utmost use of intelligence particularises itself into what is popularly called the scientific method' (p. 278).

[2] This is not to say that machines cannot do any work which, when done by men, is considered 'original research'. I am informed by Professor F. A. E. Pirani that a com-

can produce manuals teaching how to do original research in ten easy lessons. Hence any explicit analysis of scientific inquiry must be incomplete, at best a schematic anatomy representing a complex physiology. This limitation applies to the present discussion as to all others; but we can hope that by recognizing its limits, and separating off the unanalysable from that which is partly analysable, we may escape some of the pitfalls which have beset earlier discussion of the problems of method.

We have already discussed 'methods' of one sort, in connection with the techniques of using tools. Such methods are the most similar to those which govern handicraft work: they can be learned, at the elementary level at least, in terms of particular precepts governing particular operations. Also, in considering scientific inquiry as a sort of craft work, we saw that a master-craftsman of science has a body of methods of his own, which define his personal style of work. But scientific inquiry differs from handicraft work in using a body of methods which are sophisticated and subtle, but which are a social possession, not restricted to the private craft wisdom of a master. We encountered examples of such methods when we discussed the necessary judgements of the materials produced in each phase of the investigation of a problem, as the 'soundness' of data, and the 'relevance' and 'reliability' of information. Such judgements are of adequacy; an equally important set, invoked at different phases of the work, are those of value. An analysis of these two sets of controlling judgements will provide materials for a later discussion of the philosophical problems of the nature of achieved scientific knowledge, in relation to the social activity through which it comes to be. Also, by appreciating such judgements, and the criteria on which they are based, as methods, we can examine the status of 'methodology', and appreciate some of its special problems as a field of inquiry. Finally, we shall consider what sort of knowledge is constituted by the body of methods which govern scientific inquiry, and contrast it, and its channel of communication, to the results of scientific inquiry.

Adequacy : the Lack of Certainty

In our earlier analysis of scientific inquiry, we got as far as the completion of an individual research report. At that stage the

puter program exists whereby passable M.Sc. dissertations in certain fields of applied mathematics can be produced with only minimal human intervention.

product is a solved problem; the establishment, in the conclusions of an argument, of new properties of the classes of intellectually constructed things and events which constitute the objects of the inquiry. Up to that point it is inappropriate even to speak of 'facts', to say nothing of 'knowledge' or 'truth'; these categories belong to the later, social phase of the activity of science. A scientific problem, unlike a textbook exercise, carries with it no guarantee that there exists a 'correct' solution against which those actually achieved can be tested. When this lack of certainty is recognized as significant, it entails some very paradoxical and sometimes troubling conclusions. For the certainty and objectivity of knowledge about the world of nature, as proclaimed in an old and powerful tradition in the ideology of natural science, are called into question. The individual scientist can achieve no more than an adequate solution to his problems; and the criteria of adequacy are set by his scientific community, not by Nature itself. To be sure, in a healthy and matured field these criteria, as those of value, will be set in the light of successful experience of penetrating into the natural world; but there is no objective, certain or 'scientific' method for setting or testing them. It still remains possible that certain and objective knowledge can be attained by a community of scientists developing and testing it over generations; and we shall discuss this in detail later. But this then amounts to *veritas temporis filia*,[3] and is the opposite of what has generally been conceived as the process of scientific discovery. Indeed, a reflection on criteria of adequacy, their importance and their origins, can lead to the conclusion that the differences between work in the most matured of the natural sciences, and in the most fledgling of the human and social sciences are of degree rather than of kind.

These paradoxes are very real, and will be discussed more closely later. In brief, we may say that they result from the peculiar character of scientific disciplines, which is most easily discerned in the more 'theoretical' ones. That is, in them we have demonstrative disciplines whose conclusions cannot be certain. This uncertainty has generally been studied only in the context of abstract epistemological debate. The processes whereby relative and provisional

---

[3] 'Truth is the daughter of time'—a classical Latin epigram. This may have been the source for the title of Francis Bacon's unpublished essay 'Temporis partus masculus', 'The Masculine Birth of Time' of 1605; see B. Farrington, *The Philosophy of Francis Bacon* (Liverpool University Press, 1964).

certainty emerges from the social activity of scientific inquiry had not been a subject of close analysis until very recently. Also, since the earlier defenders of natural science were usually concerned to establish its credentials to the greatest degree possible, these radical flaws in the structure of scientific knowledge have not become widely known, either among those in other disciplines who would emulate its successes, or among those who teach it to the younger generation.

The terms 'deduce', 'verify', and 'prove' are still so widely used in science, especially in teaching, that it is worthwhile to give a brief résumé of the inconclusiveness of the basic patterns of argument used in natural science. The simplest and most common situation in the teaching of a developed experimental science is where a general theory is elaborated, experimentally testable consequences are deduced, and the relevant experiments are then performed. If the particular results agree with the predictions, the student is satisfied that he has 'confirmed', 'verified', or even 'proved' the theory. But in this stark and simple form, nothing whatever has been established. It is not merely that a particular collection of data can be interpreted as information in a variety of ways, so that five points which lie reasonably close to a straight line can (in the abstract) equally well be interpreted as lying on a polynomial of the fourth degree. Even if the information is accepted as being equivalent to a deduction from the theory, the 'support' it lends to the argument is, in strict logical terms, nil. For this pattern of argument is a fallacy known since ancient times as 'affirming the consequent'. On this simple pattern, one can 'verify' the hypothesis that the moon is made of mouldy cheese. One need only deduce that it would then have spots, and then establish that the predicted spots do exist. In descriptive sciences, the statements which are 'verified' are generally those of the properties of members of a class. A sample is studied, and if the claimed properties can be identified, the general assertion is passed as true. But all that can be claimed logically is that one more confirming instance has been found; and no matter how many more are accumulated, they cannot offer the slightest ground, in terms of any valid pattern of argument, for expecting future experience to be the same.

In both of these patterns of argument, negative or 'disconfirming' particular experiences seem to have a stronger logical force. For if the consequent of a particular hypothesis turns out to be false, then the hypothesis itself is certainly false. Thus, although the deductive pattern of argument is quite incapable of telling us when our theories are correct, we seem to have at least the gloomy consolation of knowing certainly when they are false.[4] Unfortunately, the inference back from a disconfirming instance works only in the case of very simple arguments; if more than one hypothesis is involved in an argument (as is nearly always the case in science), we can learn nothing more than that they cannot all be true together. Similarly, an assertion of particular properties of a class of things is not simply overthrown when a single contrary instance appears. The original assertion can be defended by a slight redefinition of the class, so that the offending sample is then excluded; or the sample can be dismissed as one of those 'anomalous cases' which abound in any detailed study of the workings of nature.[5]

Even this slender link with certainty is lost when probability statements are contained in an argument or a conclusion. For no set of samples can formally confirm or disconfirm a predicted probability distribution. To establish their relevance, one must invoke an extra probabilistic theory relating to the mode of production of the properties of the samples, then estimate the probability that the deviation between results and prediction arises from 'chance', and finally decide what is a 'significant' level of odds against a chance deviation in this case. The special tools, assumptions, and judgements involved in this sort of 'testing' ensure that it will remain very far indeed from the realm of certainty.

Philosophers of science have attempted, with some success, to provide a rationale for the different basic patterns of argument, showing why it is reasonable for an intelligent person to place reliance on them. A common technique is to exhibit ancillary assumptions, which when added to the bare argument do make it

[4] Karl Popper is considered with some justification as the 'philosopher of falsification'. But as his ideas on this have developed over the decades, the simple doctrine ascribed to him has been refined nearly out of existence. For a discussion of the various 'Popperian' positions, see I. Lakatos, 'Criticism and the Methodology of Scientific Research Programmes', *Proceedings of the Aristotelian Society*, 69 (1968), 149–86.

[5] Michael Polanyi describes how the theory of electrolytic dissociation proposed by Arrhenius was accepted for several decades, in spite of applying only to weak electrolytes; the strong electrolytes, as common salt or sulphuric acid, were considered as 'anomalies'; see *Personal Knowledge*, pp. 292–3. In his classic study of the Euler Polyhedron Theorem, I. Lakatos shows how 'monster-barring' strategies are also applied in pure mathematics. See his 'Proofs and Refutations', *British Journal for the Philosophy of Science*, 14 (1963–4), 1–25, 120–39, 221–43, 296–342.

valid. But as these philosophical arguments become more refined and sophisticated, they drift further and further from the practice of science. They remain as important studies in epistemology, but they offer little illumination to a scientist on the reasons why the complex patterns of argument he actually uses seem to work and are accepted as adequate by his community.

For what happens in any practical situation is that the basic invalid patterns of argument are not so much supplemented by extra assumptions as enriched by particular tests and controls appropriate to the materials of the problem and its characteristic pitfalls. For example, a deductive argument may fail the test of adequacy in a particular field, unless a set of known likely alternative hypotheses for the same tested conclusion are discussed and eliminated. This is no logical protection against the infinity of other hypotheses which might yield the same conclusions, but it is an insurance against at least the more obvious pitfalls. In an inductive argument, the quality of the sample can be scrutinized, and specialized arguments used to show that it is likely to be a good representative of its population. The making of inferences of causal connections between members of two classes can be tested through supplementary investigations, in the manner laid down in Mill's 'canons of induction'. In an elaborated argument of any sort, a piece of evidence that has a significant load put on it may be required to be tested for its strength by a particularly close examination of the conditions of the production of the data and information from which it is derived. None of these tests provide certainty, and there is not even a certainty that a particular test is the most appropriate one for the problem being investigated. But they are the best that can be obtained, for the work of seeking knowledge about the external world; and a study of their special character will provide an understanding of the ways in which that knowledge can, but need not, approach certainty.

Judgements of Adequacy

Each component of the argument of a solved problem, either an inference-link or a piece of evidence, can be no more than adequate to its function in the total structure. And what is 'adequate' will depend not merely on its context in the problem, but on the general criteria of adequacy for the class of such problems imposed by the community. In the beginning of this part I gave an example of the necessity of judgements of adequacy, in the discussions of the 'soundness' of data, and of the 'reliability' and 'relevance' of information. An appreciation of what is involved in such judgements may be gained from a consideration of a common and routine procedure in the formation of such judgements: statistical significance tests. For statisticians do not simply say that a correlation is 'significant' or 'not significant'; rather, they will speak of significance at a certain level. Those who have any craft skill in the use of such tools will appreciate that the significance level to be adopted is not assigned by God, but must be decided by the user. The decision will be based on estimates of the direct costs and the risks associated with each level. For each level of significance involves the possibility of two sorts of error: of rejecting worthwhile information, or of allowing dubious information to pass. The more stringent the test chosen, the safer; but also the more costly, because of the extra time, care, and resources required for producing material that will pass it. The choice of a particular level of significance must depend on a judgement of what degree of safety is required, for that component in its context in the total problem. And this judgement must be based on general criteria of adequacy applied to that particular situation. There can be no perfectly safe test of the quality of the material, and neither can there be a certainly correct decision on the degree of stringency of the test.

In general, we can say that imposed criteria of adequacy are necessary for scientific work because of the inconclusiveness of the arguments used in science. This inconclusiveness follows from the peculiar character of the objects of scientific inquiry: classes of intellectually constructed things and events, the evidence for whose properties is derived from particular experiences. To move from the reported properties of particular samples, to the properties of the classes they are intended to represent, a demonstration is necessary; but a formally valid argument, yielding certainty or truth, is impossible.

A scientific problem is thus incapable of having a solution which is 'true'. Rather, the solution will be assessed for adequacy; and for this every component must be so assessed. We can distinguish two sorts of criteria and judgements of adequacy: those relating to the argument, and those relating to the evidence. In the former class are the tests of the various inferences which carry the argument through the statements of the properties of its objects to the conclusions. Even a deductive argument is not exempt from testing for adequacy; as soon as one passes beyond the simple syllogism in sophistication, the obvious and intuitive tests fail to apply. Arguments cast in mathematical form must satisfy criteria of 'rigour' to the appropriate degree; and the rigour of physicists' mathematics is not, and need not be, in the same class as that of the pure mathematicians. Even in pure mathematics, disagreement over criteria of adequacy have erupted at crucial points in the development of the subject; the most recent important case occurred at the beginning of the present century, in connection with the use of actually infinite sets and constructions in mathematical proofs.[6] The criteria of adequacy relating to evidence are more varied; for they control not only the conditions of the production of data and information but also the strength and fit of the evidence in its particular context. It will frequently be necessary for some of the evidence to be explained

[6] At the end of the nineteenth century it could appear to mathematicians that they had reached evolution's end, in the perfection of criteria of adequacy. Thus Henri Poincaré: 'Today there remains in analysis only integers and finite or infinite systems of integers, inter-related by a net of relations of equality or inequality. Mathematics has been arithmetised. . . . Have we at least attained absolute rigour? At each stage of the evolution, our fathers believed that they too had attained it. If they deceived themselves, do not we deceive ourselves as they did? . . . We may say today that absolute rigour has been attained.' (Quoted from E. T. Bell, *The Development of Mathematics* (McGraw-Hill, New York and London, 1945), p. 345.) Poincaré was here reflecting on the achievements of Weierstrass and his school in establishing the very tight patterns of argument for 'potential-infinite' proofs in analysis. But the façade of perfection was already revealing cracks in two places. First, one of the most basic objects of inquiry, 'real number', had been under investigation for some decades, and this 'foundations' work was encountering anomalies and paradoxes at an increasing rate. These were related to the basic structure of logic, and to the 'actual infinite', which, after its complete banishment in the middle of the century, was creeping back into mathematical arguments. At the turn of the century there was a 'foundations crisis', which led to the inspired works of Russell, Hilbert, and Brouwer. The sequel to this was the creation of a new field of mathematics, in which many problems were set and solved, some of them very deep (Gödel's theorem being the outstanding example). But the basic problems are still far from solution; see I. Lakatos, 'Infinite Regress and the Foundations of Mathematics', *Aristotelian Society Supplementary Volume*, 36 (1962), 155–84. For a personal account of the reaction to the failure of all of the programmes to establish 'foundations', see John von Neumann, 'The Mathematician', *Collected Works*, vol. i (Pergamon Press, 1961). He remarks that after the failure had been perceived by working mathematicians, they 'decided to use the system anyway', for 'it stood on at least as sound a foundation as, for example, the existence of the electron' (pp. 5–6). Thus working mathematicians operate with respect to the 'foundations' of their subject much as engineers do with nineteenth-century rigorous analysis: their techniques are designed to avoid the known pitfalls, and no new ones have been detected. Since it is indisputable that the 'foundations' of mathematics are not established, the absence of any new 'foundations crises' in the past half-century might be considered as evidence of a lack of deep conceptual development in that period.

and defended explicitly; and these subsidiary arguments must also meet criteria of adequacy appropriate to their function. Thus the complexity of a solved problem is matched by that of the set of relevant criteria of adequacy; and that set will depend closely on the field of inquiry. Hence it is impossible to produce an explicit list of criteria of adequacy applying to a wide class of problems. This component of 'method' is thus a craft knowledge, but no less essential to successful scientific inquiry than the formal, explicit knowledge deposited in the public record.

This brings us back to the paradox of adequacy with which this discussion began. It would appear that the certainty and objectivity of scientific knowledge are dissolving in a set of intuitive judgements based on principles which frequently cannot even be made explicit, let alone defended or tested. The easiest escape from the paradox is to show that it is a false one; that the situation is not at all one of objective truth being based on subjective guesses. For we have the historical knowledge that some fields of science do achieve objectivity and near-certainty in their results, while others do not. In all of them, there is an absence of formally valid proofs, and a presence of controlling criteria of adequacy. The difference between them does not lie in this logical aspect of their arguments and conclusions, but in the particular circumstances of their development.

The criteria of adequacy associated with a field are, however, directly relevant to the strength it attains. Indeed, they perform an essential function, and a field is matured and effective only when they perform it well. There the deficiencies created by the lack of formally valid patterns of argument for basing general conclusions on particular reports are remedied for practical purposes by the accepted criteria of adequacy. For the tests of adequacy imposed on the components of a problem at each phase of its development serve for the avoidance or the timely recognition of the known pitfalls in the way. In a matured field, unknown pitfalls are relatively rare, and so the conclusions of the arguments will be well-established. Although they are still not immune from the eventual discovery of error or inadequacy, they provide a firm basis for immediately subsequent work in the field. Also, the social tests on the materials of a research report, necessary for their becoming accepted as 'fact', can be applied in a straightforward manner, and a direct transition from one status to the other can be the normal event.

Criteria of Adequacy and the Maturing of a Discipline

The function of criteria of adequacy is thus to make 'facts' possible. It is worthwhile to examine this more closely; we can do so in terms of the maturity or otherwise of a field of inquiry. The condition of ineffectiveness or severe immaturity in a field can be recognized (by those who wish to do so) through several symptoms. In such a field, controversies on results range indiscriminately and inconclusively from criticism of raw data to abstract methodology. (Whether such controversies are common, depends on the social character of the field; roughly, whether it is nascent or moribund.) In this we see an effect of the weakness of the accepted criteria of adequacy; they are not strong enough to channel debates on to well-defined problems, nor to provide an agreed foundation for the debates. Another symptom is that 'facts' do not exist. There is no cumulative development of accepted results, stable under testing and repetition, and invariant under changes in the problems in which they are used. Instead, the results of each school or tendency die with the problem around which its work was organized.

These two symptoms are related through a third, the ubiquity of pitfalls. The failure of facts to be achieved is an effect of undetected pitfalls. The conclusions of nearly all problems are soon seen to contain errors; and although this is no hindrance at all to further research in the same line, their materials are then valueless as information to be used in other work. But to controvert a result is not the same as to identify that flawed component in the argument which vitiated it. Such flaws are rarely errors in logic; for as we have seen, the basic patterns of argument in science are not formally valid logical ones. The flaws are truly pitfalls, false assessments of the quality of evidence or insufficiently strong inferences; and they are not easily identifiable even in retrospect, for the pattern of any argument in a real scientific problem is too complex to permit of an easy dissection. The presence of such pitfalls is caused by insufficiently strong criteria of adequacy. For when a particular pitfall is identified, it is signposted by a particular criterion of adequacy: at this point the work must be done rigorously, in a particular way, lest disaster result. Conversely, a weak set of criteria of adequacy fails to guard against pitfalls, and yields a situation where conclusions of problems are illusory, and facts cannot exist.

The judgements of adequacy thus perform the same function in scientific inquiry as the tests for quality control in industrial manufacture; and the criteria of adequacy stand in relation to them as the standards of acceptable quality set for the tests. The solution of a scientific problem is like a complex and delicate manufactured product, in that a single weak component can destroy the whole work. In manufacture, as in science, perfection is impossible; and the standards of quality must be determined in the light of the experience of the costs and risks associated with each level. The difference between manufacture and scientific inquiry, in this context, lies in the uniqueness of each scientific problem; even its components are not all uniform, easily tested objects. Hence the standards of quality control for scientific work are not capable of being set out in handbook form, nor the tests carried out by a routine. The criteria and judgements of adequacy are necessarily an informal, largely tacit knowledge; and in mastering and applying them the successful scientist is a sophisticated and highly skilled craftsman.

In the work of bringing a field towards maturity, an important part lies in the strengthening of the criteria of adequacy. This is not all, of course; the development of new tools, and the creation of an appropriate social environment, are equally important. Nor can the strengthening of the criteria of adequacy be done in an abstract, automatic fashion, as by the attempted imitation of a successful field. Such a strategy can produce totally misdirected criteria of adequacy, leaving a field in worse condition than it was originally. For the relevant criteria of adequacy are, as we have seen, intimately related to the characteristic pitfalls of the problems investigated in the field; and these will be very particular to it, in its objects of inquiry, its sources of data, its tools, and its patterns of argument. Moreover, each particular criterion of adequacy must, if it is to be effective, carry with it the craft knowledge of making judgements of whether it is satisfied in particular cases, as well as the tools and techniques for preparing materials of a quality sufficient to satisfy it.

Over recent centuries, the founders and leaders of ineffective disciplines have believed and proclaimed that the application of the correct 'method' would quickly and automatically bring their studies to a state of maturity and effectiveness. The conception of method which has generally prevailed has been a combination of two elements: a simplified pattern of argument (either inductive or hypothetico-deductive, depending on the current fashion), and a commitment to simple, preferably quantifiable data. Leaders of influential schools in the social sciences have sincerely believed that real science is done by putting masses of quantitative data through a statistical sausage-machine, and then observing the Laws which emerge.[7] From such caricatures of the process of scientific inquiry are derived criteria of adequacy which enforce an apparently rigorous procedure of research,[8] but whose results can but rarely escape vacuity. Such programmes of reduction and mathematization base their claims on the undoubted successes of the physical sciences since the seventeenth century; but they ignore the long series of dismal failures in applying this approach to the sciences of life, thought, and society. And the latter history, largely unwritten, is precisely as old as the success story: 'l'homme machine' of Descartes's chimerical physiology is an exact contemporary with Galileo's successful mechanics. The principle that each field of inquiry has a degree of precision of argument appropriate to its objects was laid down by Aristotle; and the great mathematician Gauss observed that 'lack of mathematical culture is revealed nowhere so conspicuously, as in meaningless precision in numerical computations'.[9] If we extend these principles to criteria of adequacy

[7] Thus, T. S. Kuhn, 'The Function of Measurement in Modern Physical Science', *Isis*, 52 (1961), 161–93, starts: 'At the University of Chicago, the façade of the Social Science Research Building bears Lord Kelvin's famous dictum: "If you cannot measure, your knowledge is meagre and unsatisfactory."' Granting the importance of measurement in physical science, he asserts, 'I feel equally convinced that our most prevalent notions both about the functions of measurement and about the source of its special efficacy are derived largely from myth' (p. 161). He argues this thesis at length; and the reader will see the influence of this essay on my own ideas. It is reprinted in H. Woolf (ed.), *Quantification: a History of the Meaning of Measurement in the Natural and Social Sciences* (Bobbs Merrill, Indianapolis and New York, 1961).

[8] Much of the classic study, O. Morgenstern, *On the Accuracy of Economic Observations* (2nd edn., Princeton University Press, 1963), can be interpreted as an analysis of misdirected criteria of adequacy. On the basis of data that is deeply flawed by inherent inaccuracies, 'applied' economists aggregate national statistics to a high degree of pseudo-precision, and 'theoretical' economists construct models in which these very dubious numbers are manipulated in sophisticated and sensitive mathematical procedures.

[9] Quoted from O. Morgenstern, *On the Accuracy of Economic Observations*, p. 99. The German original reads: 'Der Mangel an mathematischer Bildung gibt sich durch nichts so auffallend zu erkennen, wie durch masslose Scharfe im Zahlenrechnen.' My translation differs slightly from that given in the text. For Aristotle, see *Nicomachean Ethics*, tr. W. D. Ross (Oxford, 1915), ch. 3, 1094[b] 12–28: 'Our discussion will be adequate if it has as much clearness as the subject-matter admits of, for precision is not to be sought for alike in all discussion, any more than in all the products of the crafts. . . . In the same spirit, therefore, should each type of statement be received; for it is the mark of an educated man to look for precision in each class of things just so far as the

in general, we can appreciate the fallacy of naïve imitation as the road to success.

One sign of a field having achieved maturity is a certain underlying stability, which persists through all the rapid changes in results, problems, and even objects of inquiry. This stability is revealed by the absence of new pitfalls, except at the outermost frontiers of research; and, related to this, a set of appropriate and stable criteria of adequacy. When these have remained in the interpersonal channel of communication for some time, they become part of the basic unselfconscious craft knowledge of the field; and the particular judgements of adequacy which are made do not seem to depend on anything but common sense. In these conditions the very existence of criteria of adequacy can be overlooked; and philosophers of science, basing their analyses on the experience of just such fields, can remain in ignorance of their existence and importance. But, as we have seen, in less matured fields they cannot be taken for granted; and they will be of central importance in several of our later discussions of the activity of science.

Criteria of Value

The criteria of value, and the judgements based upon them, form an interesting contrast to those of adequacy. In those we saw a close relation with philosophical questions of the possibility and nature of scientific knowledge; while here we shall find ourselves involved in problems of the social activity of science. The difference arises from the functions of the judgements: the one is basically that of assessment of work done, either a whole problem or a component; and the other is an important determinant of the direction of future work. For although judgements of value are also made on completed work, their crucial role is in the choice of problems to be investigated. Thus, while the strength of the existing achievements of a field depends (as we have seen) on its criteria of adequacy, its health and future prospects are intimately related to its criteria of value. Because the criteria of value are complex, and necessarily involve predictions of the future, the achievement of appropriate judgements of value is a delicate social task, even in matured fields where the criteria of adequacy are well-established. In this respect, the social problems of scientific inquiry in a field require attention even when

nature of the subject admits; it is evidently equally foolish to accept probable reasoning from a mathematician as to demand from a rhetorician scientific proofs.'

the philosophical problems of scientific knowledge have been given a practical resolution.

The exclusion of problems of value from the traditional philosophy of science has its roots in the ideology of modern natural science as it was formed through many generations of struggle. The earliest conflict was on the philosophical plane; the pioneers of the new philosophy of the seventeenth century conceived a world of nature devoid of human properties, and accordingly rejected all explanations in terms of 'final causes', or purposes of things and events. Although they usually invoked the Deity in one way or another, to give meaning to the world of nature in general, their explanations of particular phenomena were cast in terms of purely efficient causes. As the new science became established, the struggle for the autonomy of its collective goals required an ideology which denied any direct relation between the purposes of the community at large, and the functions of the products of its work. Spokesmen of science would remind lay audiences of the benefits which ultimately flowed from their work, but would warn them that the work must be pursued strictly for its own sake, lest the free play of creativity be stifled. Thus science, conceived as a body of factual knowledge, was cut off from considerations of value in two ways: its assertions were purely descriptive, with no normative element; and the considerations of social value by which all other human activities are assessed were declared irrelevant.[10] The deeper problems in which values enter, such as those concerning ethics and morality, could for a long time be neglected because of the favourable social conditions inside science and its very slight responsibility for the social effects of industrial and military production.

Certainly the facts presented in a textbook of mathematical or physical science bear no signs of being conditioned by considerations of value of any sort. Even here, the appearance is illusory; for the selection and presentation of these facts has been strongly influenced by judgements of value. When one passes to books in the biological sciences or applied sciences, considerations of value enter through

[10] The defensive position of those who wanted to argue for the presence of values in human knowledge is indicated by the title of a book by the great psychologist W. Köhler, *The Place of Value in a World of Facts* (Liveright, New York; Kegan Paul, London, 1939). A review of the literature on this problem will be found in J. Leach, 'Explanation and Value Neutrality', *British Journal for Philosophy of Science*, 19 (1968), 93–108.

the term 'function'; devices are assessed in terms of their performance of a function, and things occurring naturally are even explained by the excellence of their performance of a particular function. This is still a long way from bringing human purposes into scientific explanation; but it is a reminder that the claim of 'science' to be value-free is based on very partial evidence.

In all the attempts to free science from every consideration of value, no one has extended the argument to the activity of research. For such an argument would not even have an initial plausibility; choices and decisions must be made at every stage of scientific inquiry, and such actions are impossible without judgements of value. The extension of the boundaries of the known into the unknown does not take place like the spreading of a wet spot on a piece of blotting paper. Without a strategy and tactics, a field of scientific inquiry has as little chance of success as an army in battle which is simply told to 'advance'.[11] The statement of a problem to be solved is analogous to an objective to be taken; and the selection of some problems for investigation, necessarily to the exclusion of others, must be governed by competent judgements based on sound principles. The judgements which determine such decisions are those of value, feasibility, and cost. It is easy to see that all three aspects of a project must be taken into consideration; producing a result of little value will be a wasteful use of resources; an unfeasible project, however great the value of its anticipated solution, can lead to disaster; and excessive cost on a particular project will starve others equally deserving. Given the feasibility of a project, the decision to invest in it, rather than in others, will be based on a balancing of value against cost, in comparison to other competing projects. We notice immediately a feature of these judgements which makes the decision process in science extremely delicate: the things being balanced, value and cost, are incommensurable. For (neglecting remote effects of the result to be achieved) the cost is of the inputs, measured in human effort and money; while the value is of the output, and is complex and basically not quantifiable. Because of this incommensurability, the 'planning' of science cannot be reduced to a mathematical exercise in the same way as the planning of a commercial venture.

For reasons I shall discuss later, the majority of competent

---

[11] See James B. Conant, *On Understanding Science*, ch. 4: 'Certain Principles of the Tactics and Strategy of Science'.

scientific workers do not frequently make the choices which involve an explicit assessment of the value of problems, and their balancing against costs. But judgements of value are commonplace in the informal, conversational communication among scientists, and are used in the assessment of colleagues and of whole fields. A man will be considered as something of a bore if he restricts himself to working, however well and industriously, on problems whose value, in terms of significance for future work, is moderate. Similarly, neighbouring fields or rival schools will be subjected to judgements of value, in terms of their potential for leading to new advances; we shall see that such judgements are a very important part of the government of science. These judgements are based on criteria of value which are accepted, usually implicitly, within a field. Work which deals with the objects proper to a particular field, but which is so novel as to be incapable of judgement by the existing criteria of value, will be harshly treated. The pioneer whose work lies outside the common experience of the field must expect not only technical incomprehension by his colleagues, but also a failure on their part to appreciate the value of his endeavours. To the extent that the initiation of research in a field requires the investment of funds, and hence social approval, then the field may find itself becoming ingrown and eventually stagnant through the entirely natural working of these social judgements of value against deeply original projects.

Components of Value

The criteria of value applied in judgements on scientific problems and results do not relate directly to the Good; but they are not thereby deprived of interest.[12] We can distinguish three independent components in these criteria, two relating to the functions of the completed piece of work, and the other to the purposes of the scientist in undertaking it. Of these 'objective' components, the first is the 'internal' one: to what extent the solved problem will, or does, advance knowledge of the objects of inquiry of the field, either directly or through suggested descendant-problems; this is the dominant criterion of value in 'pure science'; and we note that it is the least amenable to precise estimation, to say nothing of

[12] The classic paper on criteria of value is that of Alvin M. Weinberg, 'Criteria of Scientific Choice', *Minerva*, 1 (1963), 159-71, and ibid., 2 (1964), 3-14; reprinted in *Reflections on Big Science* (Pergamon Press, 1967), pp. 65-122.

reduction to quantitative measurement. The criteria of 'internal' value adopted for a field will depend both on the social experience of what has been successful in the past, and also on the general conception of where the field is, or should be, heading at the point of its development. For example, there may be a 'grand problem', inherited from the founders of the field, whose eventual solution defines the ruling criteria of value.[13] Alternatively, if the field is in a phase of consolidation after a significant breakthrough, then systematic studies of masses of particular details of a certain sort will have considerable value, whereas previously the objects of inquiry would have been changing too rapidly for such work to be worthwhile. Within such general considerations of strategy, the tactics of scientific progress are so varied as to offer a great many possible functions for the results of solved problems, corresponding to the variety of problem-situations out of which problems are born. A knowledge of these functions, and an assignment of value to each of them, is part of the craft knowledge of each field; and it would be pointless to attempt an exhaustive formal list of them covering all of science. Any attempt to make one sort of function, such as 'explanation', 'prediction', or 'unification under a general law', to be the unique and defining basis for criteria of value, is bound to lead to an unrealistic account of scientific inquiry.

Parallel to this is the 'external' component of value: the contribution that the completed project makes to the solution of problems, or the accomplishment of tasks, outside the given field. There is a considerable variety in these external domains: they may include other fields of science, or technology, or (in earlier ages) philosophy, or even the social or ideological needs of the community of science or its larger society. Hitherto I have implicitly restricted my discussion to problems whose dominant value is 'internal'. I

---

[13] The classic example of such a 'grand problem' is the development of mathematical methods to account for the perturbations and slow variations in the orbits of the moon and planets; the problem was bequeathed by Newton to his successors, and culminated in Laplace's demonstration that the variations are cyclic. Newton thought that an occasional intervention by God would be necessary to prevent the system degenerating; and this is probably the reference in Laplace's legendary remark, 'Je n'ai pas besoin de cette hypothese.' See A. Pannekoek, *A History of Astronomy* (Allen & Unwin, London, 1961), ch. 30, pp. 297–307.

The alternative phenomenon is a commonplace in contemporary science; the individual researchers who join in the exploitation of a successful new technique may be partly motivated by the following of fashion; but in a healthy field the fashion is justified by the need for detailed studies.

shall discuss the other sort in some detail later on; for the present, it will be sufficient to mention that the nature of the dominant component of value will strongly influence the criteria of adequacy which are appropriate for the solution of the problem, as well as the general path of evolution of the problem as it is conceived and investigated.

In the ideology of science which became dominant towards the end of the last century, any form of external component of value was considered as a falling away from genuine standards of scientific integrity. The corresponding ideal of 'pure science' was certainly important in the effective establishment of refined norms of behaviour; but as we face the problem of the intimate mixture of external components of value in the work of industrialized science, it is useful to see that completely 'pure' science is itself a recent creation. First, we can say in general that internal components of value can be appropriately applied only to the extent that there is something to which the results of the inquiry can be internal. When a field is still in the process of being created, purely technical studies need to be supplemented (in varying degrees) by philosophical inquiries on the one hand, and by the solution of practical problems of organization and support on the other. For when a field is too immature to yield facts from its investigations, nothing is more futile than the attempt to amass 'positive' knowledge in the absence of any reflection on the nature of the objects of that knowledge or on the methods of its achievement. Also, even the first genuine achievements can be jeopardized by the neglect of the tasks of social construction in the field. These will include justifying its existence to a lay audience; and the work done will inevitably, and correctly, be conditioned by its function in this respect.

Until the nineteenth century, no field of natural science had such internal and social strength that its leaders could, or even wished to, reject all external components of value from its work. The 'purity' of science seems to have been developed first in the university environment of nineteenth-century Germany, where the natural sciences struggled to win a place alongside the established humanistic and philological disciplines. With the eventual establishment of university science on a large scale in all advanced countries in the earlier part of this century, the ideology of purity also became a convenient means of preserving the autonomy of an increasingly expensive social activity. By the entirely natural processes of the formation of a folk-history, the tradition of the purity of science was extended back in time to the origins of modern science, and across to all fields of scientific inquiry.

Scientific activity with only internal components of value is certainly far simpler to manage and maintain than one where external components are effective. For in the distinction between these two components, we have the root of the inevitable tension between a scientific community and the lay society which supports it. Speaking very roughly, we may say that the 'internal' components of value are generally esoteric, and incomprehensible to the lay public which pays the bills; while the completely 'external' components of value in the present age are vulgar, and if dominant, can eventually debase and destroy the activity itself. In practice, the health of a scientific community is maintained by a delicate and inherently unstable compromise between the two.

An analogous compromise, effected through different institutional structures, is necessary between these objective components of value, and those relating to the purposes of the individual scientist in undertaking a particular project, the 'personal' components. It is quite conceivable for a man to pursue science with a single-minded devotion, totally oblivious to the vulgar rewards of power, prestige, and material comfort; and doubtless some do. But such saints are usually a minority, even among the most gifted scientists; and even such a person cannot totally exclude the personal components of value from his judgements of problems. For the most effective advancement of the field requires more than a single, isolated worker; and winning support requires propaganda, both to explain the work and to convince others of its value. To ignore these social aspects of the task when presenting one's results, or even when planning the strategy of a long campaign of investigation, is in effect to abdicate responsibility from the work, and to be concerned only with one's private pleasure.

The extreme case of the inclusion of personal purposes in the valuation of a problem is thus one where the motivation is quite selfless. From this one can pass to a variety of other sorts of personal components of value, all of them justifiable in terms of acceptable ethical principles. A university teacher may 'keep up' with some research simply as a form of mental gymnastics, to prevent his going stale and out of date in his field; in this case the cost of the research is justified 'externally' by the maintenance of the quality of his teaching. The problems given to research students will frequently be of very limited value by any objective standard; but their prime function is to train the student in research methods, and the feasibility of a project for the student may well be a more critical feature of it than its objective value. At the far extreme are scientists whose criteria of value are nearly entirely personal; there we can find men who follow, or who try to create, successive fashions in science for the enhancement of their careers, with the same facility as that with which many women adjust the hemline of their skirts. To demand that scientists abstain from including any personal components in their valuation of problems would be ridiculous and, to the extent that efforts were made to enforce such an impossible standard of behaviour, they would only breed hypocrisy and corruption. Hence for proper valuations to be made, there must be informal mechanisms for harmonizing objective functions and personal purposes in science. These must be analogous to the 'hidden hand' of classical economic theory, which ensured that the aggregate of private decisions, each set by enlightened self-interest, yielded (in theory) the maximum social benefit. But for scientific problems there is no simple, quantitative, automatically established measure of value analogous to the market price of commodities. Hence the mechanism for the achievement of proper valuations in science must be more complex and subtle, and will necessarily be more vulnerable to distortion.

Problems of Value in the Social Activity of Science

The argument as I have developed it so far might appear to be leading to the same sort of paradoxes as those of adequacy. In this case, they may be even more unpleasant. For there have been many traditions in the philosophy of science which have weakened or abandoned the claim to truth in natural science; while the long-standing dichotomy between objective scientific facts and subjective value-judgements has hardly been challenged at all. Even though I am not directly attacking that philosophical position here, I am clearly arguing that complex judgements of value which are in principle highly fallible, condition or even determine the selection of those facts that actually come to be. To protect the world of scientific fact from this invasion of judgements of value, one might hope for an application of 'scientific method' itself, either made unselfconsciously already, or deliberately through a 'science of science' in the future. Such a hope is a vain one, and a policy based on it would be disastrous. My discussion of 'methods' up to now has shown that there is no simple and automatic recipe for the production of factual scientific knowledge; and the difficulties and pitfalls of decision-making in the accomplishment of practical tasks are, if anything, more severe. This is not to deny the importance of social investigations into the structure of decision-making in scientific communities; indeed, the present work may serve to enhance the effectiveness of such studies. But an attempt to eliminate the elements of craft experience and personal wisdom in these judgements and substitute for them a bureaucratic routine would soon produce gross errors of planning from which committed scientists would need to protect their field by clandestine research. For the unknown is full of surprises; and to the extent that previous experience is codified and made a rigid base for criteria of value in decision-making, the further penetration into the unknown will be blunted, and eventually reduced to the routine and the trivial.

The importance of the judgements of value, and the particularly complex, delicate, and ever-changing character of the criteria of value on which they are based, raise severe practical problems for the government of science. As I showed earlier, the industrialization of science has produced a situation where decisions must be taken more formally, and more centrally, than in earlier periods: there must be a decision-machinery for channelling the capital investment in the production of scientific results, as a prerequisite to the work itself. Hence the judgements of value which are influential in setting the direction of new work are not merely the aggregate of those of workers in the field; an extra, and very weighty component consists of the judgements of those closest to the focuses of decision-making power. To do their work well, such men must have unusual wisdom and breadth of vision. They must not merely have the imagination to make judgements of value on problems in fields outside their special competence; but they must judge the value of whole fields. And since they stand between the scientists and the external agencies which are investing the funds, then the external value of fields and of problems within them must be an important element in their assessments.

Even during the 'academic' period of science, there were problems of government; but the 'aristocratic' principle of leadership by distinguished men was usually prevented from enforcing an unhealthy rigidity by the ability of the next generation to choose its own patrons and problems. The execution of this choice depended on the social character of the community; in the German culture area, one could freely move between universities; and in Great Britain a large proportion of the significant work was done by amateurs. Only in France were the sciences organized in a centralized, hierarchical, State-subsidized community; and it is significant that after the single generation of the pupils of the great founders of the French academic system, in the physical sciences at least the scene was barren.[14]

It would be a mistake to imagine the acts of value judgement to be a constant and dominating feature of the work of science. Scientists do not generally live in an existentialist agony of balancing imponderable components of value of problems. Indeed, most scientific workers need make few explicit assessments of value when planning their work. But this does not mean that the choice of problems is random or arbitrary. Rather, the majority are content to accept the criteria of value which are currently dominant in their community. But this makes the problems of government in science more, rather than less acute. For the crucial tasks of determining the directions of advance are concentrated among the aristocrats of the community; these have experience and wisdom, but they may also have a personal investment in work started in earlier decades, and the burden of their political roles may leave them with insufficient time for an informal, sensitive acquaintance with new developments. Hitherto the community of science has functioned

[14] The best synthetic account comparing the conditions and styles of scientific inquiry in Britain, France and Germany in the early nineteenth century is J. T. Merz, *A History of European Thought in the Nineteenth Century*, 4 vols (1904–12; reprinted Dover Publications, New York, 1965), vol. i, ch. 1–3, pp. 89–301, especially summary, pp. 298–301. The nineteenth-century historian Buckle provides a good example of the traditional English distrust of centralization in cultural matters. Proceeding from his critical analysis of the state of intellectual culture under Louis XIV of France (where, unlike all previous historians, he showed that it was abysmal), Buckle concluded that the 'protective spirit' of the State applied to science through official patronage will inevitably be destructive in its effects. He posed the problem of decision: if merit is decided by the State, it will be done incompetently; but if it is decided by the leaders of the community receiving the rewards, it will be 'disgraceful' and degrading. Citing the earlier cases of the reigns of Augustus and Pope Leo X, he said 'in each of these ages there was much apparent splendour, immediately succeeded by sudden ruin'. The editor of the later edition of the work, the freethinker John M. Robertson, disagreed violently with Buckle's whole argument. See Henry Thomas Buckle, *Introduction to the History of Civilisation in England* (1857–1861), new and revised edition (Routledge, London; Dutton, New York, no date). I am indebted to R. G. A. Dolby for this reference.

well without explicit institutions for the exercise of influence from below; but whether such informality in this respect is still appropriate when so much else is institutionalized is a question worth considering.

Generalizations of Method

A full discussion of 'methods' requires mention of two more senses of the term, which are important in the historical development of modern science, and which are relevant whenever a new field is brought into being. At crucial points in the development of science, there will frequently appear proclamations of 'method', including in their scope not only the sorts of methods we have already discussed, but the objects of inquiry as well. Thus they serve to give a complete definition of a field of inquiry, or of a whole discipline. The classic statements on the new philosophy of nature by Galileo, Descartes, and Bacon fall into this class; and the major works of the latter two were explicitly treatises on 'method'. Such proclamations form an exception to the general rule that the principles and precepts of method are transmitted through the interpersonal channel of communication. The reasons for this lie in the exceptional character of their audience and their message. For they are not intended primarily (if at all) for pupils in a field, serving to supplement the craft knowledge they have received informally. Rather, their audience is a wider one, including a part of the general public; and their function is to explain and justify their subject to this wider audience, and to draw support and recruits from it. Similarly, the message will have a minimum of technical detail and precepts for solving particular problems; rather there will be general principles, supposedly applicable to a wide (or even universal) class of problems.[15]

The special character of these proclamations of method also

---

[15] Although the three 'essays' accompanying Descartes's *Discours de la Methode* are on mathematics and natural philosophy, Descartes's programme was much more extensive. The first title for the *Discours* itself was *Le Projet d'une science universelle qui puisse elever notre nature a son plus haut degré de perfection*. See R. Descartes, *Discours de la Methode*, p. xii. The Utopian and millennarian aspects of Francis Bacon's thought are fairly well known, as well as his concern for manners and morals. Although his *Novum Organum* is primarily concerned with natural philosophy, Bacon elsewhere expressed his belief that the 'primary axioms' of philosophy apply equally to all aspects of God's creation, including the social and moral as well as the natural. See the section on 'Philosophia prima' in the *De Augmentis Scientiarum*, Book III, ch. 1; *Works*, iv, 336–40 (translation).

derives from their particular concerns. The technical tasks of investigating problems in a field are subsidiary to a larger philosophical mission; and this may be imbued with a touch of the prophetic as well. The experience on which these proclamations are based will correspond to these deeper concerns: not so much doing a well-defined job and then reflecting on the way it was done, as attempting to create a new sort of work altogether, and mastering its basic principles along with its practice. The investment of one's life and talents in such hazardous work demands a deep commitment to the success of its results, both in the achievement of new knowledge and in the diffusion of the new approach. Hence (unless the knowledge is esoteric, a rarity in our present civilization and automatically excluded from our natural science), there will be a natural tendency for the author to believe that the 'method' characterizing the new approach is as easily communicated to a wider audience as are the scientific results of the work. The subtle, particular, tacit component of the work will accordingly be neglected in the reflection on its principles, since this would obscure the message and decrease its prophetic effect. Because of this, the proclamations on method have strengths and limitations opposite to the accounts of the 'art of inquiry' in particular fields. For the latter sort is strongest on precepts of craft practice and tends to confusion or vacuity when the author philosophizes; while the great proclamations provide general illumination and philosophical depth, but their explicit prescriptions on method are sterile or banal when applied to the craft work of investigating scientific problems.[16]

Such general proclamations of method are usually intended as clear calls to essentially straightforward action; but when they are received by an audience, they are rightly interpreted as philosophical theses. For their nature demands that they include an analysis of the knowledge of their domain of inquiry, and also that they have clarity and coherence as well as depth of insight. Hence they are judged by the criteria of adequacy and value prevailing in the relevant branch of philosophy. Their authors may well be dismayed by

[16] The 'four rules' of method given by Descartes in his *Discours de la Methode* (Second Part) were probably a fair sketch of the personal style of problem-solving he had developed as a precocious child; but as announced they were of little use to anyone else engaging on detailed philosophical or experimental work. By contrast, Bacon attempted, in the Second Book of the *Novum Organum* (translation in *Works*, iv, 119–263), to provide a complete list of the types of judgements involved in the various stages of inquiry in his style; but here the edifice of distinctions collapsed under its own weight.

such a reception, especially since there is a tendency among deeply original philosophers to see their work as finishing off that unsatisfactory discipline once and for all, and laying the foundations for solid scientific work in its place. But philosophy always refuses to lie down and die, and the classic proclamations of method take on a philosophical career of their own, producing a crop of descendant-problems within the ancient field of methodology.[17]

As a field of philosophy, this is a most difficult one in which to maintain traditions of work of high quality. For the criteria of adequacy are necessarily of two conflicting sorts; both a philosophically coherent analysis, and a perceptive insight into the actual practice of science, are required. The criteria of value for problems in methodology are correspondingly twofold. If there is no continuing delicate balance between these two aspects of the inquiry, or even worse if there is no awareness of the distinction between them, discussions of methodology will be largely sterile in practice, or conceptually muddled, or both. Over recent years, most discussions of 'scientific method' have suffered from a confusion over the methods by which they themselves are governed. In particular, the relation between their discourse, and the practice of science, is but rarely made explicit, and the ambiguities in the assertion, 'The scientist does . . .' are never clarified.[18] For this might refer to what scientists actually do and know that they do, most of the time; or it might refer to principles which are implicit in their work (as syllogistic reasoning in ordinary argument); or again it might refer only to a 'rational reconstruction' of their activity, for revealing the logical structure of scientific results. Further, the assertion might be prescriptive rather than descriptive, and would thus indicate what scientists should do in one of the above senses. Evidence for the presence of these confusions is provided by some interpretations

[17] For histories of this field of inquiry, see K. McKeon, 'Philosophy and Scientific Methods', *Journal of the History of Ideas*, 27 (1966), 1–22; and also L. Laudan, 'Theories of Scientific Method from Plato to Mach', *History of Science*, 7 (1968), 1–65.

[18] For strict 'logical empiricism', the question of what it is talking about, has been effectively probed by Stephen Toulmin, in 'Are the Principles of Logical Empiricism Relevant to the Actual Work of Science?', *Scientific American*, 214 (February 1966), 129–33, a review of C. G. Hempel, *Aspects of Scientific Explanation and Other Essays in the Philosophy of Science* (The Free Press, Glencoe, 1965). See also the correspondence between E. Nagel and Toulmin, in the April 1966 issue, pp. 8–11. The sharp dichotomy between the 'context of discovery' and the 'context of justification' is now coming under attack even by philosophers of science raised in the tradition in which it was established. See P. Caws, 'The Structure of Discovery', *Science*, 166 (1969), 1375–80.

of Popper's philosophy of science; in these, it is asserted that scientists normally do, and always should, make the refutation of an initial hypothesis the goal of their work as in a research project.[19]

Yet although the field of methodology is difficult, its problems are not vacuous. As we have seen, even though scientific work may appear to be nothing more than common sense to the majority of its practitioners, it is a highly sophisticated activity dealing with very artificial objects. At any point of genuine innovation (without which science becomes stagnant and dies) the inherited unselfconscious craft knowledge of objects and methods is inadequate to the task. A deep innovation in science necessarily involves an innovation in methods; and if the innovation in methods is itself deep, its explanation and justification will involve arguments in methodology.

Taken all together, the 'methods' of a field of enquiry govern its work, from the most particular techniques of tool-using, through the criteria by which the quality of work is assessed, to the most general principles underlying the practice of the field.[20] They thus form a natural correlative to the objects of inquiry; the set of these latter defines the subject-matter of the field, while the former define the ways in which this subject-matter becomes a part of human knowledge.

Without a body of methods of this sort, the social activity of science could not exist. It is possible for a man of genius, working single-handed and in isolation, to create a new field of study, developing his methods intuitively or even unselfconsciously as he defines and solves his new problems. But for a field to reach a matured and effective state, there must exist in the interpersonal

---

[19] In his *Towards an Historiography of Science* (The Hague, Mouton, 1963) (*History and Theory*, Beiheft 2), J. Agassi proposes that historians of science should proceed by testing and refuting each other's bold conjectures (p. 74); and his examples indicate that great scientific discoveries are in fact made by refutation of previously accepted hypotheses.

[20] To cite a very simple example, in 'rational mechanics' the objects are those of Euclidean geometry as enriched by the concepts of velocity, acceleration, force, mass, and energy, with others depending on the special problems, and all of them linked by definitions. The methods include in the first place the accepted procedures for using certain mathematical tools (historically, differential equations but now including other parts of mathematics), where the procedures include accepted interpretations of the mathematical symbols, and appropriate standards of mathematical rigour. Other methods include the more subtle criteria of adequacy regarding the relation of the results of an interpreted mathematical argument, with experiment or experience. See C. Truesdell, *Essays in the History of Mechanics* (Springer-Verlag, 1968), pp. 334–40.

channel of communication a large body of knowledge of the different sorts of methods; for men of genius are too rare, and too individual in their style, to be sufficient for the production of that collection of solid facts without which a field cannot steadily increase and deepen knowledge. This phase of the development of a field requires numerous craftsmen who are competent to operate effectively with inherited skills. Of these, only a minority can have the talent and commitment necessary for teaching themselves such skills from their own successes and failures. Without a period of training for research, where they can absorb an existing body of methods, they would only blunder through ill-conceived and misdirected projects, and disrupt the delicate system of social behaviour which enables a field to exist. But with an informal instruction in methods, they can do their essential work. In this way, an established body of methods has a function analogous to one of those of tools and their associated techniques. They make possible a downgrading of tasks, so that there can be a division of labour, and the possibility of that complex social organization which is necessary for the effective advancement of knowledge on a broad front.

Methods as Craft Knowledge

Our study of 'methods' has shown that they are a very heterogeneous collection of things, governing different aspects of the complex social activity of scientific inquiry. But they have one feature in common: except for straightforward techniques, they are all largely informal or even tacit knowledge; and they are transmitted through the interpersonal channel of communication, rather than through the public channel of printed reports. In these respects they form a contrast to the objects of inquiry of a field; and a study of the full set of contrasts and relations between these two opposite components will enhance our understanding both of the activity of science and of its products.

I have already argued that the objects of scientific inquiry are classes of intellectually constructed things and events. They must be mainly public and explicit, rather than informal and tacit, if they are to perform their function in the achievement of this particular sort of knowledge. For scientific knowledge, as it has been understood since Aristotle, seeks a rational understanding of its objects rather than a dialogue or communion between them and the knower. Even if the objects are human, or are conceived as endowed with some human properties, the contact between knower and object is not an empathetic one of shared inner experience, but an indirect one, achieved through the conclusion of an argument. Although scientific knowledge is ultimately based on particular experiences of interactions with an external world, any report of a single such experience is only a fragment, nearly meaningless and useless in itself. For the significance of such a report is not what happened to the personal agent, but the clues it can give to the constitution and workings of the external world. Similarly, research reports are not designed for the communication of the personal experience of the agent, but to provide materials for others working in a similar fashion. Hence the objects of discourse must be so explicit that there is a sufficient overlap among the private understandings of them for them to be taken over, used, and adapted, with success. In this sort of work, ambiguities and a variety of interpretations are not a source of richness and depth, but of waste and error.

The line of demarcation between 'science' in this extended sense, and other sorts of knowledge, is not between the human and non-human worlds as domains of inquiry. For until a few centuries ago, the dominant traditions in the study of the natural world saw it as human in character and suffused with divinity; the alchemists and magicians tried to establish a dialogue or communion with it. And in the disciplines studying man and his creations, including psychology, history, and even theology and literary criticism, are 'sciences' in this sense. Any sort of scholar argues from evidence to a conclusion; and even if his purpose is to help his audience achieve a sympathetic understanding of a personal experience, his particular task is accomplished with an argument rather than an incantation.

The objects of scientific inquiry cannot be perfectly public and explicit; and we should review the limits and borderline cases to their public character. For example, I remarked earlier that the first achievement of a deep insight on a scientific problem can be an intensely personal experience, sometimes nearly mystical in its quality. Also, the objects of scientific knowledge cannot be taught purely by demonstration; especially in elementary instruction, they are conveyed by indoctrination and manipulation, and become part of the unreflective craft knowledge of the trained person. Even in advanced research, each scientist has his own private understanding of his objects of inquiry, deriving from his unique craftsman's experience of working with them. Finally, those objects which become established as concepts of genuine scientific knowledge never cease to undergo change, and they carry with them a history of ambiguity and concealed contradiction.

Nor is the contrast to other sorts of knowledge an absolute one. Even, for example, most poetry and the reports of mystical experiences are intended as social possessions. Also they exist in a language which inevitably is culturally conditioned, and in part intellectually constructed. But the knowledge which they yield is a fabric of interwoven private experiences, and the special, unique character of each of them, and the ambiguity in their messages, is a contribution to the richness of the whole. We may illuminate the contrast between such 'personal' knowledge and 'scientific' knowledge by considering a report of an inspired insight into a scientific problem. If this is well written, it may stir the imagination and the emotions, and enhance the reader's appreciation of the workings of the mind of the scientist. It may also yield evidence for studies in the history of science or in the psychology of invention. But as material for the further development of knowledge in that field of science, the report is of only speculative value. For it is the content of the insight, not its quality as an experience, that is relevant to this function; and until the insight is proved to have a content which can be understood and applied by others independently of any sympathy with the author's experience, it remains in the realm of poetry rather than science.

The sort of practical craft knowledge which is embodied in the methods of scientific work is midway between the 'scientific' and the 'personal' knowledge which I have put into sharp contrast. For it is incapable of explicit specification, and yet it is a social possession; although the principles and precepts of method cannot be established by an argument of the sort characteristic of scientific knowledge, there must be a uniformity of understanding of them, as reflected in practice, for a field to exist. Methods are midway in another sense; for the very private experience of a working scientist with his materials is converted into a public, social possession through operations governed by the body of informal, interpersonally communicated methods.

The Two Channels of Communication

It is useful to compare the channels of communication whereby these two quite different sorts of knowledge, the properties of the objects of inquiry, and the methods, are transmitted. Research reports, embodying new conclusions about the objects of inquiry, go mainly through a 'public' channel of papers published in recognized journals. This provides an impersonal dissemination of tested materials to an unrestricted audience; anyone with the technical competence to understand a paper is entitled to read and use it. Although the procedure of such publication may seem simple, the public channel actually accomplishes a variety of tasks, each serving a different social function. The most obvious work of the public channel is the widespread and rapid dissemination of new results; this is not only facilitates the advance of the field, but also exposes the work of every individual and school to the critical scrutiny of colleagues. Supplementary to this is the role of the journals as an archive, a permanently open repository of the past achievements of the field, where they are available for later use. The system of publication in journals also serves to protect the personal interests of individuals; for a published paper is the property of its author, and the system of citations is designed to ensure that his property is recognized. Finally, none of these functions could be performed were it not for one other: that of quality control. If there were not a test of each paper before its acceptance by a journal, then every intending user would be forced to examine it at length before investing any of his resources in work which relied on it. Under such circumstances, the co-operative work of science as we know it could not take place. Similarly, the archive would consist of unauthenticated claims to results, and as such would be virtually useless. Moreover, the intellectual property of the author which is embodied in a published paper would be nearly worthless, since the status of publication would carry with it no indication that the result was genuine.

These various functions of the public channel are quite different in character, and each relates to the private purposes of scientists in its own way. Nor is it guaranteed that excellence in the performance of one such function will produce excellence in the others. We shall later see how the public channel is sensitive and vulnerable, and how the performance of its various functions, and the nature of its contents, are influenced by the social practices and attitudes of the community it serves. But provided that it works well, as it has generally done in the past, its contents comprise both news and archive, and are the materials from which genuine scientific knowledge is eventually achieved and recognized.

For the body of methods of a field, such a public channel, with its controls, is simply not available. There can be a public transmission of results in methodology; but as we have seen, this is not at all the same as the craft knowledge which governs the work being done. Also, criticisms of results and of methods will appear in the public channel, partly in the body of the argument of research reports, and partly in the 'correspondence' section of a few journals. But, even granting the importance of such channels (especially for the discussion of the political and social problems facing contemporary science),[21] such materials are not research reports like the principal contents of the public channel and they cannot achieve the subtlety and directness which is possible within the interpersonal channel. Hence, even with these exceptions, we can say that the transmission of methods is accomplished almost entirely within the interpersonal channel, requiring personal contact and a measure of personal sympathy between the parties. What is transmitted will be partly explicit, but partly tacit; principle, precept, and example are all mixed together. There is no substitute for such personal communication; messages whose transmission requires a prior formulation and clarification of ideas (as even in a letter to a colleague), will necessarily be impoverished in their content of private craft knowledge. A period of intense interpersonal communication between members is necessary for the formation of a school; only in this way can a full body of methods, including the personal style of masters, be communicated in the context of frequent informal discussions of work in progress. Gatherings of scientists perform a similar function of interpersonal communication of methods, although in a diluted form; but even a brief remark assessing some recent work can convey significant information on developments in the ruling criteria of adequacy and value.

The two channels of communication are not entirely separate in

[21] The contrast in this respect between *Nature* and *Science*, the leading weekly general scientific journals in Britain and America repectively, is instructive. *Science* has a lively correspondence section at the front of the issue, while *Nature* prints only a few letters, commenting on specific points in articles, tucked away at the back. The 'academic market place' thus has certain advantages over the 'club' for the public discussion of common problems.

their contents. The public account of a piece of work, embodied in its research report, must include enough description of the tools and techniques to enable a competent person to reproduce the argument and re-create the evidence, even if he could not have conceived the original problem on his own. For without the explicit communication of this component of method, the materials of the report would not be capable of becoming facts. Conversely, the gap between the private investigation of a problem and the public description of its solution is frequently bridged by a communication in the interpersonal channel: verbal or even written descriptions of the work for colleagues at all levels of formality below that of submission to a journal for refereeing and publication. Some measure of such interpersonal communication of achieved results is necessary for efficient co-operative work in a field; but an important part of the 'information crisis' in contemporary industrialized science is the substitution of this hybrid channel for the transmission of cumulative series of results at the expense of the public channel. The absence of controls in this channel can lead to a deterioration of quality, as I have discussed already. Similarly, the public certification of the intellectual property embodied in a research report is not achieved unambiguously and automatically; and if results become public in the interpersonal channel only, the social task of the protection of this property is deprived of its institutional support, and depends more heavily on the integrity and skill of all the members of the community.

The relation of the two channels of transmission is even more intimate than the overlapping of contents would indicate. For the materials which appear in the public channel are strongly conditioned, in their selection, by judgements made in accordance with principles transmitted in the private channel. To enter the public channel, a research report must be certified by a referee as passing, in respect of both adequacy and value. Yet the criteria on whose basis editors and referees assess papers appear nowhere in the public channel itself. They are a part, a most sophisticated and subtle part, of the craft knowledge which constitutes the methods of a field. If they are inappropriate for the development of a field, the material which appears in its public channel will be correspondingly distorted.

On its side, the interpersonal channel of communication is not independent of the public channel; but the two patterns of influence are not symmetrical. If the public channel, considered as the archive of results, shows scanty progress in a particular field, then critics may call into question some part of the body of methods governing the work in the field. But such criticisms of methods cannot be derived as conclusions to arguments resting on controlled experience; if nothing else, the assessment of the quality of progress is itself a judgement based on criteria of adequacy and value. Hence the mechanisms for the control of methods, and of their interpersonal channel of communication, use materials characteristic of that channel itself, and not those of the public channel. This holds even when some criticism and debate is formally published; for it appears as such, and not as research reports. Thus the results which appear in the public channel are passive. They can be used for a variety of functions, including the provision of evidence for the criticism of themselves. But by the nature of its contents, the public channel cannot directly 'influence' anything; and the relation between the two channels is not one of automatic mutual controls.

This asymmetry in the relations between the two channels of communication has important consequences for the government of science. For it is possible for a field to be diseased; for the ruling methods to be such that ineffective or shoddy work is accepted or even enforced. But the establishment of the existence of such a condition, and even more the attempts to reform it, are problems of a different sort from those which the field is devoted to investigating. Moreover, while gross errors in technique can be detected by the discovery of pitfalls in published work, the more subtle components of method are not easily accessible to anyone who has not already invested a part of his life in working in the field. Thus, reforming a diseased field, or arresting the incipient decline of a healthy one, is a task of great delicacy. It requires a sense of integrity, and a commitment to good work, among a significant section of the members of the field; and committed leaders with scientific ability and political skill. No quantity of published research reports, nor even an apparatus of institutional structures, can do anything to maintain or restore the health of a field in the absence of this essential ethical element operating through the interpersonal channel of communication.

In conclusion, we may consider the two channels of communication and their contents as a pair of interpenetrating opposites. The one distributes and preserves the results of the work, while the other governs the work itself; one is public and explicit, while the other is informal and interpersonal. The contents of the public channel are in principle permanent, and exist independently of the circumstances or ultimate fate of the work which produced them; while the body of methods, bound to very particular personal experience (both technical and social) directly control the future contents of the public channel. The results of scientific inquiry are in principle based on controlled experience and rigorous argument; but the methods governing the inquiry itself are a particularly subtle craft knowledge, different in nature from scientific knowledge. Thus we have elaborated the apparent paradox that scientific knowledge is achieved through an activity based on knowledge which is not scientific in character. If we restrict our attention to the investigation of particular problems, the paradox has no resolution; but I will later show how the complex social processes whereby the contents of research reports enter a new phase of development can yield a practical resolution of this paradox along with the others.

6

FACTS AND THEIR EVOLUTION

Once we have achieved the slightest historical perspective on science, and cease to view it as the steady and certain accumulation of increasing heaps of true facts about the natural world, a number of paradoxes present themselves to us. First, there is the contrast between the highly personal endeavour required for a penetrating inquiry into the natural world, always fallible and governed by mainly tacit craft methods; and the public, objective, impersonal knowledge which eventually issues from that work. Similar to this contrast is that between the ephemeral structures in which scientific inquiry is undertaken, problems succeeding each other with great rapidity and theories and concepts being ploughed under at a perceptible rate; against the steadily increasing depth and power of the knowledge which somehow remains and grows amidst the swirling currents of the day-to-day work. And, finally, we have already observed the impossibility of 'proving' a single result in science, or even of establishing its 'probability' of being true; and yet the stock of permanent knowledge, absorbing all reinterpretations and surviving all attempts at refutation, survives and grows.

There is no magic formula to explain the success of natural science; neither some special property of natural things which makes them uniquely accessible to human reason; nor a victorious 'method' whose application has laid bare the secrets of the natural world, and which can now be turned to other fields at will. Rather, the scientific knowledge we possess is the result of a social endeavour, which over the centuries has developed an approach appropriate to its limited goals, and where the work of each individual is informed and controlled by that of his colleagues in this endeavour, of the past, the present and the future. In the previous chapters I discussed the phase of the achievement of knowledge which is largely the work of an individual or a small group: the craftsman's work of investigating problems. The resolution of the paradoxes in the character of achieved scientific knowledge will come from an analysis of the further development of a solved scientific problem, when it emerges from the workshop of its creator, and takes on a life of its own in the community of science.

In this chapter I shall first discuss the processes and tests whereby a component of a scientific result comes to be recognized and accepted as a 'fact'; from this analysis we may achieve a clearer understanding of that basic but difficult concept.[1] Not all facts are, or become, genuine scientific knowledge; they must survive lengthy and rigorous processes of testing and transformation. These take place in the course of the evolution of the different components of a solved problem. From a comparison of the different ways in which the different components evolve, we will be able to identify an essential feature of those facts which survive to become scientific knowledge.

The Research Report and its Tests

The achievement of genuine scientific knowledge is the ultimate goal of the activity of scientific enquiry; but does not come quickly or easily. The completion of the work of investigating a scientific problem, even when the criteria of value and adequacy are satisfied, does not necessarily yield knowledge. Indeed, it does not even necessarily yield facts. For the product of a completed scientific investigation, we should best use a term as neutral as 'research report'. As it stands, the report is literally not to be trusted; and in practice it is not. So great are the possibilities of error, especially in work which is so complex and subtle, and whose acceptable outcome is so important to the individual who performs it, that the research report must be checked by an impartial assessor before it is released to the community. This is commonly done by a referee for a scientific journal; only after the referee has certified the problem as being of sufficient value, and the work as having been performed to the ruling

---

[1] I should make it clear at the outset that the term 'fact' as I use it refers to *assertions* of a particular sort, and *not* to a state of affairs in the external world. Much of the recent philosophy of science is governed by an implicit assumption that there is generally a perfect correspondence between non-trivial assertions about the world, and their real objects; so that 'fact' is taken as referring primarily to the world and then to its description. For me an important problem is how such a correspondence, even an imperfect one, ever comes to be achieved and also recognized; and so I must use the term 'fact' with greater precision.

criteria of adequacy, can it appear in the literature. A research report which is distributed without this certificate is, or traditionally has been, considered as nearly worthless. I have already discussed the significance of the partial breakdown of the system of certification, as a symptom of the new social conditions of industrialized science, and as a danger in itself.

Refereeing a paper is a special skill of a matured scientist. It is hardly ever possible to reproduce the data on which the evidence is based, and it is frequently onerous to follow a complex argument through all its fine structure. So the referee must use his personal knowledge of the craft to form a judgement on the adequacy of the work, using as evidence the reports and descriptions in the text before him. In this work the scientist develops skills which are much closer to those of an historian than is generally realized. Also, the filter applied by the refereeing system is inevitably conservative in its effects. Since really pioneering work involves a recasting of criteria of adequacy and value, the referee will generally find it below standard in some respect in the accepted system. Since the style of such work is radically different from that of straightforward mediocrity, it will be at risk of assessment as belonging to a much more perjorative category: that of the crank.

It might be thought that a research report, duly certified and published in a recognized journal, can be accepted as embodying a unit of scientific knowledge. Indeed, the belief in such a rapid and straightforward process of the achievement of scientific knowledge was dominant until recently, and is still widely held. But my analysis of the referee's work indicates at least that great things are not won so cheaply. For all the referee can assure is that in his judgement the problem is of value and the work adequately performed. The true value of the problem can be determined only through the further development of the field; and the real adequacy of the work can be properly assessed only by the more demanding tests of repetition and application.

Hence there is another phase of the testing of a research report, performed in the natural course of scientific work, outside of any institutional framework. The test of value is the most crucial; and it will generally be applied more rigorously by the community of scientists than by an individual referee. Fot the rejection of a paper by a referee on grounds of insufficient interest is a positive act, requiring a justification. But for a scientist to include a particular paper in the overwhelmingly large class of those which do not interest him, requires no apology nor indeed much cogitation. Almost every paper is uninteresting to almost all those who happen to see it; and the automatic, communal test of value yields a negative result when a particular paper happens to be of interest to absolutely no one. As a candidate for the status of scientific knowledge, it has then been killed. It may continue its career in a political function, as an entry in the scientist's list of publications, but as a contribution to the advancement of knowledge it has died, precisely because it is unwept, unhonoured, and unsung.

For a research report that passes the communal test of value, there is another test of adequacy, again more rigorously applied than by the referee. This is in its use. Here, the judgement is not whether it seems all right, but whether the results can be reproduced and built upon. If the test is passed the research report continues in being, as a citation in later papers, so long as the relevant component of the report continues to be of significance. If the research report fails this test there are two possible consequences. In the case of a problem of considerable significance, the inadequate performance of the work may require a public notice, so that others are not led astray by its claims. But this is always an unpleasant business, for the good faith of the author and the competence of the referees who passed it are inevitably called into question. Hence it is easier, and more peaceful, for each private user of the result simply to cut his losses, and re-build his own work around information derived from other sources. Formally, then, the inadequately performed work remains in the literature, presenting a pitfall for all those who might rely on it in the future. But the very pace of advance and change in a healthy field of science serves to neutralize such a harmful element; if it is not carried forward as a citation in later papers, it is soon buried in the mass of literature, and by its title and date announces itself as obsolete or irrelevant to later work. Thus the communal tests of value and of adequacy by a sort of natural selection; survival is achieved by the production of progeny, and a research report doomed to sterility by its weaknesses is soon forgotten.

Facts: Introduction

We can now consider those research reports which have passed these communal tests, in respect of their claims to be bearers of knowledge or, more modestly, of facts. Thus, after this lengthy preparation, we arrive at the point which many philosophies of science take for granted as the elementary, not to say trivial, component of elaborated scientific knowledge: facts. It is not easy to define 'fact', and this may be because the concept is so fundamental for the sort of knowledge of which natural science is an outstanding example. For the cumulative, progressive, impersonal, and permanent character of achieved scientific knowledge is intimately related to the special properties of the products of successful scientific inquiry. As we have seen from all our preceding analysis, science does not begin with facts; indeed, when facts are achieved (and recognized as such by the community) the process of achievement of knowledge in that particular area of research has passed its most challenging phase and is settling down to a routine. Hence if we are to explain the special character of achieved scientific knowledge, we must consider the concept of 'fact' more closely, and see how it can best be understood.

There are several sorts of human experience of natural science, in whose terms facts appear most naturally as the 'hard, massy, and impenetrable' atomic units of scientific knowledge. Facts are the best candidates as the bearers of certainty and truth in science; they are exhibited to apprentice-scientists as the foundation and proof of scientific theories; in the short term development of science they are the court of final appeal in the judgement of theories; and in a long term retrospective survey of a matured field of science there will seem to be a few salient facts which stand out as survivors from a wreckage of the erroneous theories of the past. Although the most magnificent achievements of natural science are in its great theoretical syntheses, these must be tied down to brute facts in several ways. Their foundations in experience must rest on attested facts, they must predict new experiences which prove to be real and factual, and they must survive collision with contrary facts. Finally, no matter how hard we try to illustrate the human, creative endeavour of scientific inquiry, or to show how bold theorizing and even speculation are deeply woven into any great advance in science, we must acknowledge that what counts in the long run in science is the increase in factual knowledge of the world around us.

Yet, in spite of all this, the atomicity of facts cannot be absolute. There has been much philosophical debate on whether any significant assertions can be entirely free of theoretical assumptions and entanglements; and I have already argued that the objects of the assertions in science are classes of intellectually constructed things and events. This does not quite destroy the epistemological primacy of facts in science, for one can merely require that the theory associated with a particular fact be independent of the relevant aspects of the theory being tested through an encounter with that fact. Epistemology aside, practical experience and history show that even the hardest facts are not quite impenetrable. For my analysis of the achievement of scientific knowledge up to this point has shown how much skill and judgement is involved in the work of creation and assessment at every stage. It would require a repeated miracle for those particular research reports which contain genuine facts suddenly to identify themselves clearly and indubitably before the scientific community. Hence the judgement of what is a real fact must be a social one, subject to the same sorts of influences, and subject to the same sorts of errors, as the other judgements I have discussed. And the errors are there to be found in the archive of scientific publication of the past. What is accepted as a 'fact' today is at risk of being dismissed in the near future as an erroneous conclusion based on crude data, which was interpreted by an incorrect theory.[2] Thus the objectivity and certainty of science seem to dissolve before our eyes, and we are brought back to the paradoxes with which this chapter began. For their resolution, an abstract epistemological analysis will be insufficient; we must continue with our description of the social processes of the achievement of scientific knowledge.

[2] Although particular results in research reports are frequently controverted, the destruction of a 'fact', as I shall soon define the term, is relatively rare in matured sciences. The more common fate is for an old fact to fade away; and those facts which are sufficiently basic to merit refutation will usually be statements functioning as principles, rather than directly generalizing particular experimental results. The classic example of a refuted fact is the principle that the orbital motions of the planets are compounded of uniform circular motions, the so-called 'Ptolemaic' astronomy. This fitted common sense, as well as the accepted physics of the heavens, and was the natural physical interpretation of the mathematical tools used in any calculation (ancient or modern) of a cyclic process. In more recent times, the 'conservation of mass' was a basic fact of physical science, from its enunciation by Lavoisier until its refutation by Einstein. A more particular fact, only recently refuted, is the defining property of the 'inert gases'. Their significance is shown by this description, of 1950: 'The inert gases occupy a peculiar position in Chemistry. They are practically devoid of chemical properties, and yet for that very reason they have provided the key to the whole problem of valency and the interpretation of the Periodic Classification.' See N. V. Sidgwick, *The Chemical Elements and their Compounds* (The Clarendon Press, Oxford, 1950), p. 1. In 1962 there appeared the first reports of compounds of these gases. (I am indebted to Mrs. Rachel Countryman for this example.)

Facts: a Comparative Analysis

We may gain insight into the special character of 'facts' by examining the principles underlying their distinction from other sorts of assertion about the external world. Consider, for example, the set of terms: 'fact', 'law', 'theory', 'hypothesis', and 'model'. Working our way back from the end of the list, we see that we start with very tentative assertions; an hypothesis is something which we may fully expect to be falsified, and a model is claimed only to have some analogical relation to reality. At the beginning of the list are those assertions which we make with great confidence, as 'law' and 'fact'. Proceeding along the list in the other direction, we are starting with assertions tied more closely to particular experience, and ending with those which are remote from experience and are only guesses or constructs to aid in its explanation. The most privileged position is that held by 'law', which enjoys both a firm ground in experience and a power of unifying and explaining. Hence it is no surprise that for many generations the goal of natural science was to discover the 'laws of nature', even after the theological overtones of 'law', inherited from the earlier natural philosophy, had become muted. Nor is it surprising that the two grounds of differentiation give the same ordering of labels; for in general, the more closely an assertion is tied to experience, the less likely it is to be exposed as false. However, the two schemes are not identical, and a discussion of their differences will bring me directly to my own thesis.

In the attempts to exhibit the logical structure of completed scientific knowledge there has been much analysis of the different status of facts, laws, theories, hypotheses, and models in their relations to particular experience. Such an analysis is not merely an epistemological exercise, for unless the student or the practitioner of science is aware of such differences among the assertions in his field he will blunder into pitfalls as soon as he proceeds to the use of existing results; I have already discussed this in connection with 'evidence'. But it has proved impossible to produce a set of formal criteria by which all the different types of assertions are neatly sorted out, and each assigned its appropriate epistemological status. The reason for this is that these different forms of assertion are taken as the species of completed, static scientific knowledge; and as we shall soon see, such a simple schematism is inadequate for a description of this complex and ever-changing product of the social activity of science.

Continuing for the moment with this traditional scheme of classification of assertions, let us return to the first ground of differentiation, the confidence in its survival. For our consideration of the problems of permanence and change in scientific knowledge, this is the more fundamental one. And we see that relation between expected survival and closeness to particular experience is not very tight at all. Some of the classic 'laws' of nature have been derived from particular experience by pure or nearly pure induction; such are Kepler's laws of planetary motion, and Boyle's law of the pressure and density of gases.[3] Others have been produced as necessary axioms for an elaborated theoretical science, quite remote from particular experience either in their discovery or in their testing; and yet by their solidity they have earned the title of 'law' rather than 'hypothesis' or even 'theory'. Examples of this class are the laws of dynamics and of thermodynamics. And it may easily happen that a particular set of accepted 'laws' turns out to be more solid and enduring than those 'facts' which were adduced as evidence in its favour when it was first announced; a famous case in point is that of Mendel's laws of particulate inheritance.

Hence for our present purposes we do best to neglect the traditional divisions of the species of the units of scientific knowledge and concentrate our attention on those features of an assertion about the external world which mark its capacity for survival. In this way we will approach the definition of scientific knowledge; for the essence of this is permanence, rather than a particular logical structure.

Facts: the Defining Property

In discussing the tests on published research reports, we have already described two properties which are necessary for an assertion to become a fact. These may be called 'significance', in that it must be noticed by someone if it is not to fall into oblivion; and 'stability', in that it must be capable of reproduction and use by others, if it is not to be rejected as spurious. These two are necessary for an assertion to become a fact, but in themselves they are not quite sufficient. In addition, it must have a property of 'invariance'. We have seen how science advances through the investigation of problems; how each

[3] Kepler's painful path to discovery is an oft-told tale; for the complexities hidden behind Boyle's Law, see C. Webster, 'The Discovery of Boyle's Law, and the Concept of the Elasticity of Air in the Seventeenth Century', *Archive for History of Exact Sciences*, 2 (1965), 441–502.

problem is concerned with the drawing of conclusions about certain classes of intellectually constructed things and events; and how these classes, existing only through the determination of their properties, change even in the course of the investigation of a single problem. When a solved problem has been presented to the community, and new work is done on its basis, then the objects of investigation will necessarily change, sometimes only slightly, but sometimes drastically. In a retrospect on the original problem, even after a brief period of development, its argument will be seen as concerning objects which no longer exist. There is then the question of whether it can be translated or recast so as to relate to the newer objects descended from the original ones, and still be an adequate foundation for a conclusion. If not, then the original conclusion is rejected as dealing with non-objects, or as ascribing false properties to real objects. But if such a translation or recasting is possible, then the original solved problem is seen to have contained some element which is invariant with respect to the changes in the objects of investigation.

It would be tempting to try to make this notion of invariance more precise, by constructing a formal scheme in which certain elements remained invariant in the accepted mathematical sense. But because of the subtlety of scientific arguments as well as the variety of elements in a problem which may remain invariant through later developments, it is doubtful that such an effort would be fruitful. At one extreme we have those assertions which are capable (in retrospect at least) of being derived directly from experimental data which is cast in terms of very simple and fundamental objects of inquiry, and so which can weather all the storms of reinterpretation. Such primitive specimens include Boyle's law, measured by a static force, and Black's law of the partition of heat, measured by a temperature. At the other extreme are assertions which are not capable of a simple test, and whose form underwent multiple transformations and differentiation while they were still being established; the second law of thermodynamics is an example of this class.

It is possible for the transformations under which a fact remains invariant to be very sudden and deep, to the point where a scientist can recognize a 'fact' in a piece of work whose objects and arguments he otherwise completely rejects. This occurs most commonly in mathematical sciences, when a mathematical formalism is produced which is universally accepted as true just because it is so beautiful, but whose interpretation is the subject of violent disagreement. One example of this is Fourier's equation for the propagation of heat in solid bodies, whose correctness was accepted, but for which Poisson worked for some thirty years to provide an alternative derivation.[4] Another example is Maxwell's equations of the electromagnetic field, whose derivation was unacceptable to any of his Continental colleagues, and whose conceptual objects were transformed out of recognition within twenty years.[5]

Thus 'invariance', along with significance for further work and stability under repetition and application is a necessary condition for a component of a solved problem to be accepted as a fact; and all three together are sufficient. I shall use this term to apply to any statement of the properties of the objects of inquiry (thus excluding tools) which meets these three conditions; for these are the statements intended as assertions about the external world. Two sorts of variety are thus coalesced into a single category. The first is the different components of a solved problem which may acquire the status of fact; not merely conclusions, but information, and perhaps even axiomatic statements not derived from data, but serving as part of the foundations for a theoretical argument. Also, under 'fact' I will include statements which are qualified in a variety of ways, falling under such categories as 'law', 'theory', 'principle', or 'model'. These distinctions are extremely important in some contexts. But in the present one, of the further development of the materials of a solved problem, they are not relevant. For the acceptance of an assertion as factual in this sense does not depend systematically on how it is qualified when it first appears in its research report, or on the pattern of argument within which it is established. Again, the varied functions which facts can perform, as descriptions,

---

[4] Fourier's work was submitted to the Institut de France in 1807, and received a brief derogatory notice by Poisson in the *Bulletin de la Société Philomatique* (see Fourier, *Oeuvres*, ed. Darboux (Paris, 1890), i, 215–21). The life-long rivalry between the two men led Poisson to attempt to re-derive Fourier's equations from sound Laplacian principles. In 1835 he published his *Théorie Mathematique de la Chaleur*, where this was done; by this time both Fourier and Laplacian physics were dead, and Poisson's laboured treatment of the derivation had no issue. See G. Bachelard, *Etude sur l'évolution d'une problème de physique* (Vrin, Paris, 1928).

[5] The classic statement is by Hertz: 'To the question "What is Maxwell's theory?" I know of no shorter or more definite answer than the following: Maxwell's theory is Maxwell's system of equations.' See H. Hertz, *Electric Waves*, tr. D. E. Jones (London, 1893), p. 21.

explanations, challenges, or heuristic guides to further work, are also independent of the traditional categories.

We see that facts do not come at the beginning of a project of scientific inquiry, but are the outcome of a lengthy process of work, with the individual phase of the investigation of a problem followed by the social phase of the testing, through use, of its solution.[6] Just as the achievement of an adequately solved problem is the closest that an individual can come to discovering the truth about the natural world, so the achievement of facts as I have defined them is the closest that a scientific field, at any point in its development, can come to the same goal. The difficulty of achieving genuine facts will vary enormously over different fields of study. In fields which are well matured in their methods of work and controlling judgements, and which are descriptive rather than theoretical in character, facts are easy to come by; indeed, the transition from refined data to facts may be hardly noticeable. At the other extreme are fields which are both immature and strongly theoretical, so that the objects of investigation are replaced with dizzying speed. In such cases, one can find libraries full of research reports, all of it the product of hard work and some of it of inspiration as well; but no facts.

The Evolution of Facts: Descendant Problems

To see how some facts do survive to yield scientific knowledge, we must first study the patterns of evolution of the various components of a successfully solved scientific problem. When we look at a completed research report, we find first of all a conclusion, and with it the information which serves as evidence in its argument. To the extent that the tools and techniques used in the work are other than completely standard, they too will be described; for the strength of the evidence depends on the means of production of the data and information from which it is derived. The problem, too,

[6] In framing my definition of 'facts' in terms of the survival of tests, I am of course developing insights first argued by Karl Popper. At this point of my 'rational reconstruction' of the process of achieving scientific knowledge, the tests do not involve any metaphysical or even ethical commitment on the part of those imposing them, for they are applied to the works of others. But I have already mentioned, in connection with pitfalls, that a self-critical attitude is necessary, even on narrowly prudential grounds, if worthwhile results are to be achieved. We will later see, in connection with quality control and ethics, that even such prudence will not be sustained in the absence of a genuinely ethical commitment shared by the members of a scientific community. Hence, as we shall see, Popper's deeper message, as distinct from his particular solution of an epistemological problem, is woven into my own argument.

will receive some discussion, to a degree which depends strongly on the literary conventions within the field. The objects of the inquiry will be mentioned, and their names used throughout the report; but they will be discussed explicitly only if they have suffered some significant change in the course of the work. Along with these explicit components of the research report there are those which remain implicit in all but the most deep and revolutionary works. Among these are the patterns of argument, and the criteria of adequacy and value used in the solution and the choice of the problem. These different components of a research report (which are of course only an abstraction from a complex and subtle reality) go their separate ways once the report enters the public domain. They provide different sorts of stimuli to later workers, serve different functions, and vary in their permanence. It is through the interweaving of the lines of descent of these various components of a succession of works that scientific knowledge develops its tough fabric, so resistant to trivial refutations and simple revolutions.

The problem itself is the most ephemeral part of the whole work. Setting and solving a problem is a creative act, whose accomplishment depends in many subtle ways on the person doing the work, his background and his environment; all contributing to what I have called his style. This is the aspect of scientific work which is most similar to that of aesthetic creation. Although we may know far more about mechanics than, say, Sir Isaac Newton, we cannot (even if we wished to) reproduce his particular achievement any more than we can that of Rembrandt. Of course, there are now some scientific fields which are so highly developed, particularly on the experimental side, that a problem-situation can crystallize in only one way, and then we must speak not only of simultaneous discoveries, but of simultaneous working on the same problem. But these are as yet the exception, and the uniqueness of deep scientific creation will be the rule so long as science admits of genuine growth. Even in routine scientific work, a problem known to be solved is a dead problem; the task cannot be repeated even if one might want to, for the world is a bit different from what it was when the work began. Yet the problem itself may continue to live through its descendants, in a variety of ways which we shall now discuss.

If we have an atomistic view of the growth of science, then we have no means of understanding the lines of descent between problems. Once a discovery has been made, or an hypothesis confirmed or refuted, that unit of work is completed. One may lay one's brick of knowledge upon the top of the pyramid, and then look for something else to do. But in our discussion of adequacy we saw that the practice of science is otherwise. For a solved scientific problem is not, and cannot be, a closed and perfect structure. Rather, it is the imperfect outcome of a particular project, where the conclusion which was drawn depended on the data which could be produced, the tools which could be deployed, the special skills and tastes of the craftsman who did the work, and even the first formulation of the problem with which the intensive work began. As the final conclusion takes shape in the later, convergent stages of the work, the scientist is concerned to have an argument which is adequate for the needs of that conclusion; anything more is a wasteful luxury. Hence, when the conclusion, with its argument and supporting evidence, is presented to the community, it is necessarily rough-hewn. Those who find the report significant will discover a variety of ways on which the work can be carried on, by modifications of the tools or of the argument, or by reorganizing or redirecting the work. From these possibilities, a new set of problems can be set and investigated. If they are taken up, then the original problem can be said to have direct descendants.

These descendant problems soon mix with problems derived from other sources; and so we speak more properly of a 'lattice' of descent rather than a 'sequence'. The later work produces changes in the objects of investigation and the tools; and so the original problem, itself an element of a descendant-lattice, usually sinks into obscurity and oblivion. There are two main classes of exceptional cases. The first is the 'classic' problem, where new objects and tools are created, with such depth and power that a long sequence of descendant problems can coast on the insights contained there.[7] Even when the tools are transformed out of all recognition, the permanence of the objects and of the statement of the overall problem, usually insoluble as it stands, justify the retention of the name of the ancestor problem and its author for the description of the field. It is in this sense that Newton was the father of classical dynamical astronomy,

---

[7] In some of the mathematical sciences, it is possible to have a 'perfect' problem. C. Truesdell describes it as follows: 'Every now and then something is done in a way that is final; problem and solution become one, so that soon nobody looks at the matter in any other way, and after a little it is taught to beginners as a matter of course.' See 'Rational Mechanics of Deformation and Flow', *Proceedings of the 4th International Congress on Rheology* (John Wiley & Sons, 1965).

and Einstein of twentieth-century physics. The other exceptional case lies, in a sense, at the other extreme. In fields where data is found rather than manufactured, so that the development of powerful abstract theories is impossible, a grand problem can seem to remain unsolved, and even undeveloped, after generations of work and debate. This is most noticeable in fields which are 'history', either of nature or of society. Of course, sometimes the appearance is true; and a great synthetic insight, riding on the prestige of a distinguished scholar, can entrap generations of successors who try to isolate the hard core of knowledge in it, and eventually realize that there is nothing there.[8] But this need not be so; and the eventual rejection of a grand problem can also come about through the growth of knowledge which was fostered by the study of it and of its descendants.

The Evolution of Tools

At the opposite extreme from the problem and its descendants, are the tools developed in the course of its investigation. In themselves, they do not consist of assertions about the external world, unlike problems or facts; and a good tool has a life (in itself or through its descendants) independent of, and longer than, the problem which gave rise to it. But similarity in function between tools and facts will enable us to draw parallels between their patterns of evolution. To be successful, a tool must first satisfy conditions analogous to those defining a fact: it must be significant and stable, and also applicable to problems and materials other than those

[8] Two such 'grand problems' in the history of early modern Europe are commonly associated with the name of Max Weber; they concern the relation of 'Protestantism' or 'Puritanism' with 'capitalism' and with 'science'. On the former, a possible way out of the morass has been shown by H. Trevor-Roper in the title essay of *Religion, the Reformation and Social Change* (MacMillan, London, 1967), pp. 1–45. The latter thesis has always had to contend with the contrary evidence of the Catholicism of such important figures as Copernicus, Galileo, Descartes, and Gassendi; and the focus for study has shifted to England of the middle of the seventeenth century. Even there, the 'Puritan' influence now seems to be dissolving into 'sectarians' on the one hand, and upper-class, moderate 'Latitudinarians' on the other. For these, see P. M. Rattansi, 'Paracelsus and the Puritan Revolution', *Ambix*, 11 (1963), 24–32, and B. J. Shapiro, 'Latitudinarianism and Science', *Past and Present*, 40 (1968). 16–41. A clue to the vacuity of such 'grand problems' in history is their association of two very general entities, each of them, to the extent that they exist at all, undergoing a complex development in time and place. One might equally well argue for a fundamental antipathy between the 'spirit' of 'Judaism' and those of 'capitalism' and 'science', since during the crucial three centuries from 1500 to 1800, the Jewish contribution to either is small.

associated with its first creation. Once established as a genuine tool, it undergoes evolution along several independent paths.

The most straightforward path of evolution is the refinement of the tools used in a particular problem, for the more easy and effective production of the same sort of data and information. The descendant lattices of such refined tools will be closely related to that of descendant problems. Indeed, the technical problems of refining tools have a direct influence on the patterns of evolution of their associated problems. For the choice of problems is governed by judgements of feasibility and cost, along with value; and improvements in tools, real or potential, will affect those judgements. There is also a more subtle influence; for the available tools can even alter the objects of investigation, shaping them towards a conformity with the data and information which the tools can produce.[9] But the tools are not an unconditioned, constant external force in this process; they too suffer changes in the process of their evolution. For their refinement makes them more powerful, but also more sophisticated and frequently more specialized. Hence they require a deeper technician's knowledge for their mastery and use, and may call into being a separate class of tool-experts, with their ambiguous relationship with their clients. Also, as the tools become more specialized, the understanding of them becomes restricted to a decreasing group of scientists or tool-experts, and they are in danger of becoming esoteric and hence sterile.

This danger is generally averted through the realization of another basic path of evolution of tools: their being applied to fields completely outside the descendant-lattice of their original problem.[10] For a good tool is capable of extension to a wider class of problems and objects, unanticipated at the time of its first devising. We can see how this is possible by recalling our earlier discussion of the special properties of tools. They are usually applied to producing materials for the prior phases of the work on a problem, as data and information. Such materials are less specialized to the particular problem at hand than is the evidence derived from them; and the tools, even

[9] The influence of the physical equipment in determining the possibilities and direction of advance in a science is clearest in the case of astronomy; and we may note that theory and even grand speculation have not been inhibited thereby. See A. Panneboek *A History of Astronomy* (Allen & Unwin, London, 1961), especially Part 3.

[10] At this point we may use the language of sociological theory, and say that a tool survives through the eventual discovery of 'latent functions' for which it is suited. I am indebted to Dr. J. Wootton for this point.

if designed for their function in that particular problem, will be of correspondingly wider applicability. Also, as we saw in the case of statistical tools, information which yields crucial evidence for a particular argument may well relate to some very general and abstract properties of the objects of investigation; and the tools which produce such information can be extended very widely indeed. Finally, those very special mathematical tools which provide the language in which an argument is cast can, by their complete abstractness, perform such a function in quite unrelated fields.

The process of extension of the use of tools comes about in many ways. The most dramatic is when a fully developed, sophisticated tool is seen to be capable of use at the centre of the investigation of problems in another field: either in the production of that data and information from which the essential evidence is derived, or as the language of the argument itself. In such cases we can speak of the 'invasion' of a field by a new tool, its subsequent transformation (including its objects of inquiry), and the development of new descendant-lattices of problems and their associated tools.[11] The other extreme case is one of 'evolution' rather than 'revolution': where the function of the borrowed tools is less critical in the solution of the problem, and where neither a sophisticated version of the tool nor a specialized technician is called for by the needs of the situation. Only when the tool will be involved in the production of a crucial or delicate piece of evidence is it necessary for its user to have a full appreciation of its possibilities and pitfalls. If it is applied in a routine fashion in the production of data, then a simple, rough-and-ready understanding is sufficient. What is desired in such contexts is a standardized version of the original tool, robust rather than refined in its design.

Hence we should distinguish between the extension of the use of

[11] The field of crystallography, which had evolved through the whole of the nineteenth century using precision optical instruments for angle measurements, and the abstract mathematics of group theory for theories of internal structure, was invaded in 1912 by X-ray techniques invented by von Laue: and the field of X-ray crystallography soon grew to become of the greatest significance in physics and biology. Materials for an history of this field have been assembled in an exceptionally fine collection, *Fifty Years of X-ray Diffraction*, ed. P. P. Ewald (Oosthoek, Utrecht, 1962). In this, an essay by J. D. H. Donnay, 'For Auld Lang Syne' (pp. 564–9) quotes a crystallographer of the old school, A. F. Rogers, 'Geometrical crystallography has had a glorious past, and it will have a glorious future.' (I am indebted to Mrs. Rachel Countryman for this reference.) Such invasions by tools are not always beneficial; particularly when an immature field is invaded by quite inappropriate tools, its work can be seriously disturbed. We shall discuss this in more detail below.

the tool itself and the development of standardized versions of it.[12] The two processes are not exclusive; for whenever a particular physical tool is manufactured in quantity, or an intellectual tool described in a book, it is necessarily standardized to perform a variety of functions, operating on different objects in various fields. We can see the extreme case of standardization in the development of versions of tools suitable for use in teaching. Having learned something of the craft of manipulating with this class of tools, from experience with his 'fool-proof' models, the student will in later years be better equipped to cope with more sophisticated versions in the course of his scientific or technical work.

One feature of the standardization of tools is relevant to the general problem of this chapter. This is, that a thoroughly standard-ized version of a tool will be longer-lived than a sophisticated, specialized version. This longevity arises from several causes. A tool which is designed for the most effective production of very speci-alized data or information will be more sensitive to changes in its function; while one which performs reasonably well in the pro-duction of non-critical material in one problem will be likely to do so for its descendants. Also, the standardized tool, being designed for functioning in a wide variety of fields, will be less affected in its overall use by changes in any one of them. Finally, unless fashion and prestige-replacement are dominant in a particular scientific community, such tried and true tools, whose possibilities and pitfalls are well known to the amateur craftsman users, will be replaced only reluctantly.[13] The extreme case of such permanence will be seen in

[12] The microscope provides a good example of a tool whose use has been extended over all fields of science and technology, and which has developed new specialized versions and their standardized descendants. Indeed, so pervasive is such a tool, that the writing of its history would be an enormous task, involving the close attention to experi-mental technique in a great variety of contexts. The standard recent history, S. Bradbury, *The Evolution of the Microscope* (Oxford University Press, 1967), necessarily confines itself to advances in design in a more or less linear sequence.

A picturesque but also penetrating description of the evolution of tools was given by J. Clerk Maxwell, in a review of papers by Lord Kelvin in 1872: (The reader) 'may also study, in the recorded history of electrometers, the principles of natural selection, the conditions of the permanence of species, the retention of rudimentary organs in manu-factured articles, and the tendency to reversion to older types in the absence of scientific control.' See *The Scientific Papers of James Clerk Maxwell* (New York, 1890; reprint Dover Publications, New York, no date), ii, 304.

[13] A most interesting study on the cultural influences on the extension of a mathe-matical tool is A. N. B. Garvan, 'Slide Rule and Sector: a Study in Science, Technology and Society', *Proceedings of the Tenth International Congress of History of Science*, (Hermann, Paris, 1964), i, 397–400. The historical problem is why the slide rule,

schoolteaching, where the combination of poverty and inertia can occasionally cause the retention of experimental apparatus for generations after it has ceased to have any scientific significance whatever.

We shall soon return to a discussion of standardization in a wider context; but there are some features worthy of notice in all these paths of evolution of tools. Much useful work in science is done in the development of tools, and even in the exploratory work of seeing whether they can be adapted to perform new functions. Such work rarely involves the great conceptual advances that are the most dramatic part of the growth of scientific knowledge. This is not to deny that such work can require great skill, talent, or even courage. But the production or design of things for a preassigned function, whether they be tools or particular items of information, is a problem more of a 'technical' character than scientific; and it generally calls for a different style of work. It is on such tasks that uninspired 'scientific manpower' can be usefully employed.

But it must not be thought that all work on tools is for the unsung and anonymous workers, no more than that tool-experts are always in a servant-relation to tool-users. It is not merely that the first devising and development of very powerful tools requires great talent and commitment; but the further development of such tools can give rise to coherent, self-contained scientific disciplines, whose objects of investigation are the descendants of the original tools. In such fields, the possible functions of the objects may eventually be reduced to a minor or even negligible part of the inquiry, and purely scientific problems become dominant. The tendency for tools to become the ancestors of scientific fields is most noticeable in mathematics; from geometry itself, to differential equations and statistics, the process has repeated itself many times. An analogous process can be seen in the development of chemistry, which emerged from the status of a tool-providing art for metallurgy and medicine to become an independent scientific discipline. Such an evolution of a field entails drastic changes in its methods, and in particular in the controlling judgements of adequacy and value; and with them the

invented in the early seventeenth century, remained in obscurity for 200 years before replacing the sector or 'mathematical compasses' as the standard hand-calculating instrument. There are many factors involved, including the permanence of craft tradition, the teaching of mathematics through geometry, and the retention of a 'classical' theory of architectural design, all until the later nineteenth century.

corresponding transformations in the character of the community involved.

The Evolution of Facts: Standardization

The patterns of evolution of problems and of tools present the sharpest possible contrast. A problem is ephemeral, and lives on only through its descendants, which become increasingly remote in form and content from it. If a standardized version of it survives, this will be only as a textbook exercise; transformed so as to be soluble by routine manipulations, it will be a caricature of its original. Tools, on the other hand, have not only direct descendants, but also descendants from their extensions, and finally the more permanent standardized versions of their original or later forms. The patterns of evolution of facts lie between these two extremes; and this could be expected from the position of science between philosophy at one extreme, and the crafts at the other. In the short run, an important function of facts is to pose problems or, more precisely, to create problem-situations. An interesting fact presents a challenge: if it is in conflict with other accepted results, one or the other must be rejected or modified; and if it promises further advance, it calls for its own improvement. In these ways, the established fact is embedded in the descendant-lattice of problems derived from its original; and in this path of evolution it will survive only as long as its context in the problem.

Those facts which continue in being long enough to become knowledge must do so by a process of extension analogous to that of tools: they must be seen to be relevant to problems in other fields of inquiry, and remain so through all their changes.[14] As in the case of tools, the process can take place by invasion, as when a fact developed in one field forces its way into the awareness of practitioners in another, as a significant challenge calling for a response. The more common path of extension, however, is when facts in one field are seen to be useful as information in the problems in another. Such extensions are more demanding in the case of

[14] The necessity of extension for the survival of results is shown very clearly by the different fortunes of the leading fields of mathematics in the nineteenth century. Three became extinct or nearly so: elliptic functions; invariants of algebraic forms; and synthetic geometry. On the other hand, differential equations grew and survived through its contact with physics; and the theory of limiting processes in analysis provided the basic patterns of argument, and essential results, for analysis and its generalizations into abstract mathematics of the present century.

facts than of tools. For, as we have seen, tools can operate on aspects of the materials of the problem which are not at all specialized to the problem or even its field; while a fact, consisting of assertions about particular objects of inquiry, must be capable of translation from its original field to that in which it is being applied. Given this difference, which entails a more rigorous selection of facts by the processes of extension, the function of such extended facts is similar to that of extended tools. They may be used directly as information, or as part of the explanation associated with a particular tool or tool-subject; but in any case, they perform a function as a means to the solution of the problem, and are not part of the evolving materials of the problem itself.

The process of extension of facts will then necessarily involve standardization, just as in the case of tools; and the reasons for the longevity of the standardized materials are the same in both cases. It is important to realise that standardization is not merely a sufficient condition for the survival of a fact to become knowledge; it is necessary as well. For if a fact remains tied to its original problem and its descendants, it will die with them. It will stay alive longer only if it performs a function in work on other problems in other fields; and for this it must be available in a standardized form, for use as information or in connection with a tool. Thus the materials from which scientific knowledge is achieved are necessarily remote in form from their specialized originals; they are robust, general-purpose, standardized versions. For appreciating the properties of achieved scientific knowledge, and the conditions for its achievement, it is necessary for us to study these standardized materials in some depth.

Although the functions of standardized facts are very similar to those of standardized tools, there are some significant differences between the things themselves. One way of describing the difference would be through the distinction between a certain class of technical problems (for tools), and didactic tasks. These differences are partly characteristic of the objects of the work; but they include social and institutional aspects as well. A standardized and simplified version of a physical tool is in no way 'inferior' to its original; the differences in design correspond to differences in function. But the change in content which occurs when a fact is standardized can be interpreted as a degeneration. For a standard fact is not only something which performs a function in the solution of problems; it is also an assertion about classes of things and events, intended to relate to the external world. Now, as such an assertion passes into the social phase of its development, after its original appearance in a research report, its content inevitably changes. Even in the descendant-lattice of problems, its increased precision and sophistication will frequently be achieved by the sacrifice of the deep analysis which enabled its first formulation, but which was either unnecessary or incapable of retention in subsequent work. In particular, points of obscurity and of unresolved conceptual confusion in the objects of inquiry will tend to be overlooked, in the concentration of interest on those aspects which are capable of straightforward further development. But when the fact undergoes standardization, not merely the nuances of its first intimation, but even some important but subtle aspects of the assertion or its objects, are smoothed over and forgotten. This seems, and may indeed be, a regrettable vulgarization, especially when the end-product is examined by an expert in the corresponding descendant field of research. But it is quite necessary, if the fact is to be useful to those who lack the time, skill, or inclination to master the elaborate theoretical context in which its sophisticated versions are comprehensible.

We may illuminate the distinction between the standardization of facts and of tools by considering two intermediate cases. One of these is the body of 'explanation' which accompanies any tool or piece of information. Although this will involve assertions of a factual character, it is very strictly designed for a special function: to assist in the mastery of skills for the competent use of the material. Hence the requirements of sophistication, coherence, or even faithfulness to its original, are of quite secondary importance. As we have seen previously, even a totally incorrect theory can serve as a guide to successful craft practice; and so an explanation of this type may be adequate to its function in spite of containing factual assertions which have been vulgarized out of all recognition. On the other hand, the standardized versions of intellectual tools which are taught rather than manufactured, have a natural upper limit to the entropy increase. For if they are to perform even simple functions adequately, and not yield utterly vacuous materials, they must retain some essential features of the design of their originals. We can see this 'natural minimum' of content in the case of statistics; there is a certain critical level of sophistication of the numerical manipulations, and in the craft skill of operation, below which nothing but obvious rubbish can be produced by any particular statistical tool.[15]

From these examples, two important features of the standardization of facts can be discerned. One is that the content of a standardized fact may decay, almost without limit; the degree of sophistication and of faithfulness to its original which is necessary for its adequate performance of its function will depend very strongly on its use. Also, it can be seen that a version of a standardized fact which is good enough for one function can be quite inadequate for another; and since any standardized fact performs a variety of functions, it will naturally appear in a variety of versions. These two features will be discussed at greater length, for they are directly relevant to the nature of genuine scientific knowledge.

The Evolution of the Objects of Inquiry: Entropy-Increase

The patterns of evolution of the objects of scientific inquiry are very similar to those of facts, since these are the things about which factual assertions are made. That they do evolve, there can be no doubt; what is meant by such terms as 'force', 'molecule', 'acid', or even 'iron' is in constant flux. For these are intellectual constructs; their relation to the external world is neither immediate nor certain. Philosophical reflection on this feature of the objects of scientific inquiry can lead to a denial of the possibility of real knowledge of the external world; but this consequence follows only when the analysis is abstract and over-simple. Thus, the insight that a theoretical concept has meaning only in terms of the operations yielding its measurement, was deprived of depth by the implicit assumption that there is a unique and simple relation between each concept and its associated operations.[16] When we appreciate that each object of scientific inquiry carries with it a complex burden of meaning, derived from its history of use and adaptation, the way is open for showing how genuine knowledge of the external world is possible, even when cast in the terms of such artificial objects.

[15] I believed that I had been exposed to the ultimate in absurdity of the application of statistical tools when a graduate student asked me for advice on calculating Standard Deviations; he had a sample of three readings. My naïvety was exposed by a colleague (Dr. J. Dobrzycki) who had had a similar request, where the number of readings was two. Neither of us have any evidence of this technique being applied to a single reading; but authentic instances would be welcome.

[16] This was the philosophy of the distinguished experimental physicist P. W. Bridgman, first announced in *The Logic of Modern Physics* (MacMillan, New York, 1927). There is no doubt that 'operationalism' has been beneficial as a criterion of adequacy in disciplines where concepts could otherwise be allowed to float freely.

Those objects of inquiry which survive to become the materials of scientific knowledge are contained in standardized facts; and we can discuss the phenomenon of decay of information through standardization in connection with them. As the objects of inquiry are brought down through successive transformations in descendant-lattices of problems, and by extension, the collection of accepted factual assertions about them becomes ever larger and more heterogeneous. It sometimes occurs that a change in the objects of inquiry enables a more clear and penetrating explanation of an existing standard fact; but the opposite effect may also occur. For the new objects of inquiry, organized around new experiences and problems, may be ill-suited for the explanation or even description of the inherited standard fact. For example, the partition of heat among bodies in proportion to their mass and 'capacity' is easily understood on the analogy of water reaching the same level in a set of interconnected vessels; for this was the implicit model in the original experiment. But to explain this simple experiment in terms of 'energy' or some other conceptual objects developed a century after Black's work in the course of much more sophisticated problems, would only lead to confusion. Those who teach the old facts then have the choice between a difficult and sophisticated 'correct' explanation, and a continued use of the old, obsolete and incorrect one.[17] It is because of this that the relics of old and discredited

[17] When facts are retailed in an older, incorrect version, it is necessary that the students be steered away from pitfalls which were later discovered in that version; and so the facts as purveyed become a vulgarized version of something that never existed. For example, a consideration of 'latent heat' in the theoretical context of the eighteenth-century work could lead an unwary student to repeat an error common then. Quantity of heat was measured by the temperature change resultant on its introduction to a body of given mass and heat capacity; and this measure was naturally extended to the measurement of latent heat. At that time, there was no reason to suppose that an experimental result as 'I have, in the same manner, put a lump of ice into an equal quantity of water, heated to the temperature 176, and the result was, that the fluid was no hotter than water ready to freeze' could not be transformed into a general measure of heat of fusion, as 'But this quantity of heat (which disappeared in melting the ice) would be increasing, by 143 degrees, the heat of a quantity of water, equal in weight to the ice alone'; and then applied to *any* substance, as ' . . . the heat which any given quantity of water loses upon being frozen—were it to be communicated to an equal weight of gold, at the temperature of freezing, the gold, instead of being heated 162 degrees, would be heated $140 \times 20 = 2800$ degrees or, would be raised to a *bright red heat*.' The first two quotations are from Joseph Black, *Lectures on the Elements of Chemistry* (Edinburgh, 1803) i, 124–5; the third is from Rumford, 'An Inquiry concerning the weight ascribed to Heat', *Phil. Trans.* (1799), p. 193. I am indebted to Dr. D. S. L. Cardwell for this point; he has used it in his teaching to show that even pure 'phenomenological' physics can have its pitfalls.

theories clutter the teaching of science, not only in the schools, but wherever it is being conveyed in a standardized form as a tool-subject. But to attempt to clean up all this confuséd mass, and to purify the material by deriving it directly from its basis in modern research, can lead to a syllabus which, however satisfying intellectually to the teacher, leaves students in a state of confusion worse confounded. Hence those old objects of scientific discourse which are still useful for the description of standard facts, will remain alive even after they are obsolete for the purposes of scientific inquiry. And because of the complexity of the development of those objects, as well as the variety of teaching situations, it is impossible to ensure that every student is brought forward in a neat progression from convenient old errors, towards powerful new truths. More commonly, the different sets of objects in a family will coexist, peacefully among teachers and researchers who each use only one set, and confusedly among students.[18]

It is sometimes possible for a field to be 'unified', so that its

[18] In the teaching of elementary chemistry until recently, chemical combination was explained by a theory of 'valency' descended directly from the 'dualistic' theory of Berzelius in the early nineteenth century, and given a convincing rationale in terms of electron structure and the electrically neutral state of a complete molecule. It was then difficult for students to grasp the 'co-valence' of organic compounds, to say nothing of diatomic molecules as Oxygen gas. Little did we know that the same difficulties were present in chemistry in the second quarter of the nineteenth century; that they were partly responsible for the decay in the acceptance of Avogadro's hypothesis; and that our 'basic' theory was the relic of a grand synthesis that had failed. The influence of a subsequent research tradition on the formulation of a standard fact can be seen in the case of Boyle's Law. It was originally expressed in terms of a proportionality between the 'spring' of an elastic fluid, and its density; and we may translate 'spring' into 'pressure' without too much violence to the original. But in modern times 'density' has been replaced by 'volume', which in the context of the original law is not the fundamental physical factor involved. It seems likely that when a more general gas theory developed in the nineteenth century, with volume as a significant parameter, the version of Boyle's Law used in elementary teaching was modernized, with a consequent loss of clarity. It is even necessary, on occasion, for the doctrine of a science as taught to ignore certain reliable phenomena, which may once have been of great interest, in order to maintain a tidy structure of theory and experience. It might be expected that this will occur in optics, where the student's eye is involved in the experiments. In *Optics, the Science of Vision*, V. Ronchi cites the example of a simple convex lens with an object placed at the distant focus; the observer should see virtual images behind the lens at an infinite distance. No one does, but this does not hinder the teaching of simple lens theory. In more advanced wave optics, one learns of the position of images cast by concave spherical mirrors; experimental tests of these theories lead to the most astonishing results, none fitting the theories. Yet the phenomena of highly curved reflecting surfaces were the subject of great interest from antiquity to the seventeenth century, and it appears that they had to be suppressed in the modern period in the interest of a tidy and plausible explanation of refraction and reflection. (See pp. 202, 133–41.)

various objects of discourse can be transformed into mutual coherence, or rejected. Such unification usually takes place at the highest level, either summarizing a completed research tradition, or creating a coherent field out of previously diverse problem-areas. But even in such work, one cannot escape from history. That which is being unified was previously diverse, its problems and objects deriving from different sources in experience. In the programme of unification, a small set of these experiences, and their associated objects, are taken as fundamental, and the others are exhibited as being capable of derivation from them. But it is impossible to make a perfectly neat structure out of such disparate materials; and then those who teach the 'unified' science must cope with, or conceal, the awkward unconformities at the boundaries. This phenomenon can be seen in mechanics, which of all physical sciences seems the most naturally unified, and derivable from a few basic axioms and experiences. Its sources in experience come from diversity of practical tasks, and several crucial phenomena. For the tasks, we have such as load-lifting, balancing, falling of bodies, projectile-throwing, and overcoming a resistance by impact. The phenomena include the pendulum, and the transfer of motion by collision. These latter served as the foundation of dynamics in the seventeenth century, through their use by Galileo and Descartes respectively. Newton grafted a general idea of 'force' on to the collision phenomenon for his mechanics; while Huygens and Leibniz generalized from the pendulum to '*vis viva*', the ancestor of kinetic energy. The unification of mechanics, started at the formalistic level by Legrange, and deepened in the nineteenth century by the new concept of 'energy', still leaves these separate roots in existence. Thus, Ernst Mach's programme of banishing the concept of 'force' from mechanics as an anthropomorphic relic leaves the field of statics as 'the science of nonexistent motions'.[19] The lifting of a load at constant velocity, as by a water-wheel, can be explained in terms of work and energy but it coheres not very well with the Newtonian concept of force and acceleration.[20] And certain paradoxes of impact which were common

[19] See *The Science of Mechanics*, 5th English ed., from 7th German ed. (Open Court Publishing Co., La Salle, Ill. and London, 1942), ch. 1, section v, paragraph 6 (pp. 95–6). Mach defines 'force' as 'any circumstance of which the consequence is motion'; notes that these may be conjoined as to result in no motion; and says 'statics investigates what this mode of conjunction, in general terms, is.'

[20] On the inapplicability of Newtonian dynamics to the problems of analysis of power machinery in the eighteenth century, see D. S. L. Cardwell, 'Early Development of the

in the sixteenth century, such as the radically different effect of a hammer resting on a nail and the same hammer dropped on it from a height, are omitted from modern syllabuses, doubtless because of the confusion they would cause in immature minds.[21]

In general, the standardization of the materials of a field necessarily kills them. What was created in a succession of turbulent and frequently confused waves of advance, and then abandoned before being fully clarified when the frontier of research shifted elsewhere, must now be given a tidy organization. It must be presented as if it all started with the discovery of an elementary fact by elementary techniques, and then grew by a linear, logical development with the injection of new elementary facts at the appropriate places. It matters less that such a presentation destroys the actual history, than that it implicitly purveys a false and deadening picture of scientific knowledge and of its achievement, and presents its materials in truncated and vulgarized form.[22]

Because of all these effects in this standardization of facts and objects of inquiry, it is possible to speak of a decay of information, or an increase of entropy, accompanying the process.[23] This should not be surprising; the physical transmission of any 'signal' involves an admixture of 'noise'; and an entropy-increase in the manual copying of texts was a universal phenomenon, which now enables the reconstruction of descendant-lattices of medieval manuscripts. A similar effect has been observed in encyclopedias, ancient and modern.[24] Nor should we be surprised if the increase in entropy is

Concepts of Work, Power and Energy', *British Journal for the History of Science*, 3 (1967), 209–24, especially pp. 212–14.

[21] This paradox can be put quite vividly, in terms of a bow and arrow, first put in tension with the arrow's point against a plank of wood (making only a slight mark on the surface), then withdrawn still extended, and released. The arrow will penetrate the wood; why? Attempts by pupils (and teachers) to solve the problem shows clearly how their training in elementary mechanics is confined to manipulation of standard, safe examples. This thought-experiment was cited by Leonardo da Vinci, *Selections from the Notebooks* . . . , ed. I. A. Richter (Oxford University Press, 1952), p. 66; and Galileo struggled vainly to resolve the paradoxical properties of impact forces (see S. Moscovici, 'Remarques sur la dialogue de Galilée, de la force de la percussion', *Revue d'Histoire des Sciences*, 16 (1963), 97–137.

[22] E. A. Carlson, *The Gene: a Critical History* (Saunders, Philadelphia, 1966), displays this process particularly well. In his last chapter, 'Historical Conclusions' (pp. 244–58), he asserts that the 'beads on a string' version of chromosomes was 'never more than a straw man'.

[23] I am indebted to Dr. D. S. L. Cardwell for this and many other features of the evolution of facts.

[24] On the degeneration of the encyclopedic tradition in Rome, see W. Stahl, *Roman*

frequently greater than might be considered strictly necessary for the particular function of the standardized fact; for the natural pressures of a didactic situation will produce such a tendency. Also, the person involved in the didactic task of standardizing material downwards from one level to another will frequently lack a sufficient mastery of the material at the higher level for a good presentation at the lower.

It is easy to single out schoolteachers as the prime targets for a critical analysis of standardization. Their social isolation from research activity cuts them off from the stream of sophisticated discussion of the facts and their objects; and in any event their standardized versions of facts will usually have little similarity to those used in current research. Hence they are generally forced to retail standardizations of standardizations, or vulgarizations of vulgarizations, as the case may be. These inherent limitations of the schoolteaching situation, along with its function of imparting basic craft skills rather than 'understanding', must be recognized if there is to be any fundamental improvement in its quality. However, the process of standardization of information and facts, like that of tools, goes on at all levels up to the highest; any publication other than a pure research report involves standardization. To do the work well requires a special skill, which is rewarded in the lasting success of a classic textbook and its descendants. To do the work badly is also possible and easy at all levels; vulgarization and incomprehensibility can be found even in the most advanced and specialized monograph literature.

At some time in the future, it might be possible to enliven the teaching of science by re-creating parts of the history of which the standardized materials are the relics.[25] By such a means, students could share in the excitement of discovery and conflict which attended all the great achievements, and appreciate that the apparently tidy organization of the material is only a means of making it comprehensible for its function as a tool. But the difficulties in the way of such a programme must be recognized. It is not sufficient merely to have good historical studies (of which not many exist as yet); but the reconstruction of the context of discovery and debate would involve

Science: Origins, Development and Influence to the Later Middle Ages (University of Wisconsin Press, 1962). A modern analogue is discussed by H. Einbinder, *The Myth of the Britannica* (Grove Press, New York; MacGibbon, London, 1964).

[25] A model for such an approach is V. Ronchi, *Optics, the Science of Vision*, which I have already discussed.

teaching some amount of quite obsolete and erroneous technical material, which could be laborious and also confusing to the weaker students. It would seem that such an approach to the teaching of science will be restricted to a very few convenient examples, until there are deep changes in the tools and techniques of teaching. One can imagine a separation of the two aspects of learning, with students mastering manipulative skills privately on a machine, and then having open-ended discussions with the teacher on the understanding of the material at all levels, technical, historical and philosophical. But for this, neither the machines, nor the historical and philosophical materials, nor the teachers, are likely to be available for a long time to come.

7

THE SPECIAL CHARACTER OF
SCIENTIFIC KNOWLEDGE

THE materials which comprise scientific knowledge are facts, of a certain sort: those which have survived the processes of testing and transformation, so that they remain in use, and hence alive, long after the disappearance of the problem which first gave rise to them. We have already seen that these processes involve losses as well as gains; in particular, the standardized facts and their associated objects of inquiry are liable to suffer a loss of their content of information, and to be vulgarized. This may well be regrettable, but it is inevitable. No one (except possibly the historian) has any reason for preserving a significant fact in its pristine form; the facts stay alive only as long as they are useful in new contexts, and if they cannot be put into a convenient form for their limited functions, they are soon discarded and forgotten. Although such standardized facts comprise the body of scientific knowledge, it is not simply a collection of atomic units of hard, long-lived facts. Indeed, its special character results from the complexity and interconnectedness of its materials, as they evolve through the complex and fallible social processes of their use and adaptation.[1]

Differentiation of Standardized Facts

When we examined the standardization of facts, we observed that this is done in order that the existing fact can perform one or several new functions, as a means to the solution of new problems. In general, there will be a great diversity of such functions: for the fact can be applied in a variety of different fields (each requiring

[1] E. T. Bell, *The Development of Mathematics*, comes close to my conception when he speaks of 'residues' of epochs: that which survives, having been sublimated from a mass of detailed results, to remain alive in the work of future epochs (p. 23).

its own translation of its objects of discourse); it can serve as information, or as part of the explanation of a tool; and in each possible field and type of use it is capable of being applied at different levels of sophistication. The number of conceivable functions for a standardized fact will then be very large; and although each particular version will perform well over a range of functions, there is still room for a multiplicity of versions of the same standardized fact, differing among themselves and all remote from their original. Thus standardization does not exclude a rich differentiation; and the more basic the fact, the greater the degree of differentiation. Hence when we speak of a particular fact as having survived to become a part of scientific knowledge, we should mean a family of particular versions of a fact, mutually related by a complex lattice of descent from their original, and all still in flux.

This variety escapes systematic attention because it can usually be ignored in practice. The research worker in science has a limited range of functions for any tool or fact within his competence; and the philosophers of science have hitherto generally assumed that a fact, once it has emerged from the murky, irrational phase of creation, takes on a permanent, unique, and rigid form. Only teachers find the multiplicity of doctrine to be troublesome, and then only sometimes. And in any event their experience is not considered weighty evidence in the socially more exalted realms of science. But when teachers must cope with the misunderstandings and misinformation inherited by their pupils from earlier instruction, or try to agree on the standardization of materials designed for different functions (as 'units' in physics), or even teach their own material as tool-subjects for students in other fields, the difficulties, pedagogical and sometimes social and institutional as well, can be acute. The debates which ensue in such situations are frequently vitiated by the assumption that there exists a unique, perfect, and true version of the facts in question (and even that this perfection is to be found among the standardized versions in dispute).[2] But the case

    [2] Ernst Mach detected this tendency thus: 'Until contemporary times, however, workers in this field seem more or less unconsciously to have sought for a natural measure of temperature, a real temperature, a kind of Platonic Idea of temperature, of which temperature read from thermometers is only an incomplete and imperfect expression.' See E. Mach, *Die Prinzipien der Wärmelehre* (Leipzig, 1896); English translation of this section by M. J. Scott-McTaggert and B. Ellis, in B. Ellis, *Basic Concepts of Measurement* (Cambridge University Press, 1966), p. 190. I am indebted to R. G. A. Dolby for this item.

is otherwise; we have only a variety of functions, and criteria of adequacy of the performance of the functions by the materials of the various designs.

For those who have not had this unsettling experience of disagreement over the obvious, it may be hard to imagine that the fundamental facts of science are capable of such obscurity and variety. For such facts, and their objects, imparted as part of the most elementary craft instruction, tend to become part of the unquestioned and usually unquestionable common sense of a field. Thus, when C. P. Snow bemoaned the deep ignorance of the 'arts' graduate concerning science, he used as the test of elementary scientific literacy, the question 'What do you mean by mass, or acceleration?'[3] Completely encased in the scientist's common sense of his generation, he was not to know that the answer to the second question lies deep in the mysteries of the calculus; and that on the first question there have been as many opinions as philosophers of physical science. The first, natural reaction, on the discovery of the variety and obscurity in this elementary, common sense knowledge, is to search for the true version in a sufficiently thick book; and the second is to attempt to sort it out for oneself. For, the thicker the book that one examines, the more advanced the material, and (usually) the less illumination one will find on the elementary concepts. Also, each thick book will put the elementary material to a different use, and so will offer a different version of it appropriate to its use in the discussion of a special set of advanced problems. The bold person who strikes out on his own to discover the unique true meaning of a fact can sometimes achieve new results concerning this elementary material, which then become a part of it for the future. Mach's analysis of the foundations of Newtonian dynamics is a famous case in point. But genuine scientific knowledge is too complex and rich to be comprehended in any single schematism, didactic or analytical; the task of understanding what has been achieved is as never-ending as the task of extending those achievements to new realms.[4]

[3] C. P. Snow, *The Two Cultures and the Scientific Revolution*, 'I now believe that if I had asked an even simpler question—such as, What do you mean by mass, or by acceleration, which is the scientific equivalent of saying, *Can you read?*—not more than one in ten of the highly educated would have felt that I was speaking the same language' (pp. 14–15).

[4] Studies of the variety in interpretations of the two most basic concepts in physics have been made by Max Jammer; see his *Concepts of Force* (Harvard University Press,

Obscurity at the Foundations of Scientific Knowledge

Another property of standardized facts, closely related to their differentiation, is the obscurity of their objects. Those who have had to grapple with different forms of the same standardized fact have learned that behind the facade of simple definitions and plausible experiences there is a bewildering variety of interpretations and meanings.[5] Sorting these out, either conceptually or historically, leads into a labyrinth of arguments, in which the unique, clear and distinct definitions of the objects are impossible to locate. The best that can be achieved in practice is another, more plausible or more coherent standardized version, better suited for the particular pedagogical task at hand. These failed attempts do not receive systematic attention, because hitherto there has been no philosophical thesis for which they could function as evidence. The ideal of science as demonstrative knowledge excludes the possibility of the obscurity of the very objects of the demonstration; and in the dominant traditions of the philosophy of science, such an obscurity

---

1957) and *Concepts of Mass* (Harvard University Press, 1961). The first chapter of the earlier book, and the Introduction of the later one, provide arguments and examples in agreement with the thesis advanced in this section.

[5] The obscurity at the foundations of Newtonian mechanics was recognized as a case for concern by several German scientists in the later nineteenth century. The most famous product of this concern is Ernst Mach, *The Science of Mechanics: a Critical and Historical Account of its Development* (1st German edition, Leipzig, 1883; 1st English translation, Open Court, La Salle and London, 1893; many reprints since then). Of equal significance was H. Hertz, *The Principles of Mechanics Presented in a New Form* (original German edition, 1894; English translations from 1899; reprint, Dover Publications, New York, 1956). In the first part of the Introduction (pp. 1–14) Hertz discusses the problem of obscurity, as it is glossed over in the teaching traditions, in an illuminating way.

An elementary introduction to the obscurities at the foundations of mechanics can be obtained by consideration of the five symbols: $F = m \times a$. 'Acceleration' is a concept best left to the mathematicians; the 'times' brings in the problem of the multiplication of the measures of different physical quantities; 'mass' and 'force' are capable of different, interrelated definitions; and 'equality' is capable of several very distinct meanings. An illuminating summary of these is found in N. R. Hanson, *Patterns of Discovery* (Cambridge University Press, 1958), pp. 99–100.

The obscurity in the concept of 'image', fundamental for physical optics, is exposed and discussed by V. Ronchi, *Optics, the Science of Vision*. It is not merely a question of a 'point-image' of a point-object being an idealization, since every spherical reflecting or refracting surface produces a 'caustic' curve with a cusp, and that every finite aperture produces a spread by diffraction effects. More than this, is that the fact that what is 'perceived', even by a photo-sensitive substance as well as the eye, will be an interpretation, sometimes extreme, of the pattern of radiant energy incident upon it. To speak of 'seeing the image' created by a lens system is to be immersed, unwittingly, in the obscurities responsible for the triumphs of physical optics. See pp. 265–71.

would destroy the claim of science to be knowledge of any worthwhile sort.[6]

Indeed, the claim that the foundations of science, in its conceptual objects, are swathed in obscurity, immediately raises troubling paradoxes. If, for example, the objects of mathematics are essentially obscure and incapable of definition, then what guarantee do we have that the sciences using mathematical arguments give valid conclusions? In practical terms, if the obscurities in the differential and integral calculus are ineradicable, how do we know that the bridges will not suddenly fall down because of hidden fallacies in the mathematical arguments whose conclusions have determined their design? We can remove some of the force of this paradox by observing its similarity to those arising from the inconclusiveness of scientific arguments, resulting from the absence of formally valid patterns of argument. But this aspect of scientific knowledge has somewhat different effects on its development than the necessity for socially improved criteria of adequacy on the solution of problems, and is worth considering on its own.

The present thesis can be argued by extension from one well known to philosophers: that the basic categories of our experience are incapable of precise definition and unique analysis. Concepts such as 'cause,' 'change', and the like, have in them an inexhaustible supply of subtleties and ambiguities as material for philosophical inquiry.[7] Even basic concepts which are manipulated in scientific

[6] The faith in the essential clarity of mathematics was most beautifully expressed by Fourier. Speaking of the 'analytical equations' of Descartes, he said: 'There cannot be a language more universal and more simple, more free from errors and from obscurities, that is to say more worthy to express the invariable relations of natural things.' On mathematics in general, 'Its chief attribute is clearness; it has no marks to express confused notions.' *The Analytical Theory of Heat*, 1822, tr. A. Freeman, 1878; reprinted 1955 (Dover Publications) p. 7. A contrary impression comes from Karl Menger: '"Variable" undoubtedly is among the most frequent nouns in the mathematico-scientific jargon and hence one of the most successful words ever created. . . . Actually, few scientists seem to have given any thought to the problem of what variables are, and still fewer of those who use the term formulate clear and satisfactory answers when the question arises.' See his 'Variables, Constants, Fluents', in *Current Issues in the Philosophy of Science*, eds. H. Feigl and G. Maxwell (Holt, Rinehart, & Winston, New York, 1961), p. 304. This latter example indicates that mathematics contains obscure objects other than those which have received attention in the dominant tradition in 'foundations of mathematics'.

[7] The paradoxes of change, and also of aggregation, propounded by Zeno of Elea, still exercise philosophers. See H. D. P. Lee, *Zeno of Elea* (Cambridge University Press, 1936), for texts and analysis; and for a review of some recent controversies, A. Grünbaum, *Modern Science and Zeno's Paradoxes* (Allen & Unwin, London, 1968).

argument, such as 'number', defy all attempts at conclusive explanation. For what we have for analysis are intellectual constructs derived from very deep aspects of human experience of coping with the external world. Our practical command of them has developed through millenia of experience; and no single formalization can capture that body of completely tacit inherited knowledge.

The ordinary objects of scientific discourse are more specialized and artificial, and in that measure are more susceptible of intellectual analysis. But as they appear in an argument, they are mixed with these more fundamental, unanalysable concepts, and it is not possible to give a clear specification of their meaning relative to neatly defined alternative meanings of the basic concepts of experience.[8] More important, however, the investigation of the scientific problem in which they are used or created does not have the goal of an explication of their meaning. Although the conclusion of the argument concerns properties of these intellectually constructed objects, the properties sought for are those which relate to experience of the external world; for the ultimate goal of scientific inquiry is to advance knowledge of that external world. A problem solved by conclusions about the objects as intellectual constructs is a philosophical one. The two sorts of arguments will have different methods, and admit different sorts of evidence. Although there will be borderline cases, the distinction between the two sorts of problem was well expressed, for mathematics, by C. S. Peirce: Mathematics is the science which draws necessary conclusions, while logic is the science of drawing necessary conclusions.[9]

[8] For example, 'electric current' has 'operational' definitions through a variety of effects (the two most important being electromagnetic and electrochemical), each one of them measured by devices capable of precise specification in terms of elaborated theories. Although a penetration to the foundations of any of these will reveal obscurities, these foundations are well hidden beneath the superstructure of accomplished scientific knowledge required for the framing of the definition of 'current'. It is only in the determination of the values of the 'fundamental constants' that experimental technique must reckon with conceptual difficulties.

In his classic monograph, *An Account of the Principles of Measurement and Calculation*, N. R. Campbell observes 'the most accomplished physicists are apt to flounder when plunged into a discussion arising from general principles of measurement; international committees, charged with the definition of standards, do not seem (to me at least) to display that easy mastery of their subjects which is to be expected from the contribution of their members to original learning' (p. vi). For an example of the complications from which the international committee on electrical units did not quite extricate itself, see pp. 131–3.

[9] See C. S. Peirce, 'The Essence of Mathematics', in *Essays in the Philosophy of Science*, ed. V. Tomas (Liberal Arts Press, New York, 1957), p. 266.

Corresponding to their function in the solved scientific problem, the objects of inquiry will be subject to particular criteria of adequacy in their creation and development. Provided that those properties which are used in the argument do not lead to incoherence or inconsistency in it, they are adequate to their function. Like all other components of the solved problem, they are only as refined as they need to be. Later investigation, either scientific or philosophical, may reveal hidden ambiguities or incoherences in the terms used in the argument; but whether these constitute pitfalls depends on whether the original argument is then seen to be vitiated by these defects in its objects. Such pitfalls are not inevitable; indeed, they are the exception. For the conceptual objects with which the scientist manipulates have strong analogies in their use with tools. Those properties of a tool which are used in its application to scientific inquiry will generally be crude and over-simple, compared to those which have been elaborated by experts in the practice or in the theory of the tool. Yet the tools will be adequate to their function, for their use is on materials other than those peculiar to their own constitution, and in general they will be used on the less specialized aspects of those materials. Similarly, the objects of a scientific argument, whose concealed obscurities contain a wealth of material for philosophical analysis, are used as the components of an argument intended to relate to something other than the objects themselves, namely the external world. Hence a more rough and ready control of the properties of those objects will be adequate for the argument.

It is for this reason that the obscurity in the objects of scientific discourse is generally irrelevant to scientific work of any sort. Contact with this obscurity is made in particular ways, depending on the work being done, and (in matured fields) a practical resolution can be achieved in every case. Thus, the research scientist will usually be aware of obscurities, at the technical level, in new materials in a rapidly developing field. But as the field stabilizes, there is achieved a sufficiently clear and univocal understanding of these objects for them to be manipulated in arguments with common agreement on their meaning and an absence of pitfalls. The concealed obscurities in the standard information and tools taken over for use in problems are beyond the practical concern, and hence beyond the awareness, of the research scientist. The teacher must occasionally grapple with obscurities in this standard material;

but his task is to render it plausible and easy to manipulate by routine techniques; and so he will naturally tend to minimize or gloss over unsolved problems at every level from the technical to the conceptual. It is only when a drastic novelty of some sort is intruded into the work, and existing routine craft skills disrupted, that scientists will become painfully aware of the obscurities at the foundations of their knowledge.[10]

Only a person with a keen philosophical awareness will attempt to explore and resolve the obscurity in the conceptual objects he handles. This can be of great benefit to teaching, and to research as well, as long as this sort of philosophical problem is treated as a subsidiary problem in the general task, and governed by criteria of adequacy which are not too severe. Otherwise it can have a harmful effect on the main work; for the different sorts of problem, with their very different functions, have quite different criteria of adequacy appropriate to their solutions. Thus, an exposition of material in physics or mathematics which derives it rigorously from very deep foundations can provide intellectual satisfaction to the teacher, but leaves his pupils in a state of complete bewilderment. Also, as a field of philosophical inquiry, the examination of particular scientific concepts present peculiar difficulties, analogous to those of methodology. For while the arguments must meet the standards of rigour for genuine philosophical debate, there is an embarrassment of detailed data, some of which must be used for evidence, arising from the practice of the scientists who actually use the concepts. This sort of inquiry has social difficulties as well, for its position on the borderland between science and philosophy calls

[10] A striking example of this at the present time is the problem of converting measurements in Imperial units to the *Système International*. For this to be done properly, there must be a recognition that a string of digits is only an estimate of a continuous magnitude (or of an aggregate of individuals which have not been counted individually); and that the conventions for 'significant digits' are a *code* whose meaning depends on the function of the digits. It is not surprising that mathematical social scientists are unaware of this subtlety; but it is interesting that engineers, who generally have a good craft knowledge of tolerances, find themselves in considerable confusion when this craft knowledge is required to be adapted to the new problems of metrication. On this situation, see K. J. A. Brookes, 'Can You Metricate?', *Engineering* (6 February 1970), pp. 136–8. (I am indebted to Mr. B. Hunter, of the Department of Civil Engineering, Leeds University, for this reference.) Even the author of this very perceptive paper does not recognize the deepest obscurity of all, that of the symbol o. For it functions both as a filler for an insignificant place (in a number greater than unity), and also as a symbol of a significant zero-value in a place. The ambiguity is revealed by an expression as 420,000; which is the last significant digit? This ambiguity in the o may be one reason for its very late invention.

for a double competence in its practitioners; and our educational system provides only very few who are so qualified. For all these reasons, inquiry into the obscurities in the objects of scientific discourse is a very hazardous affair. The philosophically-minded scientist or teacher might believe, at first, that a deeper understanding of a basic concept should be easy to achieve by extension of his scientific knowledge and perusal of the relevant literature. But, paradoxically, it is easier to become skilled in the techniques of discussing a question like 'What is mind?' than one such as 'What is mass?'

With this appreciation of the obscurity at the foundations of scientific knowledge, we can see how programmes for teaching abstract sciences by starting with their 'foundations' suffer from a double fallacy. The first concerns the learning process itself; it is tacitly assumed that the mastery of craft skills of manipulation is irrelevant to true understanding. And the second is that there exists a set of clear and simple foundations, whose understanding is sufficient for the mastery of all their consequences. Both these fallacies derive from a naïve faith in the nature of scientific knowledge, which can be called Cartesian since its most influential formulation derives from the early, optimistic period in Descartes's career. This is, that truth is a simple thing, derivable by logic from clear and distinct basic ideas. For Descartes, learning the truth is a simple matter of proceeding by instantaneous steps through its structure. The sort of knowledge which requires practice is that of a skill, as playing an instrument, and has nothing to do with the possession of truth.[11] Descartes's own programme derived from his desperate commitment to locate a basis of certainty for knowledge and morals. The modern descendants derive partly from a laudable desire for intellectual honesty in teaching; and also from the wish to introduce the materials and methods of sophisticated research earlier into the curriculum, for the wider diffusion of its approach and for the indoctrination of potential recruits. In fact, any

[11] This distinction between skills and knowledge was fundamental for Descartes's philosophy; it is announced at the very beginning of his first philosophical work, the *Regulae ad Directionem Ingenii*; see *The Philosophical Works of Descartes*, ed. Haldane and Ross, vol. i (Cambridge University Press, 1911), pp. 1–2. Descartes's conclusion was that to discover truth, one should not first study particular sciences, but rather 'increase the natural light of reason'. When one is so prepared, the appropriate style for the study of particular problems is to consider them in as abstract and general a fashion as is possible.

particular set of 'foundations' chosen for teaching purposes will be an arbitrary selection, and will necessarily be purveyed in extremely vulgarized form. And students who are raised on a diet of abstract arguments and necessarily trivial problems on their materials, will emerge without either the skills or the general knowledge to be anything other than research technicians in that specialized field.

If the problems of dispelling the obscurities in the objects of scientific knowledge are truly insoluble, in principle as well as in practice, we must consider two questions: first, whether there is any point whatever in making an attempt at such a task; and second, why indeed don't the bridges fall down. The second problem is easier to solve, for the materials are already to hand from our discussion of tools and the analogous properties of the objects of scientific discourse. Of course, bridges do sometimes fall down, but in nearly every case the failure can be explained by some more straightforward, or vulgar, cause. The design can be shown to have been faulty, in failing to reckon with some feature of the technical problem: or there might even have been too much sand in the concrete. The study of failure of a structure never leads to the questioning of the basic laws of mechanics that were presupposed and used in its design, and still less the conceptual objects in which the scientific facts are cast. The reason for this is that the more 'basic' materials are not used as tools in such a technical problem until they have been made accessible through standardization, and by this process their associated pitfalls identified and removed. The physics which is used by the engineer, and similarly the mathematics used by the experimental physicist, is, by the time it reaches him, a robust tool. And as inexpert tool-users, they will restrict themselves to just those properties of the tool which have been established as safe and reliable for that sort of function.

By analogy, we can say that the scientist's handling of his conceptual objects is restricted to their more gross properties, which can be manipulated safely. It is only in the early stages of work with deeply original concepts that their essential obscurities present pitfalls to the scientific investigation. These usually do not survive into the public record, for the problems are not adequately solved until ways around these pitfalls have been charted. Examples of this phenomenon are therefore quite rare; one of the more accessible is Galileo's discussion of the paradoxes of accelerated motion, which appeared in his *Dialogue* after some three decades of unsuccessful grappling with the concept. Fortunately, he achieved a practical control over the concept of instantaneous velocity in time for the publication of his *Discourses*, dealing with the sciences of mechanics; but even then his followers and admirers found the concepts very difficult to grasp; Gassendi, in particular, defended Galileo's mechanics for some years before he realized that he had quite misunderstood it.[12] The pitfalls in the handling of the objects of the differential calculus can easily be found in the early notes of Leibniz, and some remained to be resolved by his followers.[13] Even in the works of Euler, some fifty years after the calculus was put into a more or less standard form, there are few examples of the effects of the underlying obscurity;[14] but in his hands the more elementary parts of the calculus at least, were made into the robust, general-purpose tool which it has remained ever since.

Indeed, we can say that the neglect of the obscurities in its objects is one of the necessary conditions for the maturing of a discipline. So long as scientists in a field find themselves forced to engage in philosophical inquiries to locate the causes of the pitfalls they encounter, and are unable to manipulate their conceptual objects

[12] In the First Day of the *Dialogue* Galileo tried to establish the paradoxical property of naturally accelerated motion, that 'a body passes through all the infinite degrees of slowness before reaching any assigned degree of speed.' For this he used the thought-experiment of motion along a plane of arbitrarily slight inclination. But this brought him to a more serious paradox: while we know, from the principle that bodies descending through equal heights attain equal speeds, that any two paths are traversed equally quickly, we have obvious phenomenon that the incline is traversed more slowly than a perpendicular path of fall. Galileo's struggles with this paradox were not immediately successful. See *Opere* Vol. 7, pp. 42–53; Drake edition of *Dialogue*, pp. 21–28. For Gassendi, see J. T. Clark, 'Pierre Gassendi and the Physics of Galileo', *Isis*, 54 (1963), 351–70. In calculating the distances traversed through successive 'units' of time, Gassendi encountered a pitfall by working with velocities constant through each unit, so that the simple superposition of those velocities gave a sequence of distances represented by the series, 1, 3, 6, 10 rather than the square series 1, 4, 9, 16. To 'save' Galileo's law he needed to invoke an extra force; and only in 1645 did he sort out the difficulty.

[13] *The Early Mathematical Manuscripts of Leibniz*, translated by J. M. Child (Open Court, Chicago and London, 1920), although no longer accepted in all details of its interpretation, gives a vivid impression of Leibniz's explorations, and his difficulties with the operator d; see pp. 90–103.

[14] See the 'Remarks on Mr. Euler's Treatise Entitled Mechanics' in Benjamin Robins, *Mathematical Tracts*, ii (1751), 197–221. Robins was one of the most capable of British mathematicians of his period; his contributions to the controversy over Berkeley's 'Analyst' were intelligent, and his treatise on gunnery was translated into German by Euler himself. For an illuminating discussion of the obscurities in mechanics and the calculus during the first half of the eighteenth century, see T. L. Hankins, *Jean D'Alembert, Science and the Enlightenment* (Oxford, 1970), chs. 7–10.

reliably in the solution of scientific problems, the field cannot make progress towards the creation of facts and the achievement of scientific knowledge. And since these obscurities present problems which are insoluble, and also unsettling to all but a very few students, it is natural and indeed necessary to protect the great majority of recruits from being troubled by them. In this way, they will do their work of investigating limited problems, and contributing to the advance of the positive knowledge of the field, with full efficiency. However, a policy of a complete sealing-off of the philosophical end of the range of problems in a field can in the long run lead to the same diseases as the insulation of the field from influences from other sciences and technology. For a field which goes on without the injection of new experiences and ideas will eventually become stagnant; and this rejuvenation will not always come from new discoveries within the field. Stimulus will sometimes come from other parts of science or technology, but reflection on the objects of inquiry has, in history, led to some of the most radical innovations, and the most dramatic progress, of science itself.

The practitioners in a field will, when they encounter undeniable obscurities in their objects of inquiry, generally try to minimize their significance; and it is natural and necessary for them to do so. Hence those who invest their time and talent in exposing them will generally be outsiders, who may even have a prior bias against the field or its practitioners. The debate that ensues cannot possibly be a 'scientific' one: the problems are incapable of resolution by the arguments accepted for the field, and social and personal elements enter in strongly. Yet such wrangles, however inconclusive in the short run, can be of great significance for the eventual development of the field. The classic in this polemical literature is the pair of essays by the philosopher George Berkeley, Bishop of Cloyne: 'The Analyst', and the 'Defence of Freethinking in Mathematics'.[15] The first of these was ostensibly addressed to an 'infidel mathematician'; and the argument had as a conclusion that anyone who could accept the mysteries of the calculus should find nothing objectionable in those of the Christian faith. The immediate rejoinder, 'Geometry

[15] The standard modern edition of these tracts is in *The Works of George Berkeley, Bishop of Cloyne*, ed. A. A. Luce and T. E. Jessop, iv (Thomas Nelson & Sons Ltd., Edinburgh and London, 1951), 64–138. The flurry of pamphlets created by Berkeley's attack is described in F. Cajori, *A History of the Conceptions of Limits and Fluxions in Great Britain from Newton to Woodhouse* (Open Court, Chicago and London, 1919), chs. 3 and 4.

no Friend to Infidelity', is a classic of its own type of incoherent apologetic literature; it was written by James Jurin, one of the most successful of the mediocrities who dominated English science in the name of the recently deceased Sir Isaac Newton. With such a well-defined target at point-blank range, Berkeley could produce a second attack more carefully designed, and hence even more effectively argued, than the first. His basic point was that the calculus inevitably involves manipulating with quantities which cannot be distinguished from zero; and when one takes a quotient of two such quantities, one is being either self-contradictory or hopelessly obscure. Berkeley was far from being the first to realize this; but he was the first man of real philosophical penetration who used this point to question the legitimacy of the whole subject. Although his own suggested solution to the mystery was both obscure and fallacious, it would be incorrect to think of him as merely a church-man who conducted some amateur investigations in the foundations of mathematics. For the most likely motivation for the attack was in Berkeley's antagonism, as a metaphysician and as a theologian, to the Newtonian philosophy of nature which was by then a dominant orthodoxy in English thought.

As a hostile outsider, Berkeley could bring complete clarity of thought, and of expression, to his critical analysis of the orthodoxy of the calculus. He could distinguish neatly between mastery of technique, and understanding of principles.

What I insist on is, that the idea of a fluxion simply considered is not at all improved or amended by any progress, though ever so great, in the analysis: neither are the demonstrations of the general rules of that method at all cleared up by applying them. The reason of which is, because in operating or calculating men do not return to contemplate the original principles of the method, which they constantly presuppose, but are employed in working, by notes and symbols denoting the fluxions suposed to have been at first explained, and according to rules supposed to have been at first demonstrated. This I say to encourage those who are not far gone in these studies, to use intrepidly their own judgment, without a blind or a mean deference to the best of mathematicians, who are no more qualified than they are to judge of the simple apprehension, or the evidence of what is delivered in the first elements of the method; men by further and frequent use of exercise becoming only more accustomed to the symbols and rules, which doth not make either the foregoing notions more clear, or the foregoing proofs more correct.[16]

[16] 'Defence of Freethinking in Mathematics', section 20, p. 117.

Slightly further on he gave a brief résumé of the process of the implanting of an orthodoxy in science.

Men learn the elements of science from others; and every learner hath a deference more or less to authority, especially the young learners, few of that kind caring to dwell long upon principles, but inclining rather to take them upon trust: And things early admitted by repetition become familiar: And this familiarity at length passeth for evidence.[17]

The 'Analyst' controversy is also a classic in its later effects. First, there was a flurry of polemics between defenders of Sir Isaac, who accused each other of failing to understand the simple and obvious principles underlying the calculus. A British mathematician of great distinction, Colin MacLaurin, tried to answer Berkeley in mathematical terms, showing that the problems solved by calculus are equivalent to those solved by geometrical constructions, themselves capable of perfectly rigorous proof. His book was deep and also long, but failed in its goal of justifying the manipulations themselves. Thenceforth for more than half a century the problem lived on, but at the periphery of mathematical practice; and it had to share place with other 'mysteries' involved in the calculus, as negative and complex numbers. It was not until the early nineteenth century that the challenge raised by Berkeley led to deep new advances in mathematics itself; this was in the work of Cauchy, attempting to provide, once and for all, a rigorous foundation for the calculus. With this work, there began a new chapter in the history of mathematics, that of rigorous proof in analysis. Subsequent developments showed two features which might be considered paradoxical. First, the solution of problems at each stage opened the way to the discovery of obscurities at ever deeper levels; by the early twentieth century, rock bottom had been touched, temporarily at least, with the exposure of paradoxes in the notion of 'set', and in the very structure of logic. The other feature of this history, in which both mathematics and philosophy were revolutionized, is that the mysteries in the formal manipulations of the calculus were not entirely resolved. It was of course universally believed that Cauchy's work had in principle made the calculus clear and rigorous; and with that assurance, no one bothered to observe that Cauchy's attempted justification of the symbolic manipulations of the calculus is a pastiche of borrowed ideas, none

17 ibid., section 21, pp. 117–18.

of them rendered with either acknowledgment or clarity.[18] For the symbolism of the calculus, as invented by Leibniz, encompasses such deep and basic ideas of continuity and change that it would require a philosopher of the same genius to unravel some of its obscurities in their own terms.

This case study from mathematics shows in a particularly pure form the dialectic of the problems of grappling with obscurities at the foundations of a discipline; and it is only natural that the pure case should be in mathematics. The immediate response to their exposure is purely defensive; then there is some worry about the problem, at the periphery of the field; eventually a deep attack on them yields a great development of the discipline, in its objects and methods; and through it all the obscurities remain, but now safely relegated to the obscurity of the schoolteachers' world.

Influences of Methods: the School and its Cycle

The account given so far of the further evolution of the materials which may become scientific knowledge is still deficient in one respect. For the processes of transformation of content, extension of function, standardization and differentiation, do not necessarily go on with steadily increasing cumulative success. All these developments take place in the course of the investigation of particular problems, governed by the controlling judgements of those who use and adapt the existing materials in their work. The influences from these controlling judgements, and from the rest of the body of methods, are not capable of a simple listing. But it is worthwhile to discuss one particular systematic influence on the direction of research, for its prime importance in the patterns of development of scientific knowledge.

We recall from our discussion of the conclusions of problems, that each such solution is limited and fallible. And although the components of the research report are tested and transformed by the processes I have described, these processes are not like simple quality control or sifting, whereby the 'good' is preserved while the 'bad' is rejected. For this work is done in the course of the investigation of other problems; some of them are scientific, some technical, and some practical, but each of them as limited in its own way as

[18] The critical evaluation of Cauchy's achievements and limitations has only recently begun; see I. Grattan-Guiness, *The Development of the Foundations of Mathematical Analysis from Euler to Riemann* (M.I.T. Press, 1970).

those from which their materials were derived. Hence the development of scientific knowledge is never independent of the judgements which control the investigations of particular problems; and, since we are concerned with the direction of research, the judgement of value is crucial here. Since such judgements are by their nature very imperfect, and based on an estimation of the future, we can see why the advance of science necessarily takes place largely by way of detours.

To analyse the nature of these detours, as a basis for seeing how they can eventually reveal a correct avenue of advance, I shall use the concept of a 'school'. This is most easily seen in terms of the work done on a family of problems, usually descended from some şeminal ancestor-problem. The problems are related not only by descent, but also by their objects of investigation, a battery of tools, and a body of methods. In these, criteria of value demarcate schools most sharply; and a well-developed school will also have a style of work common to its members. In addition to these technical aspects, the school will also have its social side, in the network of friendships and loyalties built up in the course of co-operative work; and an institutional aspect, perhaps in the place of employment of its nucleus, or in a stable pattern of relationships with particular investing agencies, as well as some elements of the formal apparatus of a field, such as journals and societies.

I prefer to use the old-fashioned term 'school', even though it has connotations of discipleship, rather than 'invisible college' or, say, 'super-problem'. For the former alternative term focuses attention on a particular aspect of the social character of some schools, and the latter implies a degree of definition to its work which is not always present. Not all scientific work need be done within the framework of recognizable schools; indeed, the most original advances tend to come from outside the existing schools. But even in such cases, the work of extending and consolidating such advances usually involves a technical and social situation which can then be considered as a new school. The extent to which organized schools dominate the work in a field of science will naturally depend on its technical aspects and its social history; where effective attack on the problems requires large-scale investment and co-ordinated work, then those working outside schools, even if not few, will certainly be lonely.

The methods governing the accomplishment of the tasks of a

school bear the same relation to those of a single problem as strategy to tactics in military operations. They require a longer perspective into the future, and an assessment of more complex and imponderable factors; and hence extra qualities of leadership are required for the effective direction of a school. But in both cases, decisions must be made for the concentration of resources and work in a particular direction, to the exclusion of others. In the case of the strategy of a school, the carrying out of a basic decision will require more time, and will have greater consequences for the field in which it is operating. Yet there is the same inescapable element of risk in any such strategy; the assessments on which the decisions are based are conditioned by the controlling judgements of value, feasibility, and cost.

The strategy of a school will have an obvious influence on the problems which are chosen for investigation; and there is an influence, less obvious, but perhaps even stronger, on those materials that are kept alive in the public channel of communication. For, apart from those results that find a niche in a handbook collection of very standard information, the material that is chosen for inclusion in surveys of a field, and then in specialized and advanced textbooks, will reflect the judgements of value of those who make the selections. In this task, part of the value of a piece of information is its capability for fitting into a coherent account of an important topic. Stray results which did not derive from problems in the family worked by some school will be more difficult to fit in neatly, and so will tend to be neglected. Materials which are left out of such collections are still available, in principle, to those from other fields who might be interested in using them as information and tools. But they require the work of retrieval for their discovery, or even for the establishment of their existence; and the presence of a convenient collection, implicitly claiming comprehensiveness for its functions, will discourage further searching. A similar effect, in the burying of material that does not find a natural place in a coherent account, will occur even more strongly in the selections made for oral teaching. In this way, the advanced training of recruits is inevitably conservative; by these omissions, their picture of the field as a whole is restricted to those parts which have been worked over by leading schools.[19]

[19] This process of exclusion of the results of apparently marginal fields can be seen clearly in the historical literature on scientific disciplines, which reflects their

The traditions of scholarship within the history of science have hitherto inhibited the chronicling of the rise *and* decline of successful scientific schools; but materials are now becoming available for two illustrative case histories. The first is French physical science in the early nineteenth century, which provided brilliant results, an example of a style of work, and a model of excellence for contemporaries and successors all over Europe.[20] It is well known that French influence radiated outwards, first to Germany and then elsewhere; but general histories have tended to neglect what was happening in France itself after its period of greatness. It now appears that we should speak not of one cycle but of *two*; the first being 'Laplacian physics', and the second, 'anti-Laplacian physics'. Their creative periods can be roughly dated to the intervals 1800–12 and 1815–23, respectively. The strategy of the Laplacian school was to apply mathematical arguments to objects of inquiry derived from an eighteenth-century Newtonian tradition, for the explanation of old and new experimental results. Its characteristic feature was the assumption of 'imponderable (and intangible) fluids', consisting of particles related by short-range centrally-directed attractive or repulsive forces, as the agencies of heat, light, electricity and magnetism; and it was also hoped to explain the laws of chemical affinity on this basis. The leaders of the school, Laplace and Berthollet had prestige, wealth and influence; and they were able to recruit the most brilliant students of the École Polytechnique into their circle, centred on their adjacent residences in the suburban village of Arcueil. The school enjoyed some striking successes, notably

self-consciousness and teaching at any given time. Thus, E. Nordenskiöld, *The History of Biology* (Knopf, New York, 1928; original in Swedish, 1920–4), still the classic in its field, devotes 3 pages out of 616 to ecology (pp. 558–61).

[20] The fundamental study on science in Napoleonic France is M. P. Crosland, *The Society of Arcueil* (Heinemann, London, 1967). The distinction between the two schools, of the Napoleonic and Restoration periods respectively, with the comparison of their styles and fates, has derived from my reading of a draft of R. Fox, 'The Rejection of Laplacian Physics; Turning-Point in the History of the Physical Sciences in France', to appear in *Archive for the History of the Exact Sciences*. Two other important studies of the problem are J. W. Herivel, 'Aspects of French Theoretical Physics in the Nineteenth Century', *British Journal for the History of Science*, 3 (1966), 109–32, and J. Ben-David, 'The Rise and Decline of France as a Scientific Centre', *Minerva*, 8 (1970), 160–80. It is only in recent years that historians have perceived that there was a decline; general histories, either of science or of French culture, content themselves with describing the achievements and prestige of French science during its period of greatness. An enriched perspective on the Arcueil group itself is provided by O. Hannaway in his review of M. P. Crosland's book, in *Isis*, 60 (1969), 578–81.

the discovery by Malus of polarization of light on reflection; and its recruits made up most of the leadership of French science for decades to come. But all of these recruits except for the first two (Poisson and Biot) eventually deserted the school, either quietly or in open rebellion; and (not accidentally) its period of influence is nearly coterminous with that of Napoleon. Its weaknesses can be seen most clearly in the programme for chemistry. Although it stimulated some physical research (refraction in gases, as an approach to the study of the short-range forces of particles), and was the context of Gay-Lussac's discovery of the Law of Combining Volumes of gases, on its major problem of chemical theory its achievements were few and it was soon superseded. Worse yet, it was unable to comprehend the new results which were transforming chemistry: the atomic theory of Dalton, and electrochemistry. In physics, the limitations of the strategy took longer to be revealed; but the school could not successfully meet the early challenge of Fourier in the theory of heat, and the later challenge of Fresnel in the theory of diffraction. On the social side, the style of the school was authoritarian in the extreme; competitors and mere outsiders were ignored or hindered by administrative means, and recruits were expected to show complete loyalty to the doctrines.

The anti-Laplacian revolt got underway very shortly after the Hundred Days; and a group including Fourier, Fresnel, Arago, Dulong, and Petit captured the positions of prestige and influence within a short time. The results for which they are remembered belong firmly to nineteenth-century physics, and it was under their leadership that Paris was the Mecca of the scientific world in the 1820s. But their cycle of creativity was short-lived; the early deaths of Fresnel and Petit deprived them of much-needed talent; and those who remained could not, or did not, carry on with creative work. The reasons for this collapse are still obscure; two possibilities are worth considering. One is that this grouping was *not* a school with a strategy; there was no grand problem giving unity and direction to their work. Hence when the task of consolidating the initial results became protracted (as it always does) they would too easily lose their courage and conviction. In this they form a striking contrast to the hardiest Laplacians, Laplace himself and Poisson; the latter continued working in the old way for a quarter of a century after the decline of the school, and although much of his work was sterile, some of it was of great importance. The social situation of the

Restoration grouping may also have contributed to its decline, in several ways. Although they did not possess a new orthodoxy of doctrine, they did inherit the social institutions of science which (apart from an increase in paid employment through teaching and examining) were not significantly changed from those of the *Ancien Régime*. These produced the traditional tendencies to competition for place, the neglect and exclusion of outsiders who lacked a personal patron, and the shift from a career in science to a career based on scientific eminence. As a result of this style of social organization, and perhaps as a result of political considerations as well, the two most profound thinkers of the decade, Sadi Carnot and Evariste Galois, were excluded from the official community in one way or another, and their work ignored. Finally, the absence of a third cycle of creativity may be explained by a youthful revulsion against physical science, still identified with the authoritarian Laplacianism of the Napoleonic régime; so that in the 1830s there was no wave of young men of promise available to inherit or seize the positions of leadership.

Another example of a 'school' which seems to be well advanced in its cycle is the abstract style in mathematics, frequently called 'modern mathematics'. Although there is no single figure, and no single grand problem, which unifies this tendency, there is a shared style which is possessed by the members of the 'invisible colleges' occupying the leading positions in the discipline. The origins of this movement can be traced back to the earlier nineteenth century, in the study of mathematical objects which became increasingly different from straightforward generalizations of the 'quantity and magnitude' of the ancient tradition. Considerable philosophical excitement attended the invention of 'non-Euclidean geometries', for they touched directly on the Kantian tradition. A more gradual, but more significant growth was in the conception of an 'abstract group'; and it is somewhat ironical that the concept was brought to maturity and applied by Felix Klein as the key to the unification of geometry.[21] The event which in restrospect marks the establishment of the abstract style was the solution by David Hilbert in 1893 of one of the deepest problems in the theory of invariants of algebraic forms, using entirely abstract existence arguments rather than

[21] For a history of group theory, including Klein's contribution, see H. Wussing, *Die Genesis des abstrakten Gruppenbegriffes*, (V.E.B. Deutscher Verlag der Wissenschaften, Berlin, 1969).

manipulative constructive arguments. Although at the time this was seen as a result *within* invariant theory, Hilbert's pupils later considered it as the result which had killed the field.[22] In the 1930s leadership was assumed by a group of French mathematicians with the collective name 'Bourbaki'. They embarked on a programme of re-casting all worthwhile mathematics in an abstract, axiomatic form; and by the 1950s most of the prestigious pure mathematics was dominated by their style.

The criteria of value of this school are very strongly defined; and the tendency to derogate work which involves only nineteenth-century analysis, or applications, soon led to protests over its exclusiveness. As early as 1947, John von Neumann, then one of the greatest living mathematicians, issued a warning that such a tendency could lead to the subject becoming 'baroque'.[23] The social situation of science in the present period has also led to tendencies which some have considered distortions. One recent critic[24] has complained that the research scene is characterized by the frantic

[22] See C. S. Fisher, 'The Death of a Mathematical Theory: a Study in the Sociology of Knowledge', *Archive for History of Exact Sciences*, 3 (1966), 137–59.

[23] See John von Neumann, 'The Mathematician' (1947), in *Collected Works*, i (Pergamon, 1961), 1–9. As he spoke from experience of work in many fields of mathematics, from the most directly applied, through to the philosophical, his words are worth quoting in full: 'I think that it is a relatively good approximation to truth—which is much too complicated to allow anything but approximations—that mathematical ideas originate in empirics, although the genealogy is sometimes long and obscure. But, once they are so conceived, the subject begins to live a peculiar life of its own and is better compared to a creative one, governed by almost entirely aesthetical motivations, than to anything else and, in particular, to an empirical science. There is, however, a further point which, I believe, needs stressing. As a mathematical discipline travels far from its empirical source, or still more, if it is a second and third generation only indirectly inspired by ideas coming from 'reality", it is beset with very grave dangers. It becomes more and more purely aestheticizing, more and more purely *l'art pour l'art*. This need not be bad, if the field is surrounded by correlated subjects, which still have closer empirical connections, or if the discipline is under the influence of men with an exceptionally well-developed taste. But there is a grave danger that the subject will develop along the line of least resistance, that the stream, so far from its source, will separate into a multitude of insignificant branches, and that the discipline will become a disorganized mass of details and complexities. In other words, at a great distance from its empirical source, or after much "abstract" inbreeding, a mathematical subject is in danger of degeneration. At the inception the style is usually classical; when it shows signs of becoming baroque, then the danger signal is up. It would be easy to give examples, to trace specific evolutions into the baroque and the very high baroque, but this, again, would be too technical' (p. 9).

[24] See W. G. Spohn, Jr., 'Can Mathematics be Saved?' *Notices of the American Mathematical Society*, 16 (1969), 890–4. (I am indebted to H. J. M. Bos, of Utrecht for this reference.)

production of fragmented results, with no institutions for synthesizing important work or sharing it with those outside a coterie of specialists. Moreover, the concentration on abstract research, and the shaping of teaching for the training of new recruits, has had deleterious effects in several ways. The missionary zeal for the reform of American high school teaching of mathematics brought forth a carefully-worded warning signed by some seventy mathematicians, some of them eminent men in the modern style.[25] But university teaching is the province of each group of mathematicians; and here the damage may be serious. For there is a tendency for students trained by abstract mathematicians to despise the teaching of elementary and applied mathematics as tools for scientists and engineers; and either to accomplish this task in a bored and perfunctory manner, or to abandon it to the scientists themselves. The result is a deterioration of the mathematical competence of those who need it as an essential tool in their work. The relinquishing of a monopoly on the teaching of its subject within a university can have serious consequences for a discipline, even at the crude level of the provision of jobs; if mathematicians do not justify themselves by their usefulness, then they are at risk of being considered as practitioners of a particularly pure and esoteric art form. All these tendencies are of course most extreme in America, but even in England some signs of sterility can be observed. For the 'new mathematics' curricula for secondary schools were devised by schoolmasters, with little or no participation by research mathematicians. Rather than attempting to retail an axiomatic mathematics based on the idea of 'set', they bring the new ideas of 'structure' down towards the common-sense level. Although these too have been subjected to criticism,[26] they may well create a new sort of mathematical common sense, in which 'matrix' is as ordinary and ubiquitous a mathematical object as 'function'. Yet there is little sign that the university mathematics teachers recognize the importance of this development; and the teaching of the mathematical tools of structure is frequently left to self-taught practitioners in other fields.

[25] See 'On the Mathematics Curriculum of the High School', *American Mathematical Monthly*, 59 (1962), 189–93.

[26] For an extremely vigorous argument against the British 'new mathematics' of the 1960s, see J. M. Hammersley 'On the Enfeeblement of Intellectual Skills by Modern Mathematics and Similar Soft Intellectual Trash in Schools and Universities', *Bulletin of the Institute of Mathematics and its Application*, 4 (1968), 66–85. (I am indebted to Dr. I Grattan-Guinness for this reference.)

Abstract pure mathematics still attracts recruits of great talent and vigour, and so one cannot say that it has passed the peak of its cycle of creativity. Some of the social problems discussed above may be the result of the over-expansion of the community engaged on a type of work which must be done very well indeed if it is to be worth doing at all: but in the institutional setting of modern science and mathematics it is difficult in the extreme to apply a severely restrictive policy of recruitment through postgraduate study. The danger is that some really new insights deriving from the stimulus of applications, will not be incorporated into 'mathematics', but relegated to a special field of 'application' (as computers, statistics, or the mathematics of organizations) and so fail to achieve the rejuvenating effect on mathematics which will eventually be necessary.

The work of a school, like the task of investigating a single problem, involves choices and hence exclusions; and it is only in retrospect, if at all, that the correctness of these decisions can be assessed. But the difference in time-scale and social character between the investigation of an individual problem and the work of a school is the cause of significant differences in their cycle of development. As I showed earlier, a problem has a natural end, when a conclusion can be drawn. Even if this conclusion is not as extensive or deep as that which was hoped for at the inception of the work, it exists as a real thing. That task is accomplished, and the scientist can then decide whether his next task should be to improve the previous conclusion, or to turn to something else. This can also happen in the case of a school, when a grand ancestor-problem is finally solved to the satisfaction of those working on it. But it is more common for the work of a school to lack such a neat conclusion. For, even when it is organized around such a grand problem, its work will tend to be derivative on the insights of its founder. Although its tools may become more refined, it will still work on objects of investigation basically conceived by him, and use only those tools which are appropriate to them. Only when the solution of the grand problem can be achieved by more powerful descendant tools will the programme come to a natural conclusion. More commonly, there will be a steady accumulation of results, but (after a time) no significant enrichment in the knowledge of that field. The leadership has a technical, social, and emotional commitment to the established methods; and so, unless it is very enlightened, recruits who bring in fundamentally new insights will not be welcomed or absorbed.

In such situations, the result is that over the generations the work of the school becomes less original, less significant, and eventually obsolete. Its detour has become a cul-de-sac. If in spite of this it retains political power, it can then become a serious distorting influence on the progress of the field.

Two lessons can be drawn from this description of the work of schools. The first is that the continuous increase in the number of solved problems is not at all the same as the deepening and enrichment of scientific knowledge. In his contrast of the progressive practical crafts with sterile metaphysical philosophy, Francis Bacon described two extreme patterns of development: of degeneration from a single master in philosophy, and of steady improvement in the arts.[27] He believed that a true philosophy could be as progressive as the arts; but since science lies midway between the two, it is only natural that its pattern should be a mixture. For, although there is an overall cumulative progress, within each school there is a coasting on the insights of the founder, with the possibility of degeneration if the school does not dissolve when its time is up.

We also see how it is inevitable for each school to proceed along a path which can eventually be recognized, by outsiders at least, as a detour. For the achievement of genuine progress in the long run, two sorts of practical problems must be solved. First, that the results done in each special direction can make their contribution to a common stock; and second, that a field is preserved from its members marching up a variety of separate culs-de-sac. These problems are not automatically solved; they require certain conditions of a technical and social character. The transfer of results, which we have already discussed as extension of tools and standardization of facts, requires a degree of maturity in the field, as reflected mainly in the criterion of invariance by which results are accepted as facts. If each school requires only invariance under changes of problem within its narrow limits, then they will soon be talking separate, mutually incomprehensible languages, even if they share a stock of common words. The demise of a school then entails the complete

[27] See the Preface to the *Great Instauration* (Works, vol. iv), p. 14. There the mechanical arts are described as 'continually growing and becoming more perfect. As originally invented, they are commonly rude, clumsy, and shapeless; afterwards they acquire new powers and more commodious arrangements and constructions. . . .' On the other hand, 'Philosophy and the intellectual sciences . . . stand like statues, worshipped and celebrated, but not moved or advanced. Nay, they sometimes flourish most in the hands of the first author, and afterwards degenerate.'

destruction of its store of facts as living, usable materials. Only if the work of a school has enough links with that of others for its facts to be capable of some translation into statements about other objects of inquiry will its productions survive in the common stock. Even then, there will be enormous losses, since the translations cannot be perfect, and the judgements governing the selection of such material will be different from those governing its production.

The second practical problem, of achieving the replacement of leading schools when they have outlived their usefulness, depends on the social maturity of a field. It is impossible for such a process to be a continuous and painless one; and that a field is permanently free of revolutions, conceptual or political, may be a sign of stagnation. We have mentioned this problem already in connection with the judgement of value; and we shall return to it more systematically later on.

The Historical Character of Scientific Knowledge

We can now review the processes whereby a part of scientific knowledge comes to be. Its origins lie in the investigation of a problem, where an assertion about certain objects of inquiry (and those artificial objects themselves) develops and is confirmed through interaction with the external world. The problem is solved when the assertion can be obtained as the conclusion to an argument, in which the reports of particular experiences serve as the foundations for the evidence. The work as a whole is accomplished by the use of tools, applied with tacit craft skills, and is governed by methods, including the controlling judgements of adequacy and value. If the result of the problem is of great novelty or depth (and it must generally be so if it is to survive all its subsequent tests and become a part of scientific knowledge), then the problem itself is almost certainly one involving the creation of new conceptual objects, and probably new methods as well. The problem-situation which gave rise to the work was quite likely to have been one involving more than the mere extension of existing results within its field, but was stimulated by considerations of a philosophical, or perhaps technical, character. The result of the investigation is far from being a perfect and true statement about the world; its argument is not formally valid, but only adequate to its function; and its objects of inquiry are clarified only enough to function unambiguously in the argument.

On being submitted to a community, the result is tested, first formally and then informally; and if it shows significance, stability, and invariance under changes in its objects, it becomes accepted as a fact. Again, for a deeply novel result, this process is likely to involve controversy; if it challenges the programme of an existing school, it will give rise to a debate which is ultimately methodological, and hence incapable of resolution by the methods of the discipline itself. If the result survives this criticism, then it may become the material around which a new school comes to be; and its problem becomes the ancestor-problem for a lattice of descendants. Through these, the fact becomes transformed into versions ever more powerful and sophisticated; and these versions eventually become obsolete and die when the descendant-problems themselves lose significance and are forgotten. But while this is going on, the fact also produces extensions and standardized versions, performing a variety of functions in its own field and in others. Eventually a situation is reached where the original problem and its descendants are dead, but the fact lives on through a great variety of standardized versions, thrown off at different stages of its evolution, and themselves undergoing constant change in response to that of their uses. This family of versions of the facts, and of their associated objects of discourse, will show diversity in every respect: in their form, in the arguments whereby they can be related to experience, and in their logical relation to other facts. They will, however, have certain basic features in common: the assertions and their objects will be simpler than those of their original, and frequently vulgarized out of recognition. And the obscurities concealed in the original objects of inquiry will almost certainly remain, unless there has been a philosophical critique in the meantime.

Such, then, is the character of a unit of scientific knowledge. It may appear to be a very untidy and also imperfect sort of thing to enjoy such a status. But the processes which have operated to create this family have at the same time eliminated a host of competitors. The particular fact survives only because it is capable of breeding hardy descendants, who find niches in so many areas of science that the continuous displacement of problems and whole fields, while modifying its members, do not destroy the integrity of the family as a whole. And the obscurity at its foundations is only a testimony to its depth; that which is perfectly clear is, in science, likely to be perfectly banal.

To reduce the body of scientific knowledge to such elements may seem to dissolve its unity, and indeed to destroy its reality, leaving only a heap of pragmatically justified tools. But the destruction is only of certain ideal of knowledge: one which demands that for it to be real, it must be clear, distinct, and eternal. Such knowledge of the external world may exist in the mind of God; but it is clearly beyond the capacity of human beings, who derive their knowledge of the external world ultimately from particular interactions with it, observed by means of their senses. That which we can recognize as scientific knowledge has achieved its state only by surviving a long series of ruthless selections, and of drastic changes in the meaning of its objects. It thus contains within itself a segment of human history; its roots in experience of nature are tough and necessarily manifold; and it still grows through involvement with new scientific problems, new results, and new functions.

All this is necessary for such knowledge to be proved as genuine; the diversity in form and function of any piece of that knowledge and the obscurity at its foundations, are not a denial of its reality, but a necessary condition of its existence. The unity behind the diverse appearances lies not in a unique but hidden structure and meaning, but rather in their common ancestry and their continued mutual interactions. We have an aggregate, a family, of particular versions of a penetration into the external world. No single one of these can achieve a perfect, timeless contact with that reality; and each is conditioned by the history of itself and of the family as a whole. The family of particular versions stays alive, and establishes its genuineness, by being successfully used for a variety of ever-changing functions, each particular use being governed by the judgements of men. It is through this variety in the particular forms, and their individual changes, that the inner bond between them, and the genuineness of their aggregate as real knowledge of the external world, is created and proved.

There can be practical objections to this definition of scientific knowledge, as those facts which have survived the disappearance of the problems which gave rise to them, remaining alive through a variety of descendants. For it applies only to a very small proportion of the achievements of science, and in any particular case it is recognizable only in retrospect. Surely we use the term much more freely in ordinary speech. But the objects of 'knowledge' of ordinary usage are of a different character from the objects of scientific

knowledge. The things we 'know' by a nearly immediate, unreflective sensory experience, are particular, temporary and shallow.
Knowing that this book fell to the floor is not the same as knowing
that its falling was due to the mutual gravitation of itself and the
earth. The former is indubitable, but scientifically insignificant
(unless a report of it furnishes data for a problem); the latter is
profound, but obscure in its foundations, and ever subject to evolution. For many centuries it was known that the falling of the book
was due to its earthy element seeking its natural place; and even
though we now know that the cause is universal gravitation we are
as ignorant as was Newton of the nature of that force.

How widely we should apply the term 'knowledge' is of course a
matter of convention. If we cannot bear the paradox of accepting
that genuine knowledge may be fallible, then we must ban the term
altogether from productions of the human intellect. But if we extend
it to include all that which at any moment is accepted as fact, we
are left without any means of differentiating between the ephemeral
and the permanent in the achievements of science. It seems best to
restrict the term to those results which are so solid that they live on
(in the fashion I have described) as long as does the framework of
reality in whose terms they are cast.

Paradoxes of Scientific Knowledge

It is now possible for us to consider the various paradoxes that
have been accumulating through our discussion. I cannot solve them
by a philosophical analysis demonstrating their unreality; but in the
terms of the generalized description of the development of genuine
scientific knowledge we can see how they can be resolved in practice.
Generally, the paradoxes, which revolve around the contrast between
the fallibility of the endeavour and the certainty of the result, lose
their force when we appreciate that such certainty is not achieved
by a single effort. Even the productions of the greatest genius do not
survive without immediate modification, and eventual transformation, of their objects. No single result in science can be proved to
be true; and indeed, most are not merely untrue (as assertions about
the external world) but are also of a very temporary usefulness and
life. Similarly, the rapid succession of problems, and the continuous
shift in the objects of inquiry, winnows out all those results which
were bound too closely to the circumstances of their production.
And the methods of the work, including the controlling judgements

of adequacy and value, are also tested by the fruitfulness and longevity of the results whose achievement they condition. Although interpersonally communicated, and largely tacit, these methods also evolve and mature, through the testing (significantly less direct than in the case of scientific results) of their results against further experience. The personal endeavour that is necessary for worthwhile scientific work is itself a very artificial and social creation. For the individual talent and style, and the special private knowledge, are applied in a highly stylized fashion: to the investigation of problems concerning a given set of intellectually constructed objects, working up the materials derived from experience in accordance with established methods, and drawing conclusions within accepted patterns of argument. Even the greatest creative work, in which all these components may be strongly modified, must base itself on a tradition in which such modifications themselves are a natural development.

That such genuine scientific knowedge, as I have defined it, should emerge from these processes, is an entirely contingent historical fact. The paucity of the fields in which such knowledge has been achieved so far, and the brevity of the periods of human history in which such work has been successful, show that there is nothing automatic about it. It is a particularly sensitive social endeavour, requiring an appropriate philosophical and social context, and leaders and followers of dedication and integrity. But we know that it can be achieved, for in fact it has been. With an appreciation of the complexity and contingency of this achievement, we are now in a better position to study some of the practical problems involved in the continuation of this work in its new technical and social conditions.

Limits of Scientific Knowledge

To conclude our discussion of the special character of scientific knowledge, we shall consider some of its essential limitations. First, even such genuine scientific knowledge is not absolute and unconditioned. Although it has become independent of the particular circumstances of the achievement of the results from which it derives, it is still bound to the cultural milieu in which it developed. If it is to be accepted as genuine knowledge in any other culture, the objects which it describes, and the criteria of adequacy and value which it presupposes, must be coherent with the world-picture of the culture to which it is offered. In the partial or complete absence of such coherence, the material cannot pass as knowledge. In a new environment it may be adopted as a tool for the accomplishment of practical tasks or even for the solution of scientific problems; but in that case the objects of the knowledge will soon be recast so as to make the tool meaningful and effective to its new users. Moreover, only some parts of what seemed a coherent body of knowledge will survive the transfer; and it may be that those which were considered as the essential components will be rejected as false or meaningless by the borrowers.

All the sciences of the ancient civilizations whose direct descendants are still alive as sciences in European civilization have suffered such a transformation. Perhaps the best-known example is alchemy; but in this case the manual, spiritual and mystical aspects of the operation were so intertwined that much of the body of written descriptions of the work later had to be rejected as spurious or indecipherable. Astronomy shows a neater and in some ways more interesting pattern. For the techniques of tabulating and predicting the positions of the moon and planets, developed in the Mesopotamian civilizations, are capable of being translated into modern mathematical notation and even of being simulated on an electronic computer. But the objects of those position-reckonings were, for the original astronomers, not at all the lumps of matter that we see, but gods; and in their comings and goings it was hoped to discern a pattern which, through comparison with events in their visible, earthly domain, would provide a clue to their intentions.[28] Similarly, when the great astronomer Ptolemy took over data and techniques from this source and married it to a Greek tradition of theoretical astronomy, his stars, although now localized on spheres in the heavens, were still divine agents.[29] With his astronomy he hoped to lay the basis for a sound science of astrology, and also to discover the mystical mathematical harmonies of the divine creation. For many centuries the technical astronomy of Ptolemy was neglected because of its difficulty, while astrological science was a standard part of the knowledge of all educated men, and particularly of

[28] For the general world-picture of this civilization, see H. Frankfort and others, *Before Philosophy* (Penguin Books, 1949). Some detail on the astrological theory of Mesopotamia is given by R. Labat, 'La Mesopotamie', in *Histoire Generale des Sciences*, ed. R. Taton (P.U.F., Paris, 1957), i, 73-138.

[29] For a discussion of Ptolemy's astrology in relation to his astronomy, see A. Pannekoek, *A History of Astronomy*, pp. 160-1.

physicians. But in early modern times the objects of astronomical science were transformed; Ptolemy's astronomical masterpiece was considered an obsolete handbook of tools, his astrology was cast out, and no respectable European astronomer since Kepler has published investigations into the mystical harmonies of the cosmos.

In arguing the restriction of scientific knowledge to its original cultural context I have made strong use of the distinction between knowledge and tools. The distinction is not absolute; and in connection with this problem it requires a closer analysis. When we consider scientific inquiry as a task, knowledge about the external world is its ultimate purpose, while the tools are only a means for achieving that purpose. Yet tools, even those whose function is in the performance of practical tasks, do embody some sort of knowledge. And as a form of knowledge, they can be seen as superior in several important respects to verbal assertions about those intellectual objects which are hoped to represent reality. As we have just seen, they are less bound to the culture of their origin; and they have a correspondingly greater capacity for survival and continuous development. Yet the accomplishment of practical tasks with a particular tool is not at all the same as having genuine knowledge of the principles of its operation; over history, the very successful craft techniques by which civilization was built had explanations which we now consider utterly false. Nor is the capability of tools for transfer an absolute thing in itself. What we find is that those tools which perform functions related to the more universal needs of mankind, hence those which are rather more animal than specifically human needs, can be transferred and translated. This we see from the record of archaeology and anthropology. But those which are more specialized to sophisticated material or cultural functions are restricted to their original culture in the same fashion, if not to the same degree, as the scientific knowledge in whose terms they receive their explanations.

Another limitation to scientific knowledge as we conceive it in modern European civilization can be shown by a discussion of other possible sorts of knowledge. As a bridge between these different sorts, we may first consider the contrast between scientific knowledge and personal understanding. The former is the sort of material that Aristotle said can be demonstrated and taught; it is essentially explicit and public. Understanding, on the other hand, is private and largely tacit. Each person's understanding of a piece of scientific knowledge, or even a fact, will be peculiar to himself, depending on his own history of involvement with the materials, and on his special skills and tastes. One can 'know' a classic piece of scientific knowledge from an early age, but one's understanding of it can, and should, develop as one matures. Indeed, we may say that scientific knowledge, or facts, can exist only when the overlap between the private understandings of the objects among the members of the relevant community is sufficiently great for arguments to be communicated and univocally assessed.

Now we can raise the question of whether true understanding of important things can be derived otherwise than through mastering a body of scientific knowledge. Aristotle admitted one exceptional case, which he called 'practical wisdom', deriving from a sort of craft experience of the complex and ever-changing situations of practical life. He also invoked 'intuitive wisdom' to provide a guarantee for the truth of those ultimate, unproved assertions on which any demonstrative science must rest. But this seemed to be more a matter of recognition of what was self-evident rather than a qualitatively deeper understanding.[30] However, in the other great tradition in Western thought associated with the name of Plato, scientific knowledge, public and demonstrable, is only an introduction to the real thing. This is wisdom, to be derived from an intellectual and spiritual contact with reality, rather than through the grosser senses; and which in the last resort is dependent upon a measure of direct illumination. For a deep inquiry into human knowledge, especially as its conception has developed through the history of our civilization (to say nothing of others), these other sorts of knowing must be taken into consideration; but that is beyond our present purpose.

[30] See the *Ethica Nicomachea*, Book VI, ch. 5, 1140^a–1141^a8.

Part III

SOCIAL ASPECTS OF
SCIENTIFIC ACTIVITY

INTRODUCTION

IN the previous section we gave an analysis of the processes whereby scientific results are achieved and scientific knowledge can come to be. We showed how scientific inquiry is a human activity, in the short run imperfect and fallible, and in the long run conditioned by social influences acting in an extended time. For the sake of simplicity, we restricted our discussion mainly to a matured 'pure' science and also assumed the presence of the institutions and social practices necessary for the accomplishment of the various social tasks involved. On the basis of the materials developed there, we can now enrich our analysis to include the social aspects of the work and so be in a position to analyse the social problems of science created by the conditions of the present and the near future.[1]

Our starting-point for the previous analysis of scientific knowledge was the consideration of scientific inquiry as a special sort of craft work. Here we will study the social aspects of scientific inquiry by considering this work as a special sort of socially organized activity. The starting point of this present analysis is the distinction between the collective goals of that work, and the private purposes of each of the agents involved in it. For the work to be successful, there must be a harmony, or at least an accommodation, between these two sorts of 'ends' or final causes. It is naïve in the extreme to assume that they can, or should be, identical, even in cases where the work demands dedication and self-sacrifice from the agents. For the establishment of that harmony, there must be certain social mechanisms in constant operation; but for these to perform their functions, it is necessary in turn for those who are involved in their operation (both as agents and as subjects) to have attitudes appropriate to their roles in the system. In general, as the functions to be performed

---

[1] Lest I as an outsider be challenged by scientists for my presumption in analysing their social behaviour, I may quote Jerome B. Wiesner: 'The scientific community as a whole constitutes an extremely complex social system which is very little understood, least of all by the scientists themselves.' 'The Federal Role in Science and Technology', *Bulletin of the Atomic Scientists* (November 1962), p. 45.

become more sophisticated, the methods of social behaviour necessary for the accomplishment of the relevant social tasks are necessarily more subtle; and the private purposes governing the behaviour of the individual agents must be correspondingly more enlightened. There is nothing automatic about the presence of this correlation; it is possible for any socially organized activity (including fields of science) to work well in one time and place, and very badly in another. From an examination of the different social tasks involved in scientific work, we will see the ways in which an idealistic ethical commitment is involved in its social practice. It is not an automatic consequence of the study of the natural world; but it is a component of the code of behaviour of the leading men, which is necessary for the health and vitality of the whole endeavour. Its presence is not guaranteed by any institutional arrangements, but depends on accidents of history, and on the social and cultural environment in which scientific activity is conducted.

In this Part I will consider three basic social tasks in science: the protection of property, the management of novelty, and quality control. The first two of these appeared as important practical problems in the two earlier periods of great advance in science: the seventeenth and nineteenth centuries respectively. The last has become urgent only in the present period. This chronological sequence is a natural one, for each earlier problem was presented by an increase in the scale and the complexity of the social organization of science, and then receded into the background under the changed social conditions of the later periods. The progression is also of increased sophistication in the solution of the problems. As we shall see, the operation of quality control in present conditions is extremely urgent, and requires social mechanisms and personal attitudes which are as yet far from perfect for their functions.[2]

[2] This selection of social tasks is not intended to be exhaustive; other important ones include recruitment and training, decision on the allocation of resources, and the management of the variety of relations with society at large. I have chosen these particular social tasks for discussion because of their special relevance to the problem of ethics.

8

THE PROTECTION OF PROPERTY

CONSIDERING science as an organized social activity, we can say that the collective goal of the work is the advancement of knowledge. This is not at all the same thing as to say that every scientist must or does devote his labours exclusively to that end. Rather, when he is engaged on the task of investigating problems, he will be trying to achieve several different sorts of purposes. We saw this when we discussed the components of the criteria of value, in terms of the different functions that a solved scientific problem can perform. Some of these relate to the use to which the result could be put by colleagues or successors; while others relate to the personal benefit that the scientist himself would derive from the acceptance of a result as successful. This personal benefit derives from the recognition, by the relevant members of his community, of the worth of his work. This can be known only through those results which are acknowledged to be the outcome of his labours and of no one else. The research reports in which the results are contained thus embody his intellectual property; and the social protection of that personal property is necessary if each individual is to embark on his tasks in confidence that he will receive the rewards appropriate to his endeavours.

The Research Report as Property

As a piece of property, the research report is a rather unusual object. The property comes into existence only by being made available for use by others; and a research report hoarded in secret is almost certain to depreciate in value. Nor is there a market on which it can be sold for cash: any part of the result that is capable of being sold for its applicability to technical problems must be isolated from the rest and protected through the entirely different system of patents for inventions. The report which embodies the scientist's intellectual property thus hangs suspended in a world of its own: in this aspect of his work, the scientist has neither made anything of commercial profit, nor performed services for a client. All he has done is to provide materials for others who are doing just the same as himself, in a common activity whose ultimate purposes, in terms of the benefit to the lay society that supports it, are remote and diffuse.

Yet this property is none the less real and important to those who possess it. As a certification of the scientist's accomplishment, it can bring immediate rewards. And as an implicit guarantee of the quality of his future work, it brings in interest for some time after its production. Because of this predictive and fiduciary element in the scientist's intellectual property, it is impossible for it to be 'alienated' like ordinary property, without serious damage to the system of decision and control in science. If for any reason one scientist ascribes to another a larger share in a research report than is correct, he is not merely giving him more credit than he is entitled to. He is also falsifying an important part of the evidence on which the scientific community assesses the potential for future work of the other man, and thereby distorting the operation of its system of government.

The personal benefits derived from the possession of the scientist's intellectual property will be partly material, in the way of promotion and the incidental rewards of prestige, but not exclusively or even necessarily so. Even if the scientist is not particularly concerned with such vulgar things, but cares only for the recognition of his worth, or even for nothing but confirmation by competent judges of the quality of his work, the published research report is still his basic evidence for a claim on these benefits, and is thus his main property. The fact that such varied private purposes can be served by the one social mechanism for the protection of property helps to explain why the system of scientific publication could be so successful and stable until very recently.

The technique of diffusing results of recent research through published papers of authenticated quality and certified authorship is also a very effective means for the social task of advancing the work of science. In principle, the materials become public with the minimum of delay, are guaranteed to be of at least a minimum standard of quality, and can be put to use without the time-consuming process of obtaining permission, or negotiating for rights with their owner. All that is required for the protection of the personal property embodied in them is that a result which is used in another paper be cited there. In addition to its other functions, the citation thus represents a payment for use of the material; the author of the original paper derives continuing credit from this evidence of the quality of his work.

This system of publication and citation might seem to be such a natural arrangement for the harmonizing of collective goals and private purposes that if we did not already know of the strains it is now suffering, we might imagine it to be the result of an ideal social contract drawn up at the beginning of science.[1] However, like any other mode of social behaviour, it is historically conditioned in its origins and operation, has its own characteristic defects, and requires a mixture of formal rules and an informal etiquette for its effective operation.

Earlier Social Mechanisms for Property

The social endeavour of the achievement of scientific knowledge does not depend on this particular mechanism for the protection of intellectual property; indeed, the heroic age of the establishment of modern natural science, in the sixteenth and seventeenth centuries, lacked this system, and worked with several makeshifts. First, we should realize that in Europe up to the Renaissance the category of personal property embodied in an isolated piece of new knowledge simply did not exist. In the world of learning, the 'scholastic' method was of analysis and citation of authorities, so that however novel a conclusion might appear, it could be exhibited as belonging to a strong and genuine tradition. Similarly, such practical arts as alchemy considered themselves as participating in an ancient tradition; and the force of the tradition in the authentication of any result was so strong that many genuinely original works were published as the rediscovered texts of great masters. Even in the humanist tradition, whose style of scholarship was in many ways the ancestor of our own, the establishment of a pedigree for a result was an important part of the work; and in such fields as medicine debates over real practical problems were conducted as battles of books between conflicting traditions.

The conception of a single demonstrated result embodying new knowledge, and belonging as property to its author, came first in

[1] Such an assumption seems to be implicit in N. Storer, *The Social System of Science* (Holt, Rinehart, & Winston, 1966). He sets up a model of 'exchange' of rewards by the community for the 'creativity' of the scientist.

mathematics. The priority to this field might be the result of several factors: it is possible to achieve a single, brief result which is not encased in a tradition of scholarship or practice; and the problems being solved in such fields as algebra were indubitably new. Also, mathematics was a field of learning (as distinct from a craftsman's art or a learned profession) which was practised by free-lance individuals, usually (but not entirely) in conjunction with a mathematical art as architecture or engineering. Such men already had a system of intellectual property in their inventions and constructions, which they turned to material account in obtaining employment from wealthy patrons. It would be natural for the conception of intellectual property to extend first to the branch of learning with which they were naturally associated, and which was also involved in their securing of prestige.[2]

The first devices for the authentication and protection of such property were crude and of a specialized application: the statement of a mathematical problem, as a challenge to rivals to solve it;[3] it was understood that no one in his senses would make such a challenge unless he had already solved the problem himself. This system remained in use through the seventeenth century, and was even institutionalized in the requirements for tenure of the chair in mathematics at the Collège de France in Paris, which had been endowed by the educational reformer Ramus.[4] It was the occasion of the famous encounter between John Bernoulli and Newton over the curve producing the path of quickest descent for a body moving under gravity between two points.[5] But by this later time, other methods had been developed. The first was the anagram, in which a result was publicly stated in an indecipherable form, sometimes pending its

[2] The virulence of the disputes over priority in the achievement of mathematical results in the sixteenth century makes it plain that such 'inventions', along with those of mechanical devices, were recognized as an important form of property. The notorious dispute of Cardano and Ferrari (del Ferro) against Tartaglia is chronicled at length in O. Ore, *Cardano the Gambling Scholar* (Princeton University Press, 1953), pp. 61–107. The famous solution to the cubic equation was only one of the points at issue.

[3] The challenge to a public debate (probably descended from traditional university examinations) was used unsuccessfully by Tartaglia; see Ore, *op. cit.*, pp. 99–105.

[4] This unusual system of tenure (for which arguments can be advanced) required Roberval to conceal the intellectual property of his methods and even some of his results. Hence he was sometimes anticipated by others, and became embroiled in priority disputes. See C. B. Boyer, *A History of Mathematics* (Wiley, 1968), p. 389.

[5] The challenge and problem are translated in D. E. Smith, *A Source Book in Mathematics* (McGraw-Hill, 1929), pp. 644–55. The incident with Newton is described in L. T. More, *Isaac Newton* (Scribner's, New York and London, 1934), pp. 569–71.

full confirmation by the author. Galileo used this method, and Kepler tried unsuccessfully to decipher his anagrammatic announcement of the discovery of the non-spherical appearance of Saturn. Other users included Hooke, Huygens, and Newton himself. Huygens even proposed it to the Royal Society as a regular procedure for the establishment of priority of discovery and hence the securing of property rights.[6]

Such methods, involving as they do the concealment of the result, accomplish little more than staking a claim to property; in themselves they do not guarantee the authenticity of the result, nor do they contribute to its diffusion except by stimulus. The task of establishing genuine communication in 'philosophy' was first undertaken voluntarily by a small group of men, who became known as 'intelligencers'; among these were Mersenne in France, and later Henry Oldenburg and John Collins in England. Their task was extremely delicate; for they communicated results which had frequently been revealed reluctantly by their authors. They had to take care that the results were not so completely specified that the recipient could proceed too rapidly on exploitation, and yet not so vague that a reproduction of the original result could escape the charge of plagiarism. From the difficulties they had, we can see that a significant proportion of the great 'scientists' of that age were even more concerned for the protection of their intellectual property, than for an immediate realization of its value through the prestige resulting from publication, to say nothing of contributing to a co-operative endeavour.[7]

[6] For a full summary of the correspondence between Galileo and Kepler in this period (in which Galileo sent a second anagram announcing the discovery of the phases of Venus) see *Johannes Kepler Gesammelte Werke*, iv (Munich, 1941), 478–85. In a postscript to the published version of his Cutlerian Lectures, Hooke listed some of his inventions, describing those yet unpublished by anagrams. Among these was that for '*ut vis sic tensio*'—Hooke's Law. See Hooke, *Lectiones Cutlerianae* (1697), reprinted in R. T. Gunther, *Early Science in Oxford*, viii (Oxford, 1931), 151. Newton's anagram, describing his 'secret' of the method of fluxions, was sent in a letter to Oldenburg, 24 October 1676; see *The Correspondence of Isaac Newton*, ed. H. W. Turnbull, vol. ii (Cambridge University Press, 1960); text on p. 129, translation on p. 148, decipherment on p. 159. Huygens's suggestion was sent with an anagram of his own, in a letter to Oldenburg, 27 January 1668–9; text and English translation of the letter are printed in *The Correspondence of Henry Oldenburg*, ed. A. R. and M. B. Hall, v (University of Wisconsin Press, 1968), 360–3.

[7] The properties of the cycloid were a fruitful source of priority disputes for twenty years; see C. B. Boyer, *A History of Mathematics*, pp. 389–390 and p. 400. The Newton-Leibniz dispute was a scandal whose effects damaged British mathematics for a full century; see More, *Isaac Newton*, ch. XV.

By the end of the seventeenth century, the role of the intelligencers was merging into that of editors of journals; although these were not yet the specialized journals comprehensible only to the practitioners in a field. Through the eighteenth century, original work in natural philosophy and natural history was dominated by gentleman-amateurs, and their approach to publication was more leisurely and casual. A long book or a lengthy memoir would sum up years of private inquiry; those in a medical tradition (as Joseph Black) would announce results in their lectures and only much later, if at all, write up the material for printed publication.[8] Some, like the eccentric Henry Cavendish, revealed their results only in letters to friends, even while advising them to publish their own work.[9] Claim-staking through sealed notes was a recognized practice, as is well known from the history of Lavoisier's work.[10] And the book, in which an entire field was given a definitive structure, was (and remained for some time) an important medium of publication of new results.[11]

Scientific Journals and Intellectual Property

It is clear that this variety of techniques could work well only when both the set of practitioners and the flow of results in any field were small. When property rights are secured by claim-staking on unspecified results, and immediate diffusion is accomplished by discreet personal intermediaries, only a thin stream of results can be handled without inordinate delays and serious misunderstandings about contents and property rights. The transition to specialized journals as the dominant form of publication came in the half-

[8] D. S. L. Cardwell has shown me the significance of the change in style from the lecture or memoir of the eighteenth century, to the brief papers of the nineteenth. Naturally, both forms were in use in both periods; it is a matter of relative importance.

[9] Thus, in a letter to the Yorkshire scientist, Michell, Cavendish wrote: '. . . sorry however that you wish to have the principle kept secret. The surest way of securing merit to the author is to let it be known as soon as possible and those who act otherwise commonly find themselves forestalled by others.' But English natural philosophy was still sufficiently a 'club' that priority could be secured without printed publication. Thus, Cavendish says: '. . . you ought rather to wish me to show the paper to as many of your friends as are desirous of reading it.' (Cavendish to Michell, 27 May 1783). Quoted in Russell McCormmach, 'John Henry Michell and Henry Cavendish: Weighing the Stars', *British Journal for the History of Science*, 4 (1968), 147.

[10] I am indebted to R. G. A. Dolby for pointing out the significance of this practice.

[11] For a full history of the methods of publication in this period, see David A. Kronick, *A History of Scientific and Technical Periodicals: the Origins and Development of the Scientific and Technological Press, 1665–1790* (Scarecrow Press, New York, 1962).

century coinciding with the industrialization of Europe and the changing social character of science. The medium was already there, in the variety of journals for an educated lay public; and the shift to audiences which were expert rather than dilettante could well have resulted naturally from the opportunities created by the increasing scale of the activity in different fields. The social history of science in this important transitional period is yet to be written; but we can discern, as in all other aspects of science, characteristic differences between national styles in England, France, and Germany. In England, the journals served an overwhelmingly amateur audience, and would be published by independent societies or by free-lance scientific publicists. Their contents, in degree of specialization and in quality, would naturally vary along the scale from the most expert to the least. In France, where the great achievements of the revolutionary and Napoleonic periods were concentrated among a small group in Paris, the worthwhile journals were few, managed by members of the scientific élite, and sharply differentiated from the rest.[12] Rather later, in Germany, the development of great university-based science (along with all other branches of scholarship) produced the journals serving a multiplicity of specialized fields, each journal representing a school and publishing its work.[13] As in other respects, it was the German system which became dominant in the later nineteenth century all over the world of active research in science, and which with modification we have inherited.

In its classic form, this system of specialist journals automatically performs an important function in the protection of intellectual property: its authentication. Science has always had obscure journals,

[12] French chemistry had the great *Annales de chimie ;* the other natural sciences filled the *Observations sur la physique*; and the monthly *Bulletin de la Société Philomatique* took short papers by the members. But the mathematical memoirs had to wait either for the frequently delayed publications of the Académie (or Institut) or of the École Polytechnique. The lesser Parisian journals, with the provincial journals and academies, were of a lower class.

A description of the leading French journals is given as part of the account of institutional aspects of science in M. P. Crosland, *The Society of Arcueil*, chs. 3 and 4, pp.147–231.

[13] The late start of German scientific publication is well described in H. Schimank, 'Ludwig Wilhelm Gilbert und die Anfänge der "Annalen der Physik"', *Sudhoffs Archiv*, 47 (1963), 361–73. Gilbert took over the journal from Gren in 1798 and edited it until his death in 1824. By the end he had converted it to a more technical journal, but he still had to rely on translations of foreign papers to fill it. But soon after Poggendorff took over the journal, it became a leading and very specialized periodical.

252 Social Aspects of Scientific Activity

in which boring or mediocre work could find a printed page.[14] But these tended to be recognizably different from the leading specialist journals: either provincial or colonial, publishing studies by local enthusiasts, and appealing to an inexpert audience. For a result to be authenticated as a contribution to an established field, it had to be submitted for scrutiny by one of the leaders of the field or by his chosen referees; otherwise it would tend to be ignored by the community which constituted the field. Through the system of recognized journals, the leaders of a field could effectively control the creation of intellectual property by those who aspired to, or claimed, membership in the field. Such a dictatorial system is clearly open to abuse; and some of the most seminal papers of nineteenth-century physical science were printed either privately, or in less severe journals, because the leading journals would, or did, reject them. Yet the system had the great merit of establishing a clear line between the authentic and the unauthenticated in scientific results and the intellectual property they embodied. And it was an essential component of the system of quality control.

It is clear that a system which operates strict controls on the authentication of results will tend to err on the side of conservatism; genuinely novel results will involve their own novel criteria of value and of adequacy, and run the risk of being rejected as the work of a crank. Occasional works of genius have been so mistreated by the mechanism of protection of property through publication in recognized journals.[15] But no system can accommodate genius; and these occasional mistakes are not peculiar to this one. However, in the rigid ascription of property rights we can identify a defect inherent in the system, where its performance of two functions can be in conflict. In this case it is not only that works of deep originality and

[14] I am grateful to R. G. A. Dolby for reminding me of this. It was well before the war that a group of American scientists formed an 'American Institute for Useless Research', and suggested titles of journals for the publications of its members: *The Refuse of Modern Physics*, *The Nastyphysical Journal*, *Comptes Fondues*, and *Makeshift für physik*. See letter to the editor, *Review of Scientific Instruments*, 6 (1935), 208.

[15] Authors who published important works outside the recognized channels, from choice or necessity, include Helmholtz on the conservation of energy, van't Hoff on stereochemistry, Herapath and Waterston on kinetic theory, and Newlands on the 'law of octaves' of the chemical elements. The first two soon achieved respectability, but not the others. Newlands was retrospectively honoured by the Royal Society after Mendeleef established the Periodic Table, with their Davy Medal. But he was not made a Fellow; and he is *not* mentioned in the standard bio-bibliography of nineteenth century scientists, *J. C. Poggendorffs biographisch-literarisches Handwörterbuch*, in any of the editions concerning the years 1858 to 1953, vols. iii–vii, 1898–1959.

genius are not accommodated by the system; the need to secure property rights through priority of publication can actually harm and distort the social activity of scientific inquiry. A result belongs to the man who first publishes it, or whose paper first reaches the editor of a recognized journal. In exceptional cases, the property will be shared between two or more who have achieved the result independently, when it can be proved that they could not have known of each other's work. But the criterion of impossibility is extremely strict; the claim to ignorance of a result recently published in a foreign journal is not sufficient to establish independence of work. It is easy to see the benefits of a very strict rule of priority; otherwise there could be perpetual squabbles about the sharing of ownership of results. Without some such formal system of establishment of the ownership of a result, the adjudication of priority disputes would depend, in part at least, on the personal memories of the parties involved. These are always unreliable, as regards chronology and detail; and the history of earlier disputes (the best documented being those of Newton) shows how easily a memory can push back the date of a discovery beyond all objective plausibility.[16]

The inherent defects in the system are not due to its particular rules for assigning ownership of the property: they lie in the very concept of the 'result' described in the research report, as an atomic unit with no history of its own. If scientific inquiry consisted of 'making discoveries' or even 'testing hypotheses', one might justifiably imagine there to be a very brief interval of time, at the beginning of which there was ignorance, and at the end, knowledge. But a problem, especially a deep and new one, has a complex and usually long phase of gestation. The initial insight may flicker in and out of plausibility, as the developing argument for it encounters evidence which confirms or disconfirms it. The problem itself may change in mid-course, or it may lie dormant for a while, awaiting conclusive evidence. During all this time, the problem has no status as property:

[16] D. T. Whiteside has shown that the only direct evidence for Newton's famous testing of the inverse-square law of gravitation in 1666 is his testimony of half a century later. His early scientific notebooks contain no materials which relate to such a calculation; but they do establish that until the early 1680s Newton was working with dynamical models derived from a Cartesian tradition. See D. T. Whiteside 'Before the *Principia*', *Journal for the History of Astronomy*, 1 (1970), 1–19. It should be recorded that in the discussion after the presentation of part of this paper at the British Society for the History of Science in July 1969, two American scholars cited other documents in contradiction of the thesis, but Dr. Whiteside dispatched them.

it neither brings recognition to the author, nor is it protected against appropriation.

If scientists worked in a social vacuum, this frequently lengthy period during which a project is not property, would not be too significant. But scientists generally need to communicate informally through the interpersonal channel about their work; and every such communication puts one's property at risk. In fields where the pace of work is slow, and there is a small community, the danger is not great: the incentive for poaching work in progress are small, and the penalties for being caught at it are great. But in fields which are fast-moving, large, and impersonal, one must communicate informally to keep up with events, and the hazards are serious. The protection of the property of work still in progress puts one under pressure to engage in claim-staking publication, reporting all work as soon as it is of passable quality. This tends to produce hasty and slipshod work, and also discourages the investigation of those deep and complex problems where the gestation time for the achievement of publishable results may be dangerously long. In this way, there is a conflict between the functions of the publication system as a means of the protection of property, and as the means of diffusion of authenticated worthwhile materials. This must not be thought of as an evil peculiar to this system; for as a device designed for the performance of several functions, it will naturally be incapable of attaining an ideal state of excellence in all respects simultaneously.

The Operation of the System of Protection of Property

For people to operate a social mechanism, and to submit to its operations, it is necessary for them to believe that its intellectually constructed categories are at least in satisfactory correspondence with the reality they experience. In the nineteenth century, the prevailing philosophies of science assumed the simplicity of 'discovery'; and so the divergence between the unit of property, the result, and the real thing of experience, the problem investigation, was generally ignored. This simplistic view helped the system to operate, and perhaps was necessary for its maintainance; but it resulted in a very rudimentary understanding of the possible relations between two similar scientific results. Either they were independent, in which case the similarity was accidental, or there was dependence, and one side was suspect of theft. The bitterness of the classic priority disputes of the nineteenth century can be explained in terms of this simplistic picture of the process of discovery: a challenge could not go very far without the ethics of the disputants being called into question. The common practice of the disputes being opened by friends and supporters, rather than by the principals themselves, may have been the result of an unspoken etiquette adopted to contain the explosive possibilities of such disputes. There are other explanations as well: the ruling ideology of the selfless devotion to knowledge, combined with the etiquette of the gentleman, who would not advance claims against another on his own behalf.[17]

In spite of these defects, the system of protection of property does at least provide formal rules of authentication and an etiquette of dispute for the assignment of ownership of work up to the point of publication. But it provides no systematic protection for property in the next phase of the development of a problem: the subsequent exploitation of the result. Formally, any published result is available to anyone, anywhere, who wishes to use it, the only price being its citation. But the research report may be only an interim account of work in progress; does its author have any protection against being swamped by a rival whose superior facilities or skill enable him to exploit the result more rapidly than himself? If this happens, his property in the published result will rapidly lose its value; and worse yet, the investment of his time and resources in an extended research programme will be wiped out. It is clear that the protection of exploitation rights in a result cannot be operated through a set of formal rules. The circumstances of production and use are so enormously various, that any fixed moratorium on the use of results would lead to intolerable delays in some situations, and afford no real protection in others.

Perhaps because it is incapable of formalization, the principle of

[17] The most thoroughly documented study of priority conflicts, examined from a sociological point of view, is R. K. Merton, 'Priorities in Scientific Discovery', *American Sociological Review*, 22 (1957), 635–59. His analysis includes the conception of intellectual property developed here, as well as that of conflicts between various 'norms of behaviour' in science. It is noteworthy that the participants in some priority disputes would preface their claims with an assurance that they disliked such matters, and thought them irrelevant to the progress of science. See J. W. van Spronsen, 'The Priority Conflict between Mendeleef and Meyer', *Journal of Chemical Education*, 46 (1969), 136–9. In this case the dispute was over the adequacy of Mendeleev's attribution of credit to Meyer; and it is clear that the debate was further confused by the common conviction that there was a single true discovery at issue.

fairness in exploitation of results has received no mention in the classic literature on the sociology, or ethics, of science. Yet, interestingly enough, abuses of the system of property in science through the appropriation of exploitation rights were attempted, and denounced, as soon as there was a vocationally-based community of science in early nineteenth-century Paris. The distinguished physicist J. B. Biot, who might be called the world's first career scientist, was involved in a number of unpleasant disputes over property. These usually did not involve the priority of the first discovery itself; but Biot's technique, apparently, was to learn of a discovery informally from a colleague or friend, and then with his experimental skill and the superior material resources at his disposal to exploit the discovery and report on his own series of experiments, before his informant had the opportunity to develop the work.[18] The case of the great mathematician Cauchy was even more notorious. On receiving a paper for refereeing, he could not resist the temptation of recasting the proof, improving the result, developing and generalizing it in all sorts of ways, and finally publishing it in a journal to which he had rapid access. When the paper which had originally stimulated him finally appeared in print, it would seem singularly crude and pointless in comparison to the results already published by the master.[19]

The Etiquette of Citations

In the absence of a formal rule defining a method of social behaviour, there must be an informal etiquette which governs it. In the case of the exploitation rights to published results, the etiquette will depend very strongly on the particular characteristics of the field, and (as we have seen) on the style of the local community. Even when there is a formal rule, it must be supplemented by an informal etiquette, since its categories can never encompass the varieties of the practice it is designed to govern. This can be seen even in the apparently straightforward technique of citation. This has at least two quite distinct functions. The citation of the source of materials used in an argument implicitly places that source in the argument

[18] I am indebted to Mr. Eugene Frankel, of Princeton University, for this information on Biot.

[19] This and other information on Cauchy's style of work will be found in the article on him by H. Freudenthal, *Dictionary of Scientific Biography* (Scribner, New York, 1971), vol. iii, p. 134.

itself. There is no need to argue again for the adequacy of the materials; that is assumed to be accomplished in the original report. Quite independent of this is the function of dividing the property in the published report, and providing an 'income' to the owner of the property which is used, by showing that his work was fruitful. Within each of these functions, the citation can take a multitude of meanings. The materials themselves, the uses to which they can be put, and the relations between an existing result and a new one, are as complex as the history of a solved problem. The material may be crucial, or merely incidental in the argument; it may have been central to the first formulation of the problem, or merely a late addition; and it may have been used as it was published or required extensive re-working. In all these dimensions, there is a continuous and complex scale from complete dependence to near independence. By appropriate nuance of mention, one can under-cite without actually stealing results, or over-cite with the effect of inflating the value of the property of a colleague. Thus the simple system as it stands permits a considerable range of 'sharp practice' with scientific property, which if not controlled by an etiquette, can be more corrosive in its eventual effects than outright theft.

Since the citations must convey some very subtle messages by a very crude device, the etiquette of each field will impose a code for their interpretation, whereby the entries and their possible brief comments will convey the requisite meanings to those in the field. Each such code will depend on the character of the problems in the field, on the types of mutual dependence, and also on the ruling conception of the right division of intellectual property. In every case, it will be a purely informal, perhaps tacit and unselfconscious, craft knowledge shared by the members of the field. Thus, in the last resort, this aspect of the system of the protection of property depends like the others on an informal etiquette as well as on a formal system of rules. The relation between these two sorts of methods of social behaviour is analogous to that between the public and the interpersonal channels of communication. They mutually condition each other, and are together necessary for the effective operation of the social mechanism. The means of enforcement of an etiquette will in general be very different from those for formal rules; an examination of this problem will lead naturally into the question of the ethics of scientific activity.

Property in Industrialized Science

In the conditions of industrialized science, the effective property of the scientist has undergone certain changes, which we discussed earlier. For making a claim on the rewards available to him, the scientist no longer needs to rely exclusively, or even primarily, on formally published reports of significant results of research. Two other sorts of evidence can now perform the same function: the informal reports of his achievement and promise, communicated by the interpersonal channel to those who control the distribution of research funds; and formally published reports of insignificant work, either weak in itself or hasty claim-staking. Thus two new sorts of effective property have arisen: the informal property of personal contacts, and the second-class property of titles of publications. The contexts of these new forms of intellectual property are quite different, and in general there is not much overlap between the groups of scientists who rely on them. Nor have they entirely displaced the traditional form of property; their influence depends very much on subject and place.

The development of these new forms of property is probably the main cause of the welcome decline of the bitter 'priority' disputes between great scientists which so disfigured the science of earlier periods. For neither sort is so vulnerable to appropriation of things of great value. The decrease in sensitivity over 'priority' has another cause, which has contributed to the change in the conception of the unit of property itself. The common sense of the matured natural sciences has shifted away from a belief in the simple 'discovery', and towards a recognition of the complexity of a problem, both in its structure and in its historical development. Hence when rewards are distributed for an achievement of the highest quality, they are usually shared to general satisfaction, among those who have contributed in different ways to the development and solution of the problem. Thus the earlier common sense identification of the published research report with the actual achievement has been displaced; and as a result there is an awareness of the inadequacy of this simplistic conception of intellectual property for its social functions.

As a result of these recent changes, many of the defects in the older, rigid system of the protection of property have been removed. But this does not imply that in all respects the new state of affairs is simply 'better' than the old. Indeed, if we are to cope with the effects of such social changes, we should begin to search for the inevitable defects characteristic of the new system as soon as it becomes a significant phenomenon. First, there is no guarantee that the other functions performed by the older system will be performed so well by the new. For example, the earlier mechanism of quality control in science relied heavily on the leaders' control over the creation of intellectual property, which they maintained through their control of the recognized journals in each specialized field. With the by-passing of this channel, both through informal contacts and through the proliferation of uncontrolled journals, the mechanism for quality control is inevitably weakened. A second arises in the displacement of formal rules by an informal etiquette for the operation of a social mechanism. While formal rules have the characteristic defect of rigidity and conservatism, an informal etiquette is more vulnerable to a gradual distortion, in response to an aggregate of individual pressures for the achievement of private purposes. In this case, there is no institutional inertia for the preservation of methods of behaviour which contribute to the fulfilment of collective goals at the expense of the achievement of private purposes. Of course, if a set of formal rules is rejected because it is seen by the community to be hampering its collective work, then a new set of rules, with the associated new etiquette, can be a genuine improvement. But when, as in the present period of science, the change comes in an unplanned and 'natural' way, it is more likely that the situation is one of a short-term victory of private purposes over collective goals.

9

THE MANAGEMENT OF NOVELTY

THE achievement of new results is the goal of the task of scientific research; and in this respect science has characteristics, and problems, peculiar to itself and to only a few other spheres of human activity. The uniqueness of science in this respect should not be exaggerated. Any sort of work which rises above a mechanical routine involves the solution of new problems as times and circumstances change; and in matured fields of science, extending the borders of the known can be done in a very straightforward and predictable fashion. But in science, the achievement and management of novelty do present problems which are nearly unique. In other sorts of work, an organization can be quite successful even if it merely keeps pace with gradual changes in circumstances, and does its old job well; but in science, a field in that condition eventually comes to be judged as stagnant. Also, in science, the achievement of novelty of any significant sort involves a challenge to some existing intellectual property; and the task of the management of novelty involves the orderly destruction of personal property, for the fulfilment of the collective goals of the work.

The management of novelty emerged as a problem distinct from the protection of personal property, when in its scale and social organization science developed beyond a handful of isolated, individual practitioners. When one can speak of 'schools', groups of individuals whose intellectual property is bound up with a long-term programme of research, and whose leaders have some control over the authentication of results and the distribution of rewards, then the challenge of a novel result raises particular social problems. The resolution of these problems is particularly difficult to accomplish by methods of social behaviour involving formal rules and routines; but the price of a too successful defence of existing property, is the stagnation and eventual decline of the field.

The Destructiveness of Novelty

As we have seen, every solution of a scientific problem produces some change in the objects of inquiry; for the objects exist only as the collection of their known properties. These may be explicit assertions about details, or they may be techniques involved in establishing conclusions about objects, or again they may be the the more informal personal craft knowledge of subtle features of the behaviour of the objects in the course of their study. Each time a problem is solved, the result necessarily includes new properties of the objects of inquiry, and thereby the objects themselves have changed. Of course, in the majority of cases, the change is slight and not significant; a refinement of the specifications of known properties only rarely produces such surprises as to suggest the existence of hitherto unsuspected properties. (Among the exceptions to this rule, the most famous in recent times is the discovery of numerical regularities in the wave-length of the spectral lines of simple elements.) Even when the change is significant, it will not necessarily present practical problems for its acceptance and assimilation. For so long as it does not challenge the guiding strategy of the school that controls the work it can be welcomed as a further contribution to the cumulative progress of the field.

But genuine progress in science depends on the occasional injection of deep novelty, and then such a challenge is made. This need not extend to an assertion of the incorrectness of accepted results; it is sufficient that it should claim that a new set of problems, with their own objects and methods, are a more fruitful path of advance than those hitherto studied. Such a challenge, if successful, is destructive; for the intellectual property of the existing schools is thereby rendered obsolete and sterile. From the nature of the situation, such challenges do not develop gradually, by insensible degrees. Although the roots of the new approach may lie deep in the past, and its preparation requires time, there is necessarily a brief period during which the challenge is made, recognized, and fought over. If the challenge is successful, then there is a radical transformation in the conduct of the inquiry, and one which may be too sudden in its coming for those with established ways of working to adapt to themselves. These two features, depth and destructive suddenness, are generally accepted as defining a 'revolution', and so it is in such cases that, following Kuhn, we can speak of a 'scientific revolution'. The term must be used with care, for revolutions in science can vary in their extent and depth, as much as the revolutions in society which range from 'palace revolutions' between factions of an élite, to those of Russia or China.

'Scientific Revolutions' in History and Philosophy

One of the most interesting things about such scientific revolutions, as an historical phenomenon, is that their existence was not recognized in the self-consciousness of science in the nineteenth century. The experience of science during the academic period was not so much of revolutions, where a previously effective established system is overthrown, as of pioneering. For in most fields outside Newtonian mechanics, the heritage of the eighteenth century was of fields in a rather rudimentary state both in their materials and in their social organization. The achievements of the nineteenth-century pioneers could be seen as creating new and fully effective scientific disciplines where none had existed before. Where this work involved the refutation of previously accepted theories and systems, it was relatively easy to show that in each case they were patently false and misleading, at least in the light of the new discoveries of the great pioneers of the field. Folk-history was an important part of the self-consciousness of academic science; and for each field there was developed a set of legends, in which the folk-heroes vanquished the the false and reactionary old theories. Evidence for such histories was easily obtained from the polemical publications of the side which eventually won, and from recollections and eulogies of the founding fathers. Such fields as chemistry and physiology, with the bad old theories of phlogiston and vitalism, were natural cases for such a folk-history; and the approach was extended back to such ancestral figures as Galileo and Copernicus.[1]

The varieties of this folk-history of science deserve a careful study which they have not yet received; two features of it are of significance for our present discussion. First, where folk-heroes did not actually exist, they had to be invented; thus Benjamin Thomson, Count von Rumford, was retrospectively canonized as the author

---

[1] J. Z. Fullmer, in 'Davy's Biographers: Notes on Scientific Biography', *Science*, 155 (20 January 1967), 285–91, discusses the influence of prevailing conceptions of 'science' and 'the scientist' on the framing of a biography. The biographical tradition for Davy was a fairly sophisticated one; for real hagiography one can turn to the Newtonian biographies. Only with de Morgan's critical studies of the 1840s did the warts begin to show. See A. de Morgan, *Essays on the Life and Work of Newton* (Open Court, Chicago and London, 1914).

of the 'crucial experiment' which destroyed the theory of a heat as a material substance.[2] Second, it could never be accepted, in the folk-history, that a refuted theory could have been a good one, scientifically justifiable in its own times. For this would have entailed the possibility of some serious errors in science against which integrity and discipline offered no certain protection. And the admission of the possibility of honest error would have undermined the propaganda, and the ideology, on which the science of that period depended so heavily.

It was in this context that the work of Einstein was seen as revolutionary in its philosophical implications, for a particular generation of scientists and philosophers. It seemed that his theory of relativity exhibited an error at the foundations of Newtonian mechanics, on the question of the existence of absolute space and time. And although there had already been philosophical criticism of that very idea (starting with Berkeley and renewed by Mach), this successful scientific attack at what was commonly conceived as a system of ideas that had been unchanging and apparently true for over two centuries had a profound effect. With the assimilation of relativity and quantum theory, physics was conscious of itself as being in a revolutionary state, and as dealing with obscure and difficult objects. This condition was contrasted with that of a previously stable, and comprehensible field; and the fact that physics, and mechanics, were actually in the same unstable state during the whole of the nineteenth century as they were during the twentieth, shows how strong was the influence of the dominant ideology on the self-consciousness of science.[3]

[2] The Rumford myth is perpetuated even in recent and scholarly histories of science. Thus, in A. Wolf, *A History of Science, Technology and Philosophy in the XVIIIth Century*, 2nd ed., revised by D. McKie (Allen & Unwin, London, 1952), we have 'Rumford's experiments clearly disposed of the caloric theory of heat in favour of a mechanical interpretation. His scientific contemporaries, however, had a sufficient measure of intellectual inertia to resist the new idea. So the caloric theory survived until the middle of the 19th century' (p. 198). Rumford's cannon-boring experiment of 1798 was recognized as one of the many peculiar phenomena of heat; and in the following two decades caloric theories of heat were used with great sophistication and success by Laplace and his school, particularly in the explanation of the velocity of sound in terms of adiabatic compression of air. For an authoritative discussion of the whole issue, see R. Fox, *The Caloric Theory of Gases from Lavoisier to Regnault* (Clarendon Press, Oxford, 1971).

[3] The writings of Popper show this sense of the 'Einsteinian revolution' in science and in the philosophy of science very clearly. In his own experience Einstein's theory represented (among other things) the overthrow of a dogma in science that was the paradigm case of absolute scientific truth. See K. R. Popper, *Conjectures and Refutations*

It was in such circumstances that the problem of novelty became an important one in the philosophy of science. From the time of the assimilation of Einstein's classic results, the image of the simply cumulative character of scientific knowledge could no longer be maintained; it was contrary to the new common sense of the leading field. But the question remained: how could the development of science be truly progressive, if deep new advances required the destruction of previously accepted theories? Without some clear criterion whereby the new theory can be proved to be superior to the old, there seems to be every possibility that the change will be for the worse rather than for the better. Also, if the old theory is simply rejected as false, and the achievements of the older generation are seen vitiated by error, then those of the present one are likely to be similarly exposed, and the development of science seems to proceed less in the forward direction, than sideways like a crab.

These two philosophical problems, framed as the choice between competing theories, and the relation of the old theory to the new one, have seemed for some time to be crucial to the understanding of science. For unless there are some implicit rules for the decisions made by scientists, and some logical relation of the old theories to the new for the assimilation of the earlier material, the genuine progress of science would be merely a chance occurrence. A variety of suggestions have been offered by philosophers of science for the solution of these problems; but none of them have been able to stand up for very long against criticisms on the grounds of internal consistency and of agreement with accepted historical evidence. The reason for this is that the actual processes of decision and assimilation are too complex to be plausibly reduced to simple models; and that while genuine progress does not depend on mere chance, it does depend on judgements of men for which no simple and certain rules can be laid down.[4]

(Routledge, London, 1963), pp. 192–3; and also 'Science: Problems, Aims, Responsibilities' (presented at the Forty-seventh Annual Meeting of the Federation of American Societies for Experimental Biology, Atlantic City, 17 April 1963), *Federation Proceedings*, 22 (1963), 961–72.

[4] The very fruitful concept of 'paradigm' introduced by Kuhn in his *Structure of Scientific Revolutions* leads directly to these problems, which are not solved there. For a survey of the subsequent discussion, see Dudley Shapere, 'Meaning and Scientific Change', in *Mind and Cosmos: Essays in Contemporary Science and Philosophy* (University of Pittsburg series in the Philosophy of Science, 3 (1966), 41–85). The most recent work of Lakatos, in which the decision-problem is discussed in terms of the tendency of a 'research programme' to be involved in a 'problem-shift' that is either 'progressive' or

Even the problem of the assimilation of old materials, which seems to concern meaning rather than human action, defies reduction to a formal analysis. For the objects of inquiry are related to experience through a complex structure of argument, formed by the composition of many conclusions of separate problems, each with its own history up the point of present time. A new, challenging result introduces an alternative structure, overlapping or identical with the old one at some points, but at others radically different or simply not developed. Some facts from the old structure will be given an attempted re-interpretation, but others will be dismissed as irrelevant or ignored as anomalous. An 'inclusion' of all the materials of the old structure in the new is rarely possible; but neither is it possible for either structure to be so tight, that a conclusive refutation of one by the other is feasible. Hence the relation between the two structures will generally include elements of agreement, refutation and sheer incomprehension, all mixed together.

The problems of assimilation and choice are thus insoluble at the conceptual level; simple and general descriptions and prescriptions cannot comprehend the complexity of what necessarily occurs. The problems can be resolved at the practical level; but the terms of the solution are then social and ethical rather than philosophical. The impossibility of the philosophical solution of these problems, even to the extent of initial plausibility, reflects the historical fact that progress in science is not generally uniform and linear; each innovation in science is a hazardous enterprise for everyone concerned. And on reflection it is only natural that it should be so; otherwise scientists, and science as a whole, would be able to reap great benefits while encountering small risks, and thus would escape the human condition.

The Problem of Choice: Strategies and Schools

The problem of choice is a far deeper one than choosing which of two theories is incorrect. For the challenge posed by an important novel result does not only relate to an existing set of materials, a complex collection of results, objects and methods; it also involves a gamble on the unknowable future, in the assessment of the fruitfulness of the new result in its possible exploitation and further

'degenerate', provides an illuminating formalization of the fallible judgements made in such situations. See 'Falsification and the Methodology of Scientific Research Programmes', in I. Lakatos and A. Musgrave, *Criticism and the Growth of Knowledge* (Cambridge University Press, 1970).

development. In brief, the decision is between strategies; between the established one which produced the results being challenged, and an embryonic one, implicit in the new result (or explicit in an accompanying manifesto). Thus the problem of the management of novelty in science is part of the practical problem of replacing schools when they have outlived their usefulness.

A great deal is at stake in the decision on the future of a school, and the evidence provided by any challenging new result can never be conclusive. We recall that a problem under investigation grows in interaction with its materials, and that when it is completed it is necessarily rough-hewn. This will generally be even more so, in the case of deeply novel results; for the conquering of new pitfalls, the forging of new tools, and the establishment of new objects of inquiry, require great talent, daring, and ruthlessness, and also a complete identification of the scientist with his result. He will inevitably develop his own criteria of adequacy for the problem, which will differ, at points likely to be crucial, from those of the established school. If every anomaly in experience, and every ambiguity in concept, were completely ironed out before the work was presented to the public, nothing new would ever appear. Also, genuinely new experimental work frequently involves using tools at, or beyond, their limits of reliability, especially when they are used by untrained or unsympathetic workers.[5] Hence the data itself must frequently be supported by a certain measure of faith, if they are to serve as a foundation for the crucial evidence. The newly created objects of inquiry will be designed for their function in an argument relating to these particular experiences rather than some others. It is impossible for the author of a new and revolutionary result to include in his argument answers to all objections which might come from

---

[5] The classic case of such difficulties with a new instrument is that of Galileo's telescope. The existing 'spyglasses' that stimulated his invention were incapable of improving on the naked eye for astronomical purposes; and Galileo needed to control the quality of the fine glass produced at Venice and also devise his own system of lens-grinding. Even then, most of the lenses he produced were inferior; and for the first crucial years, he simply did not have enough copies of a really good telescope to satisfy all the demands for demonstrations. The distinguished astronomer Magini brought a committee of professors to look through Galileo's telescope, and it was not difficult for them to see nothing of what he claimed was there. See Galileo Galilei, *Dialogue on the Great World Systems* (in the Salisbury translation), ed. G. de Santillana (University of Chicago Press, 1953), p. 98. For a detailed history, there are several books by V. Ronchi; the most recent is *Il cannochiale di Galilei e la scienza del 1600* (Torino, 1958). Galileo's difficulties with lenses are described in O. Pederson, 'Sagredo's Optical Researches', *Centaurus*, 13 (1968), 139–50.

those in the established school. For to be in a position to do this, he would need to be in complete mastery of its techniques, and to think in terms of its objects with full facility; and he would then be hampered in his creative thinking about his own problem. Thus it is possible for critics to dismiss the new work on quite straightforward grounds of adequacy, even showing standard pitfalls which the author has encountered. 'Brilliant but unsound' is the natural response of honest complacency to a nascent challenge.

It is all too easy to apply historians' hindsight to the counter-revolutionary camp, and to condemn them for obtuseness and prejudice against a challenge which eventually proved successsful. Indeed, one can sometimes find cases when debate over an innovation has involved the sort of personal considerations which are rigorously excluded from the ideal of scientific practice. But this need not be the result of weakness or corruption of the proponents; for if the existing criteria of adequacy for results do not give sufficient guidance, then one must fall back on other sorts of clues to the likely quality of the challenging results. These will involve the author and his past work, and also the problem itself. If the problem has a bad history of being the province of cranks and speculators, then whoever tries to rehabilitate it will have to struggle against a natural and justified prejudice.[6] Worse, if the problem has a history of involvement in a political or professional struggle, then any advocate of it runs the danger of being dragged down with it.[7]

Independently of all these constraints on their perception and assessment, those in an established position have every motive for dismissing challenges to their orthodoxy. For, as we have seen, their property is at risk. It is at this point that there is a conflict between the collective goals of the work of science, and the private purposes of strong groups within the community. For the advancement of a field inevitably involves the destruction of older intellectual property;

---

[6] John Ziman in *Public Knowledge*, mentions the case of 'continental drift', which 'dropped into the limbo of cranky and speculative notions' for nearly a half-century (pp. 56–7).

[7] This tendency will be most marked in sciences which are not fully matured, and where the innovation is related to a folk-science hostile to an academic or 'professional' science. The persecution of those who used 'mesmerism' for the relief of pain during surgical operations in the nineteenth century, is recorded by E. Boring, *A History of Experimental Psychology*, 2nd ed. (Appleton-Century-Crofts, New York, 1950), Ch. 7, pp. 116–33.

and those whose property is destroyed, are hurt in the process. Two features of this situation make it more serious, in some respects, than the inherent conflict between general goals and private purposes which is involved in the normal maintenance of quality control. For someone who loses property through the rejection of his research report by the journals is generally at fault for having misinterpreted the ruling criteria of adequacy and value. It should be part of the scientist's craft skill to know what sort of result is good enough for authentication as genuine property. Also, in each such case, only an individual or a very small group is involved, and their investment in that particular problem is necessarily limited. Here, on the other hand, reliable criteria of assessment are not furnished by the craft knowledge of the field; and the investment that is in danger can involve the careers of many men.

Yet the continuous advance of science requires that such conflicts be resolved, and that the community should occasionally decree the destruction of the property of eminent members, in the service of its collective goals. How is this to be done? If there were some simple crucial test for deciding between competing approaches, then integrity would demand that on some occasions, an elderly scientist would have to congratulate a young man on just having ruined his own life's work. As an alternative, one might have an impartial committee of adjudication, taking evidence from both sides, and then deciding which strategy is to be supported, and which suppressed. Such a solution would have the merit of finality, but it is clearly unworkable. The criteria involved in the assessment of competing strategies, are so complex, subtle and speculative, and the property at stake is so significant, that for justice to be seen to be done in such cases, there would have to be developed a system of tribunals, with their own case-law and intellectually constructed categories of evaluation. But since there is in principle no simple criterion for choice in such practical problems, there can be no simple and certain means to their resolution. In the development of a field as a whole even more than in the investigation of individual problems, detours and their associated waste and personal tragedy are inevitable.

Social Aspects of the Decision-Problem

Thus, for the making of decisions on strategy, and on the fate of established schools, there is no substitute for experience, which is gained over time. But such experience does not cumulate automatically; and the institutions for the management of novelty can be better or worse designed for the gaining of this experience. What is mainly required, is for the new strategy of work to be given a chance. This is not at all a matter of common sense tolerance between colleagues. For the leaders of the established school may, justifiable from their point of view (and perhaps correctly) consider the challenging new work not merely as unscientific, but as anti-scientific. Clever nonsense is far more insidious, and hence more dangerous, than the production of an obvious crank or dullard. By its superficial plausibility and air of excitement, this work might easily corrupt the minds of immature scholars and students. In some styles of scientific work (depending on nation and field), these reflections will be given clear voice, and then be answered in kind.[8] Personal accusations can easily be dragged in, under the guise of reflections whether such inadequate work could have been done by an honest man. The established etiquette of scientific debate serves to prevent such struggles from going to their natural extreme; but when one's property, ideology, and personal existence are threatened by what seems to be an unfounded and unscrupulous attack (whether from a rogue innovator or a die-hard reactionary, depending on one's position in the struggle), one will use every weapon at one's disposal.[9]

What seems required, in such cases, is a place for the dangerous innovator to hide, perhaps in an obscure provincial institution, where he can get on with attempting to consolidate his work. There he may find a new and rival school, hurling anathemas at the older school through the means of publication at his disposal. I have already discussed this phenomenon in connection with personal style. In the long run, the effective decision on the relative merits

[8] A minor classic in the history of scientific parodies is S. C. H. Windler, 'Uber das Substitutionsgesetz und die Theorie des Typen' (letter to the editor), *Liebigs Annalen der Chimie*, 33 (1840), 308–10. In this, the author reported reacting manganese acetate with chlorine, and by successive 'substitutions' achieving a substance with the formula $Cl_2\ Cl_2 + Cl_8\ Cl_6\ Cl_6 + aq$. The paper was not a trick on the editor, but an unveiled attack by Liebig on the work of Dumas. See J. P. Phillips, 'Liebig and Kolbe, Critical Editors', *Chymia*, 11 (1966), 89–97.

[9] An extreme case of invective against an innovator is the review of van't Hoff, *La Chimie dans l'espace*, by Kolbe: 'This nature-philosophy, which had been put aside by exact science is, at present, being dragged out by pseudoscientists from the junk-room which harbors such failings of the human mind, and is dressed up in modern fashion and rouged freshly like a whore whom one tries to smuggle into good society where she does not belong.' Kolbe's attack was quoted by van't Hoff in his Inaugural Lecture shortly afterwards; see J. H. van't Hoff, *Imagination in Science* (Springer-Verlag, 1967). Van't Hoff went on to achieve many honours, including a Nobel prize.

of the rival schools and their strategies will be taken by another generation, for whom the personal conflicts, and the passionate commitment to a particular strategy, are ancient and dead history. For the quickest arrival at this healthy state, it is best for the field to have a loose structure, so that not every member must align himself and his students, with one side or the other. Otherwise, orthodoxies may harden to such an extent that the rival schools become completely self-enclosed, immune from all outside influences of criticism or rejuvenation and eventually splitting the field into separate, but equally stagnant sections. Alternatively, if a new school does achieve a victory, then the style of the days of bitter struggle will tend to be carried over in its success; and all those who are not enthusiastically with it will be counted against it, and subject to the same administrative procedures.

Thus in several ways, the management of novelty in a field is better accomplished if the structure of the field is 'polycentric', where there are several focuses of prestige and power, not all of them too sharply defined. But such a structure is not achieved automatically; its existence depends on the social and intellectual context of scientific and scholarly work. The history of science has provided us, as yet, with little detailed evidence on this problem. But there is a striking contrast between Britain and Germany on the one hand, and in France on the other, in the progress of the physical sciences in the nineteenth century. The network of German universities, each trying to build its prestige for scholarship, provided a stimulating environment for men of great originality in all fields. In Britain, the institutional structure for science and scholarship was weak indeed, but the tradition of independent amateur scholarship supplemented by a sprinkling of jobs, provided the base for excellent work through the century. But in France, as we have seen, the centralized scientific establishment forced outsiders to fight every inch of the way. The first generation, including Fourier and Fresnel, were ultimately successful; but it was during the Restoration period that Sadi Carnot and Evariste Galois worked and died in enforced obscurity. Even later, scientists who could not secure the patronage of a distinguished man could find themselves in an uncomfortable and invidious position; such was the lot of the chemists Gerhardt and Laurent.[10]

[10] For a brief account on Gerhardt, see A. J. Ihde, *The Development of Modern Chemistry* (Harper & Row, 1964), p. 204. For Laurent, ibid, p. 176. There is quoted a complaint by

Thus, it seems that the dangers that arise from the mismanagement of novelty are not so much that bad old theories will be retained and new ones suppressed; for in the long run, provided that there are some independent centres even in remote nations, good work will be recognized. Rather it is that wherever the protection of the established property of a school becomes too rigid, then in the area where that school holds political control the stagnation of the field will be felt by new and potential recruits, and the really gifted young men who are necessary for the rejuvenation of the field will simply not be there.

Postscript: Modern Times

Bitter resistance to novelty, like the bitter disputes over priority, seems to have declined in the recent period. This change is doubtless due in part to the changing common sense of science, with an awareness of the rapid obsolescence of theories and of whole fields. Also, the change in the location of the scientist's intellectual property, from the monument of past achievements to the contacts for future work, enables past work to be jettisoned with far less cost. In addition, there might be a cultural influence, marked by the change from the German master-scholar and his group of disciples to the more mobile and egalitarian society of the American model. Because of these influences, the practical problem of the management of novelty is less severe, and the means to its resolution are more effective. The weaknesses of the present situation are also easy to perceive. Not all that is new is good, and there are dangers in the dominance of temporary fashions on scientific inquiry. Problems which require a long time for their maturing will be less likely to be investigated, since their results may be considered obsolete by the time they are achieved. The prestige given to novelty of any sort weights the scales in favour of the younger men of brilliance, and against the older men of wisdom. In some fields, it may be quite in order for a scientist to become an elder statesman at thirty-five; but in others, where years of experience are necessary before worthwhile new

Laurent: 'I have not been able to dismiss an emotion of indignation in seeing certain chemists first call my theory absurd, then much later when they have seen that the facts are in agreement with my theory better than they are with all the others, pretend that I have taken some ideas of M. Dumas. If it fails, I shall be the author, if it succeeds another will have proposed it. M. Dumas has done much for the science; his part is sufficiently great that one should not snatch from me the fruit of my labors and present the offering to him.'

results can be achieved, the accent on youth can lead to a loss of the inherited successful craft experience of the field. These considerations are very subtle, but they are relevant to the judgements by which the leaders of fields, and of science as a whole, distribute the rewards which control quality and set the direction of research.

QUALITY CONTROL IN SCIENCE

THE social activity of science has another feature which makes it nearly unique among all sorts of work: the social task of the maintenance of the quality of the products seems to be accomplished with so little difficulty that the problem of quality control has received no more than a passing mention in any systematic discussion of science.[1] When one contrasts this to routine industrial production, where quality control requires an elaborated organization, or even to the learned professions, where tribunals stand ready to judge and punish bad work of certain sorts, the situation in science seems nearly Utopian. This happy state of affairs might be explained by the moral superiority of scientists, a result either of selection and indoctrination or perhaps even of the essential nature of the work. Indeed, the idealistic propaganda for science, characteristic of the academic period, did not dispel that impression. But in the present period, we know that scientists are, after all, human; and we are also aware of the phenomenon of 'pointless publication', better described as shoddy science. Thus the effectiveness of the inherited social mechanisms of quality control for the conditions of the present and near future is open to question. Our task is to examine these mechanisms, and then to study the conditions under which they can operate well.

Assessments of quality are an everyday part of the scientist's work: any material that he is considering using in any way, must be judged for its quality in the relevant aspects.[2] There are, in addition,

---

[1] The only historical study known to me which analyses scientific results by the different grades of quality, and discusses the growth-patterns of the different sorts of work is K. O. May, 'Growth and Quality of Scientific Literature', *Isis*, 59 (1968), 363–71. His sample of materials was some 1700 papers on determinants, from the late eighteenth century to 1920.

[2] Comprehensive reviews of literature in a field must provide clues to quality; otherwise the reader is faced with a mass of information which is useless because it is undifferentiated. Among the subtle devices used is the heading 'See also' before the mass of

several contexts in which assessments of quality are made on behalf of the whole community; these are the ones involved in quality control as such, and we shall concentrate our attention on them. At the top of the scale of quality, there are the results which bear a claim for consideration for rewards: promotion, advancement, or special marks of distinction for the man who did the work. And at the lower end of the scale, there is a cut-off point for quality, below which a result is not given the formal authentication as property by being published in a recognized journal. The criteria involved, and the individuals affected, will be very different in the two extreme cases; but we shall see that they are not totally independent of each other, either in their workings or in their effects.

The Assessment of Quality, from Best to Worst

The criteria on which assessments of quality are based can be reduced to two that we have already discussed: adequacy and value. Although these criteria can be described in general terms, their application to any particular case involves the making of a number of subtle, indeed tacit judgements, which depend on an intimate craft knowledge of the work under review. It is impossible to design a simple set of routine tests, by which one could assign some numerical marks to a solved scientific problem, and then grade it on a linear scale. Nor indeed would it be feasible to erect a formal system of categories of quality, and train up a corps of expert assessors (such as exist in many other spheres of activity) to operate in their framework. For the techniques are so subtle, the appropriate criteria of adequacy and value so specialized, and the materials so rapidly changing, that any fixed and formalized categories would be a blunt and obsolete instrument as soon as it were brought into use.

The result of this special characteristic of science is that if there are to be truly expert assessments of quality of work, they must be made by a section of those who are actually engaged upon that work. In itself, this provides no guarantee whatever that proper assessments of quality will be made, or that they will be used appropriately for the maintenance of quality. Science is not unique in this respect; in the learned professions, the assessment of quality of work, both the best and the worst, is in the hands of the community of qualified practitioners. We may note, however, that the professions do have

citations too insignificant to be discussed in the survey; the instruction means exactly the opposite, of course. (I am indebted to Professor P. Ent of Utrecht for this point.)

a formal apparatus of judgement and penalties for the worst work, while in science these are lacking.

To give a rough idea of the sorts of assessments that are made, I can distinguish several classes of scientific work, and the standards by which they are defined. The best of all scientific work is that which survives, through all the many testings and transformations, to become genuine scientific knowledge. But this can be known only in retrospect; and so a sober assessment of any new result will not place it in the 'immortal' class. However, one can reasonably predict that a given result is likely to yield 'enduring facts', which will survive the demise of the original problem; if that problem was a deep and difficult one, and the result is capable of development and extension, then it is entitled to be considered as first class. Now, even a temporary fact is no mean achievement, for its existence shows that the solution to the original problem had depth rather greater than the explicit statement of the conclusion would guarantee; work which achieves this success is certainly good. Even this grade is not the minimum quality necessary for a result to be worthwhile in the advancement of its field. So long as it meets the appropriate standards of adequacy, and shows its value by being put to use by others, however briefly, in subsequent work, the research which produced it can be considered competent.

These classes are defined in very general terms, and are based on predictions of the future career of a result. In any field, there will be symptoms whereby a skilled man can form a judgement of the prospects of a particular result. These will be enormously various; elegance of argument may be basic in one discipline but irrelevant in another; novelty of the problem or surprise of the solution may be crucial or insignificant; concern for fine detail may be the mark of the master-craftsman, or of the bore. Where masses of reliable information are required for the investigation of new problems, then work which does very little more than provide such information may be quite competent. But in more theoretical or speculative fields, where the boldness of the problem and the depth or elegance of the solution constitute the value of the work, then routine problem-solving along a narrow path, with standard techniques, may have no value other than the political one for the scientist who publishes the work. Thus quality depends not merely on the character of the work done but quite strongly on the field in which it is placed. Because of the subtlety and particularity of the criteria of quality, competent judges must be in or very close to the field, and must also be capable of producing work of a similar degree of quality to that which they are assessing. For this reason, the assessment of work of the highest quality, and the social tasks depending on such assessments, must be reserved to the aristocrats of a discipline, rather than being assigned to a democratic procedure or to the laymen who administer the funds invested in research.

The assessment of quality at the lower end of the scale does not require such subtle and speculative judgements. But it does involve questions of ethics, for a published result which is not even competent, has a false claim to authenticity; and the referee who passed it for publication was either lax or incompetent. A result may be weak in a number of ways. In experimental fields it is possible for reasonably good data to be incapable of being processed into evidence which functions adequately in an argument. The desired conclusion may be a thesis more general than the evidence available; or it may be that the argument is a retrospective, *ad hoc* attempt to explain the data that has been produced. In theoretical fields, the argument may be good in parts, but those parts may lack the necessary coherence for the conclusion to follow from them. A weak result may be capable of improvement to the point of being competent or better; and in such cases the referee can offer advice. And the borderline between weak results and those which are just competent is necessarily indistinct.

At the lower end of the class, the weak results merge into the vacuous ones: those for which nothing good can be said. In a matured field, the production of vacuous results is reserved to those scientific workers whose self-respect, if they have any, derives from other areas of their experience than the advancement of their field. For the body of craft knowledge of methods of a matured field has, among its functions, enabling a conscientious worker to proceed on limited tasks without encountering fatal pitfalls, and at least producing data and information with some content. However, this does require some patience, diligence and judgement. If one seizes the first plausible set of experimental data to appear, confects an argument into which they can somehow fit, and then writes it up for publication, one can be fairly sure of achieving vacuity. In a theoretical field, one can start with some objects of inquiry which have some distant relation to experience, concentrate attention on their mutual relations at the expense of those aspects with real content, and paste together an argument, verbal or mathematical, from which are drawn invalid conclusions about non-existent things.

Only the expert eye can distinguish between a vacuous research report and one with some real content. By the use, or abuse, of craft skills, it is possible for slipshod experimental work to be dressed up in a conventional description, for unsound data to be converted into empty information by dubious mathematical techniques, and for an incoherent argument to masquerade as close reasoning.[3] Moreover, this approach to research and publication not only saves time and energy but also pays dividends in the number of publication-points achieved. For the first vacuous research report can spawn another, as soon as the scientist has found an error in his data; and so on. In theoretical fields it is even easier, since every existing cobweb of argument invites immediate further articulation. Indeed, the question for an analysis of quality control and ethics in science, is not so much how weak and vacuous work can possibly exist, but rather how such weeds are kept from choking the very artificial growth of genuine scientific inquiry.[4] For the present, we will restrict our discussion to matured fields, where the social task of quality control has at least the technical means for its

[3] It is not common for a scientist to identify productions of vacuous research publicly; only a master of a field could, and should, do so. For an example of this type of critical analysis, see F. Yates, 'Theory and Practice in Statistics', *Journal of the Royal Statistical Society*, Series A *(General)*, 131 (1968), 463–74. This was his Presidential Address to the Royal Statistical Society. According to him, one main cause of vacuous work is the tendency of university teachers of statistics to publish 'pure' research in a subject whose origins and strength lie in its association with practice. In fairness, he also cited vacuous 'applied' research; the most famous being that of using aerial photographs for assessing the damage to German production caused by the mass air raids on the towns during World War II. Both the bombers and their statisticians concentrated on the town centres, in apparent ignorance of the fact that German industrial cities, unlike the English, have a residential centre and industrial suburbs! (I am indebted to Mr. A. B. Royse for this reference.)

[4] Here we can see how my analysis relates to that of Popper, on the question of the role of criticism in science. He used the 'critical attitude', a willingness and commitment to submit one's theories to test, as a criterion of demarcation between the practitioners of 'science' and of 'pseudo-science', and hence between those categories themselves. (See 'Science; Conjectures and Refutations', ch. 1 of *Conjectures and Refutations* (Routledge, London, 1963).) In the years since his first insight, the question of the personal motives of scientists has been seen as rather more complex, and the demarcation correspondingly less neat. But we can agree with his basic insight, and re-state the dichotomy on a practical, rather than an epistemological, basis. Then we can say that the absence of a critical attitude among the members of a scientific community is a *cause* of a degeneration into vacuity and corruption, and also a *sign* of its domination now or in the near future.

accomplishment.[5] In immature fields, the situation is otherwise; and the tragic problems encountered there require a separate discussion.

Comparative Examples of Quality Control

Although quality control in science is operated with a minimum of formality, the nature of the task is the same as in other sorts of socially organized work. By analysing the process, particularly at the lower end of the scale, we will discern similarities and differences between science and the more common situations, and so be in a position to analyse the special problems encountered in science. The function of quality control is to ensure that the users of the products can rely on their being of a certain acceptable standard. The task is divided into several phases: establishment of criteria of quality and setting standards in their terms; testing the set of products for an assessment of their satisfying the standards; and enforcing the regular adherence to the standards by a system of penalties and rewards on the operatives. In ordinary industrial production, any or all of these phases may become highly elaborated. The first phase involves decisions at the highest level, since more rigorous standards lead to higher direct and indirect costs. The task of testing has developed into an applied science in its own right, and its practice can involve specialized equipment and sophisticated statistical tools. And the enforcement phase involves politics (in the wider sense), since it is assimilated into the general system of government of a factory.

In general, the system of quality control in industrial production is run along the same lines as those of decision and command: rigidly hierarchical, with the unskilled workers at the bottom of the pyramid usually being treated as simple stimulus-response machines. Their products are tested on a routine basis to simple standards by

[5] The classification of degrees of quality given here agrees well with the criteria implicit in the 'Kendrew Report', *Report of the Working Group on Molecular Biology*, Cmnd. 3675 (H.M.S.O., London, 1968). Giving their impression that quality is very uneven, the authors say: 'Some work in the field is of the highest class but there is much that is concerned with problems irrelevant to the real growing points in the field; work which may not be downright bad but which contributes little to the main advance of the subject —which, in other words, is at the best dull and at the worst trivial. Our examination of the literature leads us to think that this is in marked contrast to the position in the U.S.A., where as well as a good share of the work of the highest class there is a great deal of the sound and competent, perhaps uninspired but at least relevant, work which is essential to sustained development' (para. 9, p. 4). (I am indebted to Dr. J. Wootton for this reference.)

another set of operatives. Rewards are for large quantity rather than for excess of quality, and the penalties for inferior quality are assimilated to those for insufficient quantity, and are generally of the crudest sort. Thus, in this situation, the technical problems of quality control are of a purely administrative sort. Where skilled work is being done, the system of quality control, although the same in its essentials, must be more subtle and refined in its operation. For it is impossible to make an adequate assessment of the quality of a craftsman's product, by simple routine tests. In such work, the worker is constantly making intuitive, perhaps tacit judgements of the quality of each operation; if these were not necessary, the manipulations could be performed equally well by unskilled hands or machines. The product of the work is thus conditioned by a multitude of judgements of its quality, relating to every phase of its making. And from the mutually reinforcing touches of excellence comes the quality of craftsman's work, which reveals itself only in its performance, and cannot be categorized in a small set of measurable attributes. Hence the control of quality of craftsman's work must to some extent be entrusted to the craftsmen themselves. This can be done, in spite of the divergence between the private purposes of the craftsmen and the goals set by the hierarchical organization employing them; the conditions under which it is possible will be discussed in connection with ethics.

A situation more like that in science can be found in the learned professions. The practitioners there are not placed in a comprehensive formal hierarchy, and the profession as a group is almost entirely self-policing. The establishment of criteria of quality, the setting of standards, and the testing of work, are all accomplished informally under the guidance of the recognized leaders, with work of high quality as the main focus of attention. The recognition and rewarding of high-quality work has important functions in the system of indoctrination of recruits, in the maintenance of morale, and in relations with the lay public. Those whose work is merely competent or even weak can labour in obscurity, merely earning a living. But a genuine profession will have formal institutions for trying cases of very bad work, and imposing punishments, even to the point of expulsion. These are necessary because of the special relations of professionals to their clients, and perhaps also because of the dangers of scandal caused by seriously aggrieved clients. Even here, it is not so much simple incompetence which attracts such penalties, but rather the abuse of the client's confidence by actions serving the private purposes of the professional. Accordingly, such professionals as solicitors and dentists are more at risk of punishment, than, say, surgeons or engineers.

The Levels of Quality Control in Science

In science, by contrast, the situation seems to be one of nearly perfect happy anarchy. There is no formal hierarchy of decision and control; and the community of science has no institutions of punishing or expelling a wayward member. A university may revoke a man's title to its degree, but in general the subject-societies (usually misdescribed as professional associations) have no such penalties at their disposal. Indeed, the system of quality control in science has only one point where formal procedures operate: the assessment of a research report by the referee of a recognized journal, for authentication as a published paper. The penalties involved in this procedure is a private affair between the referee, the editor, and the author; and no punitive action is taken against a scientist who submits one, or even many bad papers. But the penalty is none the less real for being discreet. Time and resources have been invested in the investigation of a problem, and with the rejection of the paper, no property has been created. Work has been wasted, and there will be an embarrassing gap in the time-sequence of the scientist's publications.

This is how the matter works in principle; and in situations where the leaders of a field can and do exercise tight control over the authentication of results this discreet quality control at this minimum level is quite effective. But as we have seen, in the present period the mechanism does not work perfectly; and so it is necessary to examine the ways which it is maintained to the extent that it is. The referee for a journal, and the editor acting on his advice are, as we have seen, the effective agents in the system of quality control at the lower level. The referee's task is a particularly subtle sort of craftsmanship. He must not only understand the technical material under review; he must also be able to reconstruct in his own mind something of a history of the investigation, to detect errors of commission and omission. As a conclusion of his inquiry, he must decide on a recommendation for publishing, and also offer suggestions for improvement. His criteria of quality will depend not only on the field in which the problem is located, but also on the particular style of the journal and interests and competence of its audience. Although the referee may be given general guidance by the editor on the principles of publication of that particular journal, he will still need to interpret these for the special case of each research report under review.

Because the referees cannot be segregated off from the body of working scientists in their field, they cannot be made accountable to a separate hierarchy of control. A working scientist who is acting as referee will find himself in the invidious position of being prepared, on occasion, to deny authentication to the work of colleagues and friends. In order that referees can accomplish their delicate task properly, their personal skill at this craft must be supplemented in several ways. First, they must have a commitment to the maintenance of the standards of quality in that particular journal, or in the field in general, so that they can see the damage done to the immediate interests of their unsuccessful colleagues in the light of a higher good. Also, they must be protected by a strict etiquette which makes their relation to each author as impersonal as possible, so as to minimize the risk that disagreements over assessment will become personal disputes. And, finally, they, and their editors, must themselves be subject to the operations of some mechanism of quality control.

This higher level of quality control must operate directly on the journals, and through them on their editors and then on the referees. Without this, there would be no barrier against the spreading of a decline of standards among referees, consequent on a loss of morale or commitment among some. The outcome of such a development would be a Gresham's Law for scientists' intellectual property, applying directly to work which aspires to be good or competent: if property of equal political value, embodied in formally authenticated research reports, can be purchased at a lower personal cost in time and work by submission to a lax journal, then only the saintly minority will penalize themselves and their families by continuing to work to unnecessarily high standards. The prevention of such a state of affairs cannot be accomplished by formal procedures: the community of science has no institutions for withdrawing 'recognition' from a journal which publishes inferior work. Instead, there is a very informal ranking of journals in a field by their prestige; and although such a list will never be published, it can be an effective means of control over editors, referees, and authors as well.

The mechanism here is competition for the limited supply of rewards available to individual scientists; and it can operate because the assessment of the quality of a research report is partly based, in practice if not in principle, on the journal in which it appears. A paper appearing in a journal of high prestige has a much better chance of attracting attention than one in a journal of low prestige. The 'better' journal will be more widely distributed, it will be scanned more carefully by active scientists, and a paper there will be given a greater initial credit for quality. Hence it is in the personal interest of the author to publish his papers in the journals of the highest prestige which he can reach. To give his work to a weaker journal does not merely cost him the risk of under-valuation of the result; it also associates him with men who can do no better. The editor also receives rewards for his identification with a journal of high prestige; and suffers the penalty of relegation to obscurity if his journal is weak. It is then in his interest to establish a good reputation for his journal so as to attract papers which will maintain and enhance that reputation; conversely, if he publishes papers with which a good man would not want to be associated, he runs the risk of losing his high-quality contributions.[6]

All this is a matter of degree, of course; in a large field, there may be several 'leagues' of scientists and journals, distributed naturally over ranges of quality, with the hope of promotion and the fear of relegation between leagues functioning as the immediate rewards and penalties in this system of quality control. Since the aspirants for promotion will model themselves on the style of their superiors, the whole system is ultimately governed by the distribution of rewards for work of the highest prestige. However, not all journals, or scientists, are in this system. There is the mass of those who, through geographical isolation, lack of resources, or lack of talent or commitment, cannot hope to produce intellectual property authenticated within the dominant system. They then settle down, either to working within their autonomous but isolated local system

[6] Evidence of an imperfection in this informal system of quality control on journals arises from a report on recent decisions of the American Geophysical Union concerning publications. A new policy is described as follows: 'A review of the quality as well as of the length of papers by a group which is separate from the editors. This amounts to post-factum refereeing of the papers and should provide guidance to the editors, to the publications planning committee and to the A.G.U. Council.' Letter to *Science*, 166 (1969), 43–4, from S. F. Singer. It would appear that a committee accountable to the Council and not to the editors will monitor the published journals and make assessments of quality for use in future guidance of editors.

of prestige, or to production of second-rate property in the titles of weak or vacuous reports in the many obscure journals that coexist with the few significant ones. The first of these groups, of 'colonial science', have been with us for a long time; and although it frequently involves tragedy for its members, in exile from 'metropolitan' science, it does not constitute any threat.[7] But the second group, of 'shoddy science', do represent a danger; for if their characteristic property is not kept firmly in second class status, the whole system of quality control for genuine science could easily collapse.

Leadership and Prestige

Thus the question, 'Who controls the controllers?', has brought us to the leaders of each field, who control the rewards of prestige (and the penalties of obscurity) which operate on journals as well as on individuals. At these higher levels, prestige becomes an important component of the mechanisms of control. It functions as the 'cash value' of the scientist's intellectual property as assessed by his colleagues; on the basis of his prestige, he can command personal benefits and the resources necessary for the continuation and expansion of his chosen work. Also, at this higher level, the criteria of assessment of quality are significantly different from those lower down. Adequacy can be presupposed, so the dominant component here is that of value; and since this depends on a judgement of the future, it is intimately related to the task of directing current research. As a result, quality control at the highest levels is inseparable from the general tasks of leadership of the field.

The informal hierarchy of control does not end at this point; and this is just as well, for otherwise it would be impossible to maintain any unity or coherence in the endeavour of science. Since there is no class of active scientists who operate independently of all fields, the operation of quality control on whole fields and on their leaders must be the task of a community of leading scientists of which they themselves are members. The relations between fields of science are of many sorts, and in constant change: there may be areas of overlap of problems, they may be associated through the provision

[7] For an analysis of 'colonial science', which was the condition of American science until well into the twentieth century, see Donald Fleming, 'Science in Australia, Canada and the United States: Some Comparative Remarks', *Proc. 10th International Congress of History of Science*, 1962 (1964), pp. 172–96; and also James R. Killian, Jr., 'Towards a Research-Reliant Society', in *Science as a Cultural Force*, ed. H. Woolf (Baltimore, 1964).

of tools, or one field may be the parent of another. Whatever their internal relations, they are bound together, in groupings of various sorts, by the competition for limited resources and the need to allocate them wisely. A special class of leaders, of exceptional qualities of statemanship, is required for the proper accomplishment of this latter task. For the judgements involved are concerned with quality and direction of research in areas where the person cannot have any personal craft knowledge. The men involved must have the ability to extrapolate widely from their own scientific experience in order to assess quality and promise from the evidence provided to them; and they must also be able to put in proper perspective the claims for fields close to their own research interests and those remote from them.[8] Their decisions constitute a mechanism of quality control on whole fields; for a field which does not measure up to its competitors will be deprived of resources and thereby be sent into decline.

This crucial set of tasks of ultimate decision and control has traditionally been accomplished within a particular institutional framework; in those nations where the nineteenth-century German system was followed, the universities had the central role. For, although a man's working colleagues might be scattered all over the world, his career would be within the university system; and his advancement in that career would depend in the last resort upon the judgements of colleagues in a particular university. In the universities, too, there is competition for limited resources; but there the decisions are taken by scholars from such a wide variety of disciplines that in respect of the technicalities of any one of them they are

[8] Thus, looking again at the 'Kendrew Report', we see that the criterion of quality implied in the phrase 'concerned with problems irrelevant to the real growing points of the field' is capable of more than one interpretation in practice. For although the leaders of a field may be very sensitive to new approaches coming from young men in obscure centres they may be quite convinced in their collective wisdom that their own investigations, now maturing after some years of work, constitute the true 'growing points'. No set of formal rules will distinguish in practice between the many subtle gradations between these two positions; hence there is no substitute for wise leadership.

In this context, Szilard's description of the effects of even the best-intentioned system of centralized direction of research, is worth recalling. '. . . the scientific workers in need of funds would concentrate on problems which were considered promising and were pretty certain to lead to publishable results. For a few years there might be a great increase in scientific output; but by going after the obvious, pretty soon science would dry out. Science would become something like a parlour game. Some things would be considered interesting, others not. There would be fashions. Those who followed the fashion would get grants. Those who wouldn't would not, and pretty soon they would learn to follow the fashion too.' Quoted from 'The Mark Gable Foundation', in *The Voice of the Dolphins*, p. 101.

but educated laymen. Hence, within the universities, the assessment of a man and of his work, reached either through informal consensus or by formal deliberation, will have a very different basis from that within a field of science. On the one hand, colleagues will be able to judge him from their personal experience of his work outside his speciality; and this certainly gives an indication of the capacity for leadership. But for an assessment of the man and of his field (the two are closely bound together in the case of a leading scientist), his lay colleagues must rely on evidence of the prestige he commands in the community of those competent to judge.

Thus, for advancement within the universities, prestige has a very high value indeed; it is invoked as evidence for a claim to rewards, in a situation where it cannot be rigorously tested. Also, the assignment of prestige to a man depends on more than his work within a field and on the field itself. For universities also have a prestige ranking; and to be (or even to have been) at a university of high prestige, automatically confers prestige upon the man. The two systems of prestige thus interpenetrate and automatically reinforce each other; and at the top of the scale are those scientists who, ideally, are at the universities of greatest prestige because of their enjoyment of the greatest prestige among the expert and lay colleagues they have encountered during their career. Once ensconced in the greatest universities, they will, in principle, continue to work as scientists of distinction and also function as statesmen of science. By their contacts with leading scholars of all fields, spiced by the element of competition which exists even at the rarified heights, they will master the very subtle skills, and develop the very enlightened attitudes, necessary for participation in the government of science at the very highest level.

This was the traditional institutional framework for the ultimate exercise of quality control and direction in science. Appropriately, the task of controlling the controllers reaches its final stage in a very informal community, operating through a mixture of co-operation and competition to exercise control on each special area in the interests of a very general common good. The great age of this system was that of Germany, in the middle of the nineteenth century. It may be significant that those universities were not merely research institutions, but were devoted to teaching as well. So while a man might achieve a fair position by the exclusive pursuit of his research, he could not join the company of the truly great unless he had the breadth of vision and culture to be an influential teacher as well. In this way, the fragmentation of the community of scholarship was prevented, or at least delayed.[9] The functioning of this system was also improved, and perhaps even made possible, by special features of its social and cultural setting. The heroic age of German science came after the period of great philosophical endeavour and educational reform. To have a genuine university of high prestige was considered a valuable asset by the citizens of many towns; and there was available, in secondary schools and universities, teaching which was both rigorous in intellectual discipline, and philosophical in approach. For a scientist raised in that environment the highest achievement was not the making of new technical discoveries, but participation in '*Wissenschaft*': the dedicated quest for human knowledge, wherever it could be found.[10]

Industrialization of Science and Quality Control

Some elements of the classic university system (or its analogue in the centralized national academies) have been imported into all those nations where science has flourished since the nineteenth century. But with the industrialization of science, certain changes have occurred which weaken the operation of the traditional mechanism of quality control and direction at the highest level. The concentration of the scientist's audience towards those who advise the agencies has not merely caused in him a tendency to neglect the diffuse consensus of expert colleagues; it has also made him less dependent on the judgement of his lay colleagues in the university system. Indeed, in America, where the process has gone the furthest, the term 'university' has been replaced by a new one, 'multiversity'. In these, independent research contractors and entrepreneurs grace the

[9] On the relation of the ideals of teaching and research in the German universities, see H. von Helmholtz, 'On Academic Freedom in German Universities', in *Popular Lectures on Scientific Subjects* (London, 1893), ii, 237–65; especially pp. 252–62. In this essay, Helmholtz stressed the converse aspect of the situation of university scientists: that university teachers were required to be competent scientists. For the German system was the exception among universities in the nineteenth century.

[10] For a description and explanation of the greatness of the German universities in terms of their cultural and social setting, see J. Ben-David and A. Zloczower, 'Universities and Academic Systems in Modern Societies', *European Journal of Sociology*, 3 (1962), 45–84; reprinted in part in N. Kaplan, *Science and Society*, pp. 62–85. In this as in other articles, the authors seem to have a concern to 'reduce' the greatness to its social causes, and they tend to explain the decline by its inherent features; they do not give sufficient emphasis to the variable rate of growth through the century, declining towards the end.

institution with a share of their prestige, in return for a well-paid sinecure and the provision of material facilities.[11] Under these conditions, the system of assignment of prestige, always a delicate and unstable thing at the best of times, cannot operate to any worthwhile effect. The subtle and informal assessments of quality, across fields and disciplines, and the use of these assessments in distributing rewards within the university system, are impossible. Once established as a leading Big Scientist, a man is immune from control by anyone except his intimate colleagues and friends, and those investing agencies whom they jointly advise. He can create property by the mere flourish of a pen. He may choose to take time off from organizing to produce some serious research; but otherwise he can achieve a satisfactory string of publication-points (necessary for the formalities of contract renewals) by placing hasty, shoddy papers in journals controlled by himself and his friends.[12] If such men come to dominate the government of a field, the effects, especially in times of financial stringency, are easily seen to be disastrous.

The extent to which such developments have already damaged the mechanisms of quality control in science is impossible to estimate. The evidence on which such an estimate would need to be based would have to be worked up carefully from very private and confidential testimony. For our present purposes, it is sufficient to appreciate that some such things have occurred, and that they are the effect of changed technical and social conditions on a very delicate task. In the last resort, there are no direct means whereby anyone outside the world of science can exercise quality control on science. The products of the craft work of scientists are intelligible, and valuable, only to other scientists. And although they relate to the external world, their value as well as their meaning is governed by the judgements of men, those particular men who enjoy this esoteric activity. If the government of this work were accomplished through

[11] A disturbing description of this situation, organized around the problem of student unrest rather than of the government of science, has been given by Paul Doty, 'The Academic Condition in the United States', *Nature*, 224 (1969), 1063–8. The author is a Professor of Chemistry at Harvard; what he says of that aristocratic institution can be expected to hold more strongly elsewhere.

[12] Evidence that the accumulation of the pseudo-property of publication points is not restricted to those that can do no better, comes from a letter by R. W. Whalen in *Science*, 167 (1970), 1318. He reports an author who first published a paper in *Science* (itself a mark of quality), but who proceeded to use one-half of the data, and then one-quarter of the data, for two later publications in speciality journals, with no cross-referencing between articles.

formal institutions, then its response to changing conditions would be delayed, and the work might have time to adapt itself gradually while maintaining its excellence. But the nature of the work requires a government for direction and quality control which is almost entirely informal, accomplished by a series of craft skills which become ever more refined, demanding and delicate; and the work itself is very sensitive to the quality of its government. The problem of quality control in science is thus at the centre of the social problems of the industrialized science of the present period.[13] If it fails to resolve this problem, and does not develop new techniques for restricting prestige and rewards to those who deserve them, then the immediate consequences for morale and recruitment will be serious; and those for the survival of science itself, grave.

[13] The problem of quality became a practical concern for American scientists when the Johnson administration supported the policy of establishing new 'centres of excellence' in science away from the areas which had come to monopolize prestige and funds (the North-East and California). It was appropriate for a representative of the most aristocratic discipline, mathematics, to look at this problem of 'science policy' from the point of view of the maintenance and control of quality; he showed that 'excellence' cannot be expanded merely by a shower of dollars. See S. Maclane, 'Leadership and Quality in Science', in *Basic Research and National Goals* (A Report to the Committee on Science and Astronautics, U.S. House of Representatives, by the National Academy of Sciences) (Washington, Government, Printing Office, 1965). The paper was reprinted in part as 'Can we Buy Quality in Science?', *Bulletin of the Atomic Scientists*, 21 (November 1965), 7–11. British scientists have also become concerned about the problem of quality. In addition to the 'Kendrew Report' for molecular biology, there was an article by the distinguished physicist A. B. Pippard, in which quality control had a brief but significant mention. In 'Science as a Constituent of University Education', *Nature*, 219 (1969), 1307–8, Pippard made various proposals for rationalizing research and teaching in physics; but confessed that 'the problems involved in encouraging centres of excellence and, perhaps more important, discouraging mediocre research are too complex to discuss here.'

II

ETHICS IN SCIENTIFIC ACTIVITY

ON this very important topic, we need to define our problem very carefully; otherwise we are sure to encounter some of the many pitfalls that beset any discussion of the relations between what is and what ought to be. We can start with the phenomenon of the sudden shift in the public image of the morality of science. Through the academic period, science was portrayed, and widely (although not universally) accepted as an activity good and ennobling in itself, and productive of enormous although diffuse benefits for mankind in general. In recent years, the image has been tarnished; not only is science blamed for the new horrors of war and the threat to our environment, but there is a new genre of literature exposing the human foibles of scientists.[1] It is all too easy to conclude that science as a whole has declined from a state of pristine purity; but such popular and popularizing materials are extremely unreliable evidence for the state of affairs which they are taken as describing. If nothing else, they can deal with only a very small segment of a large and variegated activity; and whether this is truly representative can be determined only by an extensive study in depth. This too has its own hazards, for any relevant historical data is inherently unreliable; and without an appreciation of the special character of the problem, the most diligent research could easily yield vacuous results. Thus, a poll of elderly scientists would be quite likely to produce a majority view that in the old days scientists were more honest and self-sacrificing, and also that summers had more sun and winters more snow.

[1] The frank display of human motives in James D. Watson, *The Double Helix: a Personal Account of the Discovery of DNA* (Atheneum, New York; Weidenfeld and Nicolson, London, 1968), was a source of revelation for many reviewers. A very different reception was given to D. S. Greenberg, *The Politics of Pure Science*: the leaders of the American scientific community were not amused by the story of the human reactions of scientists when faced with the temptations of money and power; nor was the book's message clearly perceived by reviewers in the general press.

Our present task, then, is not to decide whether the two dominant images of science are correct portrayals of the past and the present; but to establish the categories in whose terms problems relating to that question can be properly investigated. Although such a study can be only a preliminary to a substantive enquiry, it is not merely an abstract exercise, and some general conclusions about the real world of scientific activity can be derived from it. This present discussion will be organized in terms of two problems. First, in what ways, if any, does the special character of scientific inquiry necessitate the adoption of a superior ethical code by scientists, for good work to be done? Second, to the extent that this is the case, then by what social mechanisms and individual actions and commitments is this ethical code enforced and maintained? In both of these questions, we will use materials developed in the preceding chapters, and amplify certain points mentioned in passing there. Also, we will assume that it is possible for good work to be distinguished from bad. This assumption is justified by the experience of matured fields of inquiry; in this discussion we can ignore the problems of the identification of good work, and restrict ourselves to using the general principle that doing good work is more costly to the individual than doing bad work.

The 'Final Causes' of a Task

We have already mentioned that one function of the various social tasks in science is to maintain a harmony between the collective goals of the endeavour, and the private purposes of individuals. For our present discussion, this distinction between the 'objective' and 'subjective' final causes of an activity must be enriched and made more precise. For this, we will develop a set of terms, using words taken over from common speech and nearly synonymous there, but which we will henceforth use in the restricted senses to be defined. Our unit of analysis is a task on which someone is engaged; and for the sake of clarity we will take as our example a task where the various causes are very distinct.[2]

Suppose we find someone on a routine, assembly-line job, say fitting together a set of components into a sub-assembly. If we wish

[2] The literature of the sociology of organizations distinguishes among the various 'goals' involved in such work, but the present classification is not to be found there. For a recent study see E. Gross, 'The Definition of Organisational Goals', *British Journal of Sociology*, 20 (1969), 277-94.

to know the 'aim' of his task, its final cause, we find it necessary to put four quite different questions: 'What use will it be?', 'What is to be done here?', 'What are you trying to do?', and 'Why are you doing it?'[3] The first question relates to the function that will be performed by the product of the accomplished task. In a system of production with a division of labour and a hierarchical organization, this part of the final cause of the task is the most basic. For, from the pre-assigned function, the goals of the task are specified; the fulfilment of the goals ('What is to be done?') is the means to the performance of the function. These goals include the specification of the accomplished task, and criteria for its adequacy. In the assembly-line example, these will relate to the properties of the product which determine its quality (tolerances of fit, shape, etc.), and those of its making which relate to its cost (as materials, wastage, and time required). Now, the function which determines the goals of the task does not exist in a vacuum; it will generally be subsidiary to other functions in a hierarchy. The ultimate 'final cause' at the top is not a function, but a set of personal purposes, individual or collective, directly expressed or imputed; for otherwise no one would have brought the activity into being or kept it going.

We notice that these first two questions are not best answered by the operative himself, but by those superiors who define, direct and control his actions on the job. Only in the last two questions do we come to a man, as distinct from a biochemically activated machine. A machine may be designed to direct its motions and processes for the fulfilment of preassigned goals, but the man has purposes. These are what he wants to accomplish in a particular task; those things that once done, will give him satisfaction and the sense of effort well spent. The fulfilment of the assigned goals of the task may well be among his purposes, either for their own sake, or more probably as a necessary condition for the achievement of a genuine purpose as entitlement to the day's wage without penalties or deductions. This distinction between the operative's purposes on the one hand, and the goals and higher final causes of the task set him on the other, is fundamental to this argument. There are still many who believe that the subordinates in a hierarchical organization (be they workers, soldiers or students) either do, or should, or in the good old days did

[3] For a closer study of some of these 'teleological' concepts, applied to sociological theory, see Dorothy Emmett, *Function, Purpose and Powers* (MacMillan. London; St. Martin's Press, New York, 1958).

have their own purposes defined and dominated by the final causes imposed on their tasks: that they do, should, or did believe in their boss and sincerely want him to be pleased with them. Without arguing this set of views, we can agree that, as a contingent fact, many workers, soldiers and students now do not particularly believe in their boss, and want only to stay out of his way: their private purposes are not dominated by the imposed goals of their tasks. Whether such a situation is new or old, good or bad, is not of concern at the moment; it is necessary only to establish that there is in principle a difference between the objective final causes imposed on a task, and the personal, private ones of the operative. Once we accept this difference as existing somewhere, sometime, then we have to reckon with the problems of quality-control and ethics in any socially organized activity including scientific enquiry.

Coming to the last of the questions defining the final causes involved in the task, we consider why the operative is at that particular task (to the extent that his situation is the result of choice) rather than at some other. We call this set of reasons his 'motives'; and we notice that these set part of the framework in which his private purposes are specified. The motives will include some desirable and some undesirable aspects of the situation; and the purposes will correspondingly include the maximizing of certain personal benefits and the minimizing of certain personal costs, either specific to that situation or of a general nature. To take a trivial example, one may find two men next to each other on an assembly line; one may be there because of the opportunities for overtime on that job, while the other lives close to the factory and works there so as to have more time for his leisure activities. The purposes of the first will then include the achievement of a significant amount of overtime pay, and he will do what he can to influence the flow of work so as to make extra hours necessary; while the other will have a contrary purpose, and his private strategy on the job will include seeing that the week's work is done in good time.[4]

The specification and achievement of a set of personal purposes on a job is thus a complex affair, which can itself be considered as a task. The different sorts of activities and unit tasks associated with

[4] My choice of the four terms 'function', 'goal', 'purpose', 'motive' is to some extent arbitrary, and even as I have defined then here they will overlap in meaning with other common terms denoting 'ends'. I have avoided 'objective', since it generally means 'goal' with a connotation of 'function', and 'aim' or 'aims' which mix 'goal' with 'purpose'.

a job are evaluated by their function in the achievement of the personal purposes, and individual unit tasks will be specified by private goals, established in accordance with these private functions, rather than being defined by those who direct and control the work. Thus we may see someone working quite hard at the accomplishment of some assigned task, but what he is 'trying to do' is defined by his own private system of basic purposes, subsidiary functions, and individual goals. The goals set by the organization operate only as external constraints on his specification of the complex private task; and these constraints are effective only to the extent that his fulfilment of the goals can be controlled. Since quantity is easily measured and (in this situation) quality is capable of assessment by routine tests, the two systems of final causes can coexist, if not in harmony then at least in relative peace.

The difference between a man and a biochemically activated machine introduces another element in the government of his activities: one which is social and in part ethical. Since the operatives necessarily have activities other than their assigned manipulations, the organization must impose a job-discipline, governing such things as punctuality, rest-breaks, and general behaviour on and off the production line. Without this, the shop would degenerate into chaos and the imposition of controls, even on the work being done, would be severely hampered or impossible. But those on the shop floor have a society of their own, with its own code of behaviour. This has the function of protecting the common purposes of its members, and will frequently run counter both to the system of controls and to the job-discipline. It will frequently require a member to forgo personal advantage for the sake of the common good, as in refraining from exceeding the standard rate of production, or in joining a strike. In order that members should do this voluntarily, they must subscribe to an ideology from which this ethical code can be derived, and which serves to justify and guide their actions in this respect. In its terms, struggles with the organization are conceived as the protection of their rights.

In this extreme case, the radical separation of the organizational and personal components of the final causes of the task creates a situation where the worker's ethical principles always operate against the fulfilment of the organization's goals. Naïve appeals to the 'common interests' of all 'co-workers' in the firm (be it privately owned or part of a Socialist economy) will inevitably be greeted with public or private derision.[5] From management's point of view, the operative cannot be expected to show initiative or be entrusted with responsibility; his advantage over a machine in the subtlety of his manipulations is offset by his unreliability on the job and his cunning applied against the job. For the worker, convenience and right combine to encourage him to get as much as he can for as little, by any practical means at his disposal. Of course, there are many cases in which enlightened or skilful management succeeds in overcoming the harshness of this dichotomy; but the raw situation is sufficiently common to serve as an example of a real case. And taking this extreme case as a real and in many ways natural situation, we can ask how it is that scientists can generally strive for the achievement of the results of the highest quality of which they are capable, in the absence of a formal system of quality control or of any imposed job discipline. Either they are drawn from an élite group of recruits of great moral superiority; or some aspects of their task, and its inherited code of behaviour, enables them to develop very enlightened private purposes which harmonize with the common good.

Quality Control of Skilled Tasks : the Insufficiency of Self-Interest

We have seen that in the case of the unskilled production-line worker, his private purposes and his ethical code are basically in

[5] This initial naïveté about the behaviour of the working class under socialism may well have been responsible for some of the worst excesses of Stalinism. A theme which runs through A. Koestler, *Darkness at Noon*, is the contradiction between the aspirations of Socialism and the intellectual and political backwardness of the Russian masses; indeed, this provided the rationale for the 'confession' by the bourgeois-intellectual Rubashov. The peasant interrogator Gletkin provides the evidence for the problem, first in telling of his discovery of the necessity for coercive methods in Socialist construction; and at the end, with the question 'Were you given a watch as a boy ?' (Penguin Books, 1947), pp. 80–8 and p. 180. At least some of the more recent revolutionaries have publicly appreciated this problem; thus Che Guevara discussed the difficulties caused by the ideological backwardness of the Cuban urban working class. See 'On Sacrifice and Dedication' (1960) in *Venceremos*, ed. J. Gerassi (MacMillan, New York, 1968). Lenin's own teaching on the problem was not very helpful. In *State and Revolution* he gave the classic formula: that the repressive state would gradually wither away spontaneously, when 'freed from capitalist slavery, from the untold horrors, savagery, absurdities and infamies of capitalist exploitation, people will gradually *become accustomed* to observing the elementary rules of social intercourse' (ch. 5, section 2, p. 462). But shortly afterwards he recognizes the necessity for maintaining 'the *strictest* control by *society and by the state*' (italics original) so long as the 'ordinary run of people, who like seminary students in Pomyalovsky's stories, are capable of damaging the stocks of public wealth "just for fun" and of demanding the impossible'. (*Ibid.*, ch. 5, section 4, pp. 469–70; the order of the passages quoted here is inverted from the original.) Page references are to the *Collected Works*, vol. xxv (Lawrence and Wishart, London, 1964).

antagonism to the fulfilment of the goals set by the organization which employs him; and he fulfils these goals adequately only because his work and activities can be controlled and disciplined. If we now consider skilled work, we find a radically different situation. For the quality of work cannot be controlled by routine tests; and if good work is to be done, it is necessary for the operatives to want to do it. And although, as we will show, maintaining the quality of work performs a valuable function for any group of skilled workers, this in itself is not a sufficient reason for any one of them to exert himself to that end. Hence something more is necessary, as a component of the personal purposes of the craftsman on his job.

I mentioned earlier, in connection with quality control on skilled production work, that the assessment of the quality of such products cannot be done on a routine basis. This is only a special case of a more general principle, which has direct relevance to scientific work. In the measure that goals of a task are complex, sophisticated, or subtle, then the assessment of the adequacy of the fulfilment of those goals must be correspondingly so. For if the tests of adequacy are crude in comparison, then a person possessing the skills to do the work properly will be able to find ways of formally satisfying the imposed tests while achieving his own purposes to the detriment of the assigned goals. This holds for the production and management of material things; and when the product of a task is immaterial, as in organizational or managerial tasks, evasion of the intended function of any system of controls is child's play. This phenomenon has been recognized in nations with centrally planned economies, where individual factory managers make certain to fulfil the bureaucrats' goals for their establishments, and yet the 'plan' never achieves the desired co-ordination of its parts. And from the advanced society at the other extreme of social organization, America, comes the aphorism, 'Wherever there's a system, there's a racket to beat it.'[6] This problem is not resolved by imposing more complex sets of

[6] For this aphorism I am indebted to an anonymous tram-driver in Atlantic City, New Jersey, in 1952. There was system which made it impossible for him to cheat on the collection of fares: passengers put their fares into a machine, and he handled money only to make change when necessary. But the fare was nine cents *or* a token worth one-third of twenty-five cents. On a Saturday evening he had many couples boarding the tram for this occasion only; the man would offer a twenty-five cent piece. The driver would give change of seven cents, as from two full fares, and drop two tokens into the machine. The profit was only one and a third cents each time, but several hundred such fares amounted to a tidy bonus for an evening's work. The driver was euphoric over his discovery and shared the secret with nearby passengers, along with the valuable aphorism.

controls on the work. For the system of control may itself become so cumbersome as to invite close analysis and really determined attempts at complete evasion; and the problem of controlling the controllers can also become acute.

It seems that this problem, which is closely related to the general social problem of corruption, can be resolved only if the agents being controlled acquiesce in the operation of the system and also participate in it. How this can be done with a general public is a most difficult problem; but we have many examples of special groups which do in fact police themselves to a large extent. In the case of an occupational group, such as skilled workers or professionals, the mechanism seems straightforward. All the members of the group enjoy special rewards in virtue of their membership; but these exist and are maintained only because membership in the group is a guarantee, to employers or clients, of the high quality of the work which will be performed by anyone so designated. If this guarantee is betrayed, the reputation of the group will be lost, and it will lose its special rewards, either gradually or with catastrophic suddenness. Thus in this case, unlike that of the unskilled worker, each task accomplished serves as a precedent for the support of the future claims of the individual worker, as a member of his group, to special rewards. Hence, we might reason, each member will naturally strive to maintain that common reputation, out of simple self-interest.

Arguments of this character, showing how a common good can derive from an aggregate of selfish decisions, are as old as the science of political economy. They all share the same logical fallacy, and all neglect the incommensurability of social cost and private benefit which can frequently occur in human actions.[7] The fallacy lies in assuming that 'all' and 'each' are completely synonymous and interchangeable logical operators. An argument, like the above, which starts with 'all' and ends with 'each', is not necessarily valid. To show this, I will use an example involving the incommensurability principle, and suppose that I discover a way of stopping my electricity meter. If my manipulation does not leave any traces, I cut it off between visits by the meter-reader, and I reduce my bill by an

[7] An argument similar to mine has been made by M. Olson, in *The Logic of Collective Action* (Harvard University Press, 1965). He shows there that individual actions which are 'rational' in the economists' sense of the term will, in general, be counter to the achievement of their share of the common purposes of their group. He is mainly concerned with the implications of his conclusions for public policy; the ethical aspects of the problem are eliminated by the concept of 'rational'.

amount which would not attract suspicion, why should I abstain? I can argue that even if my gain were recouped by the company through slightly increased charges to the other consumers, the loss to any one of them would be quite negligible; and so the total sum of 'happiness' or 'utility', measured by cash in hand, would experience a net increase! In opposition, it could be argued that if all consumers tapped the meter in this fashion, then the rates would rise until the effective tariff to each would be the same as previously, and the community would be the worse for the universal cheating. But this reasoning fails to be binding on each of the consumers, and in particular on this one. In answer, I could take two cases: where the dishonest consumers are a negligible fraction of the total, and where they are a significant fraction. The former case is the same as that where I am the only beneficiary; and in the latter case, I could argue that I would be a fool to pay significantly more than my due, to reward those other consumers for their dishonesty. Nor is the situation changed if there is a risk of detection; I could simply work out a probabilistic cost-benefit function involving the penalty, proceed somewhat more circumspectly, and argue the principle of the matter as before. And although it is clear that if each member of the group yields to temptation in this way, all will soon suffer, still there is no argument from utility to all, to prove to each man that he should take the straight and narrow path.

It seems, then, that situations where private benefit from an action is incommensurably larger than its social cost, are not controlled by the application of a 'utilitarian' ethic, nor by the principle of aggregation of enlightened self-interest. The existence of this class of apparently harmless acts of dishonesty is recognized in the English vernacular, in the term 'fiddle', as distinct from theft or embezzlement. What ethical grounds, if any, are there to deter me from tapping my electricity meter? There seem to be two. The first is the very abstract form of utilitarian doctrine, deriving from Kant: that I should judge my actions by the consequences that would result if everyone else were doing the same. And the second is that certain things are simply not done, regardless of whether anyone else knows of them or imitates them, and independently of their actual or hypothetical consequences.

This ethical problem, and its possible solutions, are directly relevant to the maintenance of good work in skilled and sophisticated tasks, including science. If I am the only effective judge of how well I am fulfilling the goals of my task, why should I refrain from cutting corners here and there, and helping myself to a bit of personal benefit within my reach, when I am quite sure that no one would notice or even particularly care? On what basis should I decide to make the personal sacrifice involved in doing my work better than I need to? Any imposed criteria of adequacy on my task and job discipline, are quite irrelevant to my decision on this matter; nor am I constrained by a group code of behaviour designed to maintain the reputation of the work of all. For me to include among my personal and private purposes, the best possible accomplishment of my assigned tasks, I must be conditioned by two closely related principles: loyalty to my group, so that I would not let the side down; and pride in my own work, so that I would not betray myself.

These two principles can be effective on individuals, only when there is good morale in the group; indeed, the presence of such principles is an essential component of good morale. In its absence, work of excellence will not be done by all; and it is hard for any one individual to maintain such work for any time, in isolation from and in opposition to his colleagues. The achievement of good morale is not a simple task; imposed systems of rewards can help, but if they do not perform their intended function, they merely become the occasion for a new system of cheating. It is an error to believe that the commitment to good work, and the associated good morale, can arise automatically because the social system of the accomplishment of the tasks will thereby provide an absolute maximum of benefit to all. This is the fallacy of 'functionalist' explanations of human behaviour, which (unlike Aristotle) assume that the 'best' final cause is simply identical to the efficient cause. It is exceedingly common for such social systems to achieve a relative maximum of general benefit by an accommodation rather than a harmony of collective goals and private purposes: where the private purposes of all the agents (including those exercising decision and control) are dominant over the collective goals, and where these latter are fulfilled in a perfunctory or spurious fashion. To bring group activity from such a slack and demoralized state, to one of excellence and commitment, requires leadership capable of establishing an ethic, based on an ideology, to which the members adhere. For the transition is a highly unnatural one, and even the maintenance of the state of excellence requires the continuous exercise of leadership.

The Management of Property: the 'Professional Etiquette'

It might appear that the ethical problems identified in the case of skilled work are not directly relevant to the case of scientific inquiry. For the tasks of the skilled worker differ from those of the unskilled only in the degree of difficulty of their control; the ultimate purposes from which they are derived are still irrelevant or antagonistic to his own. Hence it is only natural for the normal state to be one of conflict between organizational goals and private purposes, and for any true harmony to be the exception. In science, by contrast, there is no such basic antagonism; for there the situation is one of a group of colleagues freely co-operating in a common endeavour, and where the social tasks of control and even of direction are undertaken informally by a group within the community, differentiated only by their talent and commitment. A mark of the fundamental difference between the two cases is the absence of an imposed job discipline, and the refinement of the code of behaviour into an elaborate etiquette whereby the distribution of rewards for achievement is accomplished with honesty and fairness.

Yet these particular differences do not necessarily result from a uniqueness of scientific endeavour, whereby it is essentially and automatically governed by the highest ethical principles. As we have already seen in our discussions both of the protection of property and of the management of novelty, scientists have private purposes which are not always necessarily in harmony with the collective goals of the work. It is true that autonomous research scientists are not in the work-situation of an employee. The product of the task is not a precedent for the assessment of quality made by an external agent of control; it belongs to the scientist himself, as a piece of intellectual property. But there are other spheres of activity where a similar sort of property is created, and where an etiquette can function quite independently of refined ethical principles.

One essential feature of the scientist's intellectual property, as distinguished from the 'real' property of commerce, is that it exists only by being available for use by others. Even if the scientist does his research in physical isolation from colleagues, his tasks are eminently social in nature. He depends on others, not merely for the training in his craft skills, but also for the information without which he could not do any work at all; and his property takes on value only through the explicit assessments of colleagues, rather than through the workings of an anonymous market. Now, there are other sorts of work, which have similar features of essential co-operation and mutual assessment between independent practitioners, although distinguished from science in that the individual task usually brings a cash return on its own. The traditional professions are examples of this class of work, and in them the 'professional etiquette' performs a very straightforward function in the harmonizing of private purposes and collective goals.[8]

In professionals' work, the assignment of shares of personal property is done more in the definition and division of the task, than in the acknowledgement of previous work used in the current project. But in spite of this difference from science, the function of the etiquette, and the reasons for its necessity, are the same. In the learned professions in particular, the task presented by a particular 'case' may require a variety of skills for its accomplishment. The practitioner who receives such a case must decide whether to try to do the job on his own, or whether to call in colleagues expert in particular aspects, sharing the work and the payment with them. It is clearly in the clients' interest to have competent handling in all respects; and the collective purposes of the profession are served by ensuring that this will always occur. Hence the task will be parcelled out among various experts, each sharing in responsibility for its accomplishment, and enjoying a part of the reward. But this division of the task is a highly sophisticated expertise; the analysis of the task itself constitutes a problem like a scientific one and then the choice of fields of expertise and the division of responsibility among them requires a skilled knowledge of both the technical and social aspects of the task. The control of this division of the professional's task, presents the same difficulties as the control of quality of the work: no formal set of rules could encompass the variety of situations encountered, and a machinery of decision or adjudication would be impossibly cumbersome. Hence for this division to be accomplished quickly and efficiently, it is necessary for the practitioners to adhere to an informal code of behaviour, the mastery of which is one of the more subtle skills necessary for competent practice in the field. This, then, is the essence of the professional etiquette: the methods of

---

[8] For a recent summary of the literature on 'professionalization', including classic sources and modern research, see *Professionalization*, eds. H. M. Vollmer and D. L. Mills (Prentice-Hall, 1966).

social behaviour necessary for the orderly division of the property created in the particularly sophisticated co-operative work by independent agents.

Although a professional etiquette will necessarily be more elaborate and intellectually subtle than the code of behaviour of skilled manual craftsmen, it is not thereby endowed with a higher ethical content. Indeed, an aggregation of self-interested policies can suffice to maintain such an etiquette. For if one member of the group violates the etiquette by failing to call in a colleague from the appropriate group of experts, he is not merely endangering the reputation of all the members of the profession; he is also appropriating some of the payment which that group would expect to flow to a member. By a combination of collective goals and private purposes, they will be motivated to impose penalties on him; and so in this respect the system provides an automatic mechanism for enforcement. Also, it is possible to identify codes of behaviour which function as professional etiquettes in situations where refined ethics are definitely absent. As an extreme example, the code of the American Mafia is said to consist of two maxims: don't squeal to the cops, and keep your hands off other members' wives. Although the truly learned professions would not have such crude etiquettes as these, there too they can operate exclusively for the division of property, to the detriment of the interests of clients. American medicine occasionally has scandals from revelations of unnecessary consultations with kick-backs of fees; and in any profession the etiquette operates to close ranks against aggrieved clients. Even in science itself, one can find stagnant fields, where the creation and division of intellectual property is governed by the most rigorous etiquette except in those cases where genuine work challenges the existing social structure of the field.

By itself, then, a professional etiquette does not establish a harmony between the private purposes of practitioners and the professed ultimate purposes of the work. If the collective goals of the activity are contrary to those ultimate purposes, that is, serving an aggregate of interests of members instead of those of clients, then the etiquette can operate quite directly against the achievement of those ultimate purposes. For these ultimate purposes to become relevant to the definition of the final causes of the professional's task, the etiquette must be supplemented by a genuine professional ethic.

Professional Ethics and their Enforcement

A true learned profession can be distinguished from an occupational group which provides a particular service by four criteria. The client is dependent for his welfare on the accomplishment of the task; but he is not competent to assess the adequacy of the work done; recognized competence in the set of tasks is legally restricted to those certified to have completed a training of a scientific character; and in exchange for the monopoly of practice the group accepts responsibility for the achievement of the purposes of clients. The situation of the professional thus involves an essential fiduciary element; incompetence or malfeasance constitutes a betrayal of the client's trust. Should this occur, there is a risk of scandal, and the erosion or loss of the legally enforced monopoly enjoyed by the professional group.[9] It is thus in the long term collective interest of the profession to maintain standards of work and to protect the interests of clients; but this does not automatically extend to the immediate private purposes of the members of the group, either by self-interest or by the operation of a professional etiquette for the orderly division of property. A more refined code of behaviour is necessary to prevent the relaxing of standards and the neglect of the interests of clients, whereby the profession would degenerate into a group of businessmen temporarily enjoying a collective monopoly on a particular class of trade.

This social function of maintaining the health of a profession is performed by a professional ethic, as distinct from an etiquette.[10] This ethic must be based on the absolute primacy of the true ultimate purpose of the task: serving the purposes of the clients. This need not mean that the private and collective purposes of the profession should invariably be sacrificed in the interests of each

[9] In the earlier nineteenth century American orthodox medicine suffered from such competition from sectarians and quacks, partly because the professional physicians were given to 'heroic' treatments (the heroism being that of the patient enduring copious blood-letting or poisonous purgatives). Their reputation sank so low that some states revoked their monopolies. Most of the sects combined a 'democratic' style with mild remedies, thereby achieving lasting success; but the commercial 'patent medicines' were sometimes even more deadly than the physicians' prescribed remedies. See R. H. Shryock *The Development of Modern Medicine* (Knopf, New York, 1936; Gollancz, London, 1948), ch. XIII, 'Public Confidence Lost', pp. 224–48.

[10] The distinction between professional etiquette and professional ethics is clear in the paradigm code of medical ethics published by Thomas Percival of Manchester in 1803. The latter was 'reviewed' in the code, while the former was given detailed treatment. See R. H. Shryock, *The Development of Modern Medicine*, pp. 128–29.

individual client; the protection of the interests of clients as a class may well involve giving less than full satisfaction to some particular case. But as a general guiding principle, to be interpreted with skill and subtlety in each particular instance, the professional ethic is necessary if the fiduciary relationship is to be justified and preserved. The problems of establishing and maintaining such an ethic are similar to those encountered in quality control of skilled work, but even more delicate and demanding. For the control is not merely on the tasks accomplished, but also on the subtle social mechanisms whereby the tasks are organized and rewards distributed. The professional etiquette itself must be governed in its operation, and constantly modified as circumstances change. The assessment of adequacy can relate as much to the way things are done, as to the fulfilment of stated goals; and the relevant criteria can be so informal as to be incapable of verbal argument. The area of ambiguity between proper and improper conduct is broad and ill-defined, so that an individual can claim (perhaps with sincerity) that his particular bit of self-serving is neutral or even beneficial with respect to the collective goals. And when all the prior phases of the task of control are necessarily so extremely informal, a system of crude, quantitative penalties would find no middle ground between those that are ineffectually light, and those that are harshly severe.

For the effective enforcement of a professional ethic, it is not necessary that all members of the profession adhere to it with a total commitment. But even more than in the case of skilled tasks, there must be excellent leadership and good morale. The leadership is necessary to provide definition and application of the professional ethic in constantly changing circumstances; and morale is necessary if this leadership is to be effective. For the ordinary members of the profession must respect the leaders in their own terms of skill and success, and be prepared to accept their guidance and control. Also, they must have some degree of commitment to doing good work, so that a consensus can operate against those who betray the good name of the profession. Otherwise the enforcement of a professional ethic is impossible. Even the formal institutions of statute and judicial law cannot operate effectively (except in conditions of police terror) in areas where the population at large does not accept its categories; and in this very informal situation, where the policing must be done by the members themselves, general acquiescence in the fairness and necessity of the penalties is absolutely essential. Also, since the whole operation of control, including penalties, is generally conducted completely informally except in the most extreme cases, the subtle penalties of disgrace and even disapproval must be real for the ordinary members of the group, if control is to be at all effective. And for this, there must be good morale: a consensus, applied to oneself as to others, that good work and a good reputation are important.

Thus the professional ethic can operate only within certain constraints. If the ideology on which it is based (a conception of the profession and its service to its clientele and the community) loses contact with the reality experienced by members of the group, the basis of respect for the ethic will erode, the leadership which adheres to the ideology will be isolated, and the task of policing will become onerous and eventually impossible. Hence the professional ethic and its ideology cannot rise too high above the working etiquette, and the experience and values it reflects. On the other hand, the situation of the professional makes it natural, in some respects, for the professional ethic to be comprehended and accepted. The basic task is related to particular clients whose needs are visible and real; and the trust of those clients, and the advantages of a protected monopoly of practice, is part of the daily experience of the practitioner.

Levels of Ethical Commitment in Science

We are now in a position to analyse the problems of the nature of the ethics and ideology necessary for the maintenance of a healthy state in science, and of the conditions under which that ethics and ideology can be effective. It would be unrealistic to assume that a man automatically becomes a paragon of virtue on receiving his Ph.D.; and yet the delicate and subtle character of scientific work makes it particularly susceptible to the degeneration of quality in the absence of controls. We have already seen that in spite of the informal structure of the scientific community, there are several quite definite social functions to be performed, with some differentiation within the community on the basis of the accomplishment of the relevant individual tasks. We can use this rough division to explore the degree of ethical commitment which is necessary for their proper accomplishment, keeping in mind that each scientist is not included in one single group associated with a particular task.

At the lowest level of the structure of control, we have the investigation of a particular scientific problem. Considering this as a task, we can say that its goal is the achievement of a result. The function of the accomplished task is to provide materials for the further investigation of other problems; this function may have any degree of specification from the closest to the most general, depending on the context of the investigation. Since scientific inquiry is very skilled and sophisticated work, a scientist who will do only what he is told, and that not very well, will produce nothing much better than vacuity. Even a technician assisting on scientific work must have some personal commitment to the quality of his products if they are not to be utterly unreliable, even if he is closely supervised by his superior. Of course, the social context of scientific inquiry is such as to tend to eliminate in advance those who make the most simple and short-term calculation of personal cost and benefit; the lengthy process of obtaining the necessary skills and qualifications requires an attitude which, if not necessarily more enlightened, does require a more sophisticated and long term calculation. However, this process does not exclude a man who supplements his research skills with those of getting by with least work. In principle, the quality of his work is controlled by the referees of the papers he submits to journals; and they should have the special skills for detection of slipshod work. Thus, for the maintenance of a minimum standard of competence in the productions of ordinary scientific workers, it is not necessary for them as a body to have a degree of commitment to good work any greater than that of skilled manual workers, provided that they can be effectively policed by referees.

Higher demands are made on the referees, if they are to do their work properly. If, out of indolence or personal interests, they all pass everything that looks plausible, then the public channel of scientific communication in their field would quickly become choked with an undifferentiated mass of results of every degree of quality, mostly bad. For their work, which is frequently onerous and which brings no tangible rewards, the referees must be committed at least to the maintenance of the standard of quality in that field with which they are associated. But for this degree of discipline, it is sufficient in the short run that the referees adhere to a code of behaviour equivalent to a professional etiquette. The value of the referee's own intellectual property is tied up with the reputation of his field, and of those journals with which he is most closely associated. His own purposes are served by helping to ensure that the creation of property in his field is accomplished in an orderly manner, that the shares in property are allotted fairly through acknowledgement of previous work, and that the value of the property of all the members of the group is not allowed to deteriorate through admission of shoddy work. Thus, provided that he is protected from undue pressures from the other producers, and given some direction and control from above, the referee need be concerned only with the protection of his group's intellectual property, to accomplish his tasks adequately.

Unless the referees are indeed committed to something more than the protection of property, control and direction from above is very necessary for the prevention of a rapid decline in standards. As we have seen, effective control on the product of such a skilled task as scientific research, is possible only when those exercising the control are themselves expert craftsmen. But then they will be working on two sorts of tasks, involving the different private purposes associated with production and control respectively. Those whose short-term private purposes are served by the lowering of standards for the easier production of intellectual property, will not be a separate and inferior group; they include colleagues and protégés of the referees, and the referees themselves. It is even possible, in small and specialized fields, for two good friends simultaneously to referee each other's papers. In such an extreme case, it is necessary for the etiquette of refereeing to be adhered to very rigidly, if mutual convenience is not to dominate over the collective goal of the production of genuine results. But here, unlike in the case of the professionals' division of a complex task, the etiquette is not automatically enforced by the aggregate of self-interests of practitioners. And this very subtle method of social behaviour is very vulnerable to the sort of modification at the fringes, by which no apparent damage is done in the short run, but whose cumulative effects are quite destructive.

Hence, as we discussed earlier in the case of quality control, the role of the leaders of the field is crucial. If their concern is no more than the creation of intellectual property which can be cashed for material and social benefits, then there are no internal barriers to the rapid degeneration and corruption of a field at all levels. Aspirants to leadership will inevitably have a complex mixture of motives, involving both personal ambition and scientific zeal. They will take the style of behaviour of their superiors and patrons as more significant than their public pronouncements for shaping their own behaviour and purposes. In a single generation, the style of leadership can change from that of genuine scientists who are greatly enjoying unexpected material benefits, to entrepreneurs in the business of the production of results on contract. Hence if the leaders are to maintain the health and vitality of their field, their private purposes must be at least as refined as those of the leaders of a profession who are committed to a genuine professional ethic.

The Social Context of 'the Scientific Ethic'

The 'scientific ethic', analogous to a 'professional ethic' and the ideology which necessarily serves as its basis, lacks the direct contact with experience of the professional ethic, and is correspondingly more sensitive to the social and cultural context of the work. For the professional's task is defined by the needs of a particular client; and the ultimate purposes of the task, in serving the group of clients, are present every time the work is done. By contrast, the higher final causes which condition the scientist's task are remote and diffuse. The collective goal of scientific endeavour is the increase of knowledge; but it is impossible to identify a group in society whose purposes are so directly served by the continued fulfilment of that goal, that they can be said to determine it and all the final causes down to the specification of the goals of a particular task. It is true that the work and the achievements of science do perform various functions outside scientific enquiry itself, through contributions to technique, through the maintenance of the quality of high-level teaching, and in enhancing the prestige of a local or national institution or community associated with it. But in the tradition of science and scholarship that we have inherited, these external functions do not, and must not, provide ultimate purposes which dominate the final causes whereby scientific work is directed and controlled. In this tradition, the collective goals of scientific inquiry are autonomous; this not only creates an inherent tension between the community of science and the lay society which supports it, but also deprives the ideology and ethic of science of a firm basis in experience. Assuredly, the search for knowledge has long been recognized in our civilization as one of the most noble of human endeavours. But lasting success in this search is reserved for only the very few; and no one in science can be sure that he will be other than a humble participant in a task whose proper accomplishment may require the work of generations yet unborn.

The 'scientific ethic' to which the leaders of science must subscribe if they are to do their work properly is thus so refined and subtle as to rest ultimately on philosophy and faith. Francis Bacon defined the attitudes which the ideal natural philosopher should bring to his work: he should be 'sober, chaste and severe', and should approach his task in spirit of humility and innocence. He was reacting against the various corruptions of knowledge which he saw in his own time, principally the vacuous disputations of academics and the boastful, spurious claims of the natural magicians and alchemists; and for him the reformation of natural philosophy was intimately related to a true reformation of morals and religion.[11] Although his personal conception of the search for knowledge was soon lost, his definition of its goals, 'to establish and extend the power and dominion of the human race itself over the universe', remained influential for generations, for it cohered with the experiences and values of natural philosophers in many ways.[12] Among the English until nearly a century ago, the study of God's creation was (as Bacon himself believed it should be) parallel to the study of His word, and it was widely believed that this study could not but increase reverence for the Creator.[13] Elsewhere this 'natural theology' was not so strong, but the close relation of scientific inquiry to metaphysical concerns, especially among the greatest men as in Germany, provided the work with an ultimate purpose harmonizing with its immediate goals. Also, the general social benefits, in the way of material progress, thought to be largely a result of the applications of science, provided a justification of the work and an added component of human value; although (as we have seen in the

[11] This interpretation of Bacon is developed in Benjamin Farrington, *The Philosophy of Francis Bacon*.

[12] This formulation of the goals of natural philosophy is found in the *Novum Organum* I, Aphorism 129, *Works*, vol. 4, p. 114. Bacon's essential message has been variously seen as 'inductivism', as 'power over nature', and as 'the sacred mission of scientific inquiry'. The first interpretation was characteristic of the nineteenth-century British; the second has a history from the Enlightenment through to the Marxists, and the third is recent. The writings of Benjamin Farrington show a transition from the second to the third interpretation; see his *Francis Bacon, Philosopher of Industrial Science* (New York, Abelard-Schuman, 1947) and *The Philosophy of Francis Bacon*.

[13] The concluding paragraph of Newton, *Opticks*, clearly expresses this higher function of natural knowledge; and this tradition continued until well into the nineteenth century. For the continuity between Newton's fundamental natural philosophy and his theological concerns, see J. E. McGuire and P. M. Rattansi, 'Newton and the Pipes of Pan', *Notes and Records of the Royal Society*, 21 (1966), 108–43, and the later work of J. E. McGuire, such as 'Force, Active Principles, and Newton's Invisible Realm', *Ambix*, 15 (1968), 154–208.

case of Helmholtz) the greatest spokesman of science were careful to warn that this external function should be kept out of the hierarchy of final causes in science.

These very diverse strands of ideology could serve as an intellectual basis for a refined scientific ethics, because the social position of scientists, either by birth or acculturation, provided a natural basis in experience for such an ethic. For, until quite recently, one could generally not engage in free scientific inquiry if one had to earn a living. Even where there was a developed institutional structure, as in Germany, a 'career' in science involved years or decades of living on independent means or on nearly none. Hence the vocation of natural philosophy required, for all but brilliant recruits from the lower orders, the background of a gentleman.[14] Those of this class who were concerned with ordinary comforts or social position would not devote themselves to philosophy; and those who made the sacrifice would do so in the pursuit of more refined satisfactions. Among them in particular, the code of a gentleman, in which a man's most valuable property is his good name, would be powerful. In scholarship, a good name is achieved by talent and protected by integrity; and an overriding commitment to the quality and trustworthiness of work with which one is personally associated in any way is the essence of a genuine scientific ethic.[15]

[14] Thus Helmholtz, in his autobiography: 'And now I was to go to the university. Physics was at that time (in the late 1830s) looked upon as an art in which a living could not be made. My parents were compelled to be very economical, and my father explained to me that he knew of no other way of helping me to the study of Physics, than by taking up the study of medicine into the bargain' (p. 274). *Popular Lectures on Scientific Subjects*, second series. The weakness of the position of the *Privatdozenten* in the German universities of the nineteenth century, which eventually became intolerable when the professoriate did not increase in proportion to student numbers and research work, is discussed in J. Ben-David, 'The Universities and the Growth of Science in Germany and the United States', *Minerva*, 7 (1968–9), 1–35. He finds the cause of the success of American universities in their structure, while it seems more likely to have been in their growth-rate; and his chosen example of statistics as a 'growth-point' is ill-suited to his argument, since it flourished in England long before it did in America.

[15] A particularly stern and perceptive analysis of the scientific ethic was given in the nineteenth century by Fresenius, the founder of analytical chemistry: 'Knowledge and ability must be combined with ambition as well as with a sense of honesty and a severe conscience. Every analyst occasionally has doubts about the accuracy of his results, and also there are times when he knows his results to be incorrect. Sometimes a few drops of the solution were spilt, or some other slight mistake made. In those cases it requires a strong conscience to repeat the analysis and not to make a rough estimate of the loss or apply a correction. Anyone not having sufficient will-power to do this is unsuited to analysis no matter how great his technical ability or knowledge. A chemist who would not take an oath guaranteeing the authenticity, as well as the accuracy of his work,

In the conditions of the industrialized science of the present period, all these bases for the maintenance of a scientific ethic have been eroded: philosophy and religion have become irrelevant; the applications of science are not a matter of simple benefit; and one can earn a living and indeed become rich in a career in science.[16] The great pronouncements of the noble ideology of science of the earlier period would sound false or hypocritical in this new context. These changes produce many difficulties in the preservation of an enlightened ethical code in science.[17] The leaders themselves are deprived of a secure basis in their personal culture and experience for the ideology and ethics of autonomous, dedicated scientific inquiry to be natural. Should they parade such old-fashioned beliefs as arguments for adherence to the demanding ethics of science, they would risk isolation from a large section of the community they lead. Moreover, even these leaders do not exist as superiors in a formal hierarchy of direction and control; the governing of their community can be accomplished only with the active participation of a large part of it,

should never publish his results, for if he were to do so, then the result would be detrimental, not only to himself, but to the whole science.' Quotation from F. Szabadvary, *History of Analytical Chemistry* (Pergamon Press, 1966), p. 176. I am indebted to Dr. W. H. Brock for this reference. It would be naïve to believe that all gentleman have always been gentlemen. In his discussion of observations, in ch. 5 of *Reflections on the Decline of Science in England* (London, 1830), Charles Babbage devoted a section to frauds; he defined hoaxing, forging, and trimming, and then, with fine irony, described the varieties of 'cooking'; see pp. 174–83.

[16] Another danger, which until very recently was so remote as to be inconceivable, is that scientists might resort to the law of libel to protect their results from published criticism. On this, see J. R. Lewis, J. S. Gray, L. W. Pollock, and P. G. Moore, 'Can a Scientific Article be Libellous?' (letter), *Nature*, 225 (1970), 1081. The Master of the Rolls, quoted in *The Times*, was reported as commenting 'it would be a sorry day if scientists were to be deterred from publishing their findings for fear of libel actions. So long as they refrained from personal attacks they should be free to criticise the findings and techniques of others. It was in the interest of truth itself. Otherwise, no scientific journal would be safe.' As the matter is still *sub judice*, the merits of the case cannot be here considered; but, since the original report and the criticism appeared in a 'scientific' periodical (the *British Medical Journal*), the outcome will naturally be awaited with anxious interest by those concerned to protect genuine scientific criticism.

[17] An echo of the old ethical code of science is found in the review, by Hedwig and Max Born, of *The Scientist Speculates: an Anthology of Partly-Baked Ideas*, ed. I. J. Good (Basic Books, Inc., New York, 1963), in *Bulletin of the Atomic Scientists*, 19 (May 1963), 30–3. The review starts: 'The publication of [this book] is nothing less than a crime against the ethical code, unwritten but vital, of the community of scientists.' Later: 'Uncompromising, indefatigable pursuit of truth, then, is the hallmark that distinguishes the scientist from the charlatan. It constitutes the indispensable ethic of science.' They object particularly strongly to the making of priority claims on the basis of half-baked speculations. The review concludes with a list of blunders and absurdities contained in the book.

and the willing acquiescence of the rest. Should morale decline among the ordinary members of a scientific community, then it becomes impossible, either by sanctions or rewards, to restore it through actions taken at the top. And without good morale, what good work is done will fairly soon be driven out by the bad. There will inevitably be some leaders who will strive only for instant prestige; the referees associated with them will take the hint and authenticate any property which will enhance or share in that prestige; and ordinary scientists will be under pressure to achieve their private purposes most cheaply by identifying the current fashions and producing passable results in their image. If this degeneration occurs in isolated fields, it can still be checked through the system of controls across fields. But if corruption spreads to this highest level as well, then whole areas of science can become gigantic confidence-games, producing pseudo-property at a feverish pace, and resembling a stock exchange in a bull market rather than a collective endeavour on behalf of the highest human goals.

Ethics in Science: Old Pronouncements and New Problems

The most eloquent pronouncements of the ethical basis of science came at the end of the academic period, when their basis in social reality was on the point of extinction. In retrospect, they can be seen to suffer from the common fallacy of defining a social activity in terms of the idealized attributes of its perfect state. As materials for an analysis of the ordinary practice of scientific research, they have been recognized as gravely defective. But when they are recognized for what they really were, statements of a credo and warnings of danger, their true worth can be realized. They need not apply in detail or even in principle to every scientific worker who publishes a paper; but if the spirit embodied in their separate insights is lost to the leaders of science and to those who aspire to that position, then the outlook for science is grim.

The three most influential spokesmen for the ethical commitment of science in the 1930s approached the problem from different and sometimes contradictory points of view. The physical chemist Michael Polanyi concentrated on the intense personal experience of scientific inquiry, which was necessarily undertaken by a community of free and dedicated men. The philosopher Karl Popper saw science as the embodiment of intellectual honesty, realized through the principle and practice of criticism. And the sociologist Robert K.

Merton saw in the 'ethos' of science the realization of the highest standards of civilized human behaviour.[18] All three identified dangers to science: for Polanyi they lay in bureaucratic direction and control; for Popper they lay in the dishonesty latent in the denial of criticism; and for Merton they lay in the inherent conflict between the norms of co-operative scientific endeavour and those of lay society and the State. The paths which their thoughts took in later years do not concern us here; nor need we attempt a detailed assessment of their conclusions. It is, however, worth recalling that they were engaged, each in his own way, in a complex debate centring around the Marxist view of society and of science.[19] The Marxist writers, of whom J. D. Bernal was the most influential, were primarily concerned with the external, social functions of science. They believed that science had an obligation to contribute in a conscious and planned way to the benefit of mankind; and also that this was being done, with no loss to the integrity and freedom of science, in the Soviet Union.[20]

Looking back from the other side of the watershed of the Second World War and the subsequent industrialization of science, it is easy to see that the proponents of an idealistic ethic were unrealistic about the attitudes necessary for the ordinary practice of science to be successful in the short run; and that the Marxists were grossly ignorant of the problems of science under Soviet conditions. But

[18] For K. R. Popper, see the revealing autobiographical account in 'Science: Conjectures and Refutations', ch. 1 of *Conjectures and Refutations* (Routledge, London, 1963). He describes how, in 1919, he came to see that *pseudo-science* (astrology, Freud's psychoanalysis, Adler's psychology, and Marxism) survives by seeking (and finding) confirmations, while real science (Newton, Einstein) advances by making testable conjectures. For Polanyi, see *The Logic of Liberty* (Routledge, London, 1951), for example p. 51: 'The coherence of science must be regarded as an expression of the common rootedness of scientists in the same spiritual reality. . . . Only then are the conditions for the spontaneous co-ordination of scientists properly observed.' For R. Merton, see 'Science and the Social Order' (1938) and 'Science and Democratic Social Structure' (1942); reprinted in *Social Theory and Social Structure* (The Free Press, Glencoe, Collier and MacMillan; revised edition, 1957), 537–61. The later paper gives the classic formulation of the 'ethics of science' with the four 'norms' of 'universalism', 'communism', 'disinterestedness', and 'organized scepticism'.

[19] Popper had adhered to Marxism, to some extent, before his philosophical conversion; but he later devoted much time to its refutation. Polanyi was consistently anti-Marxist and anti-collectivist. Merton took Marxism seriously and did a study of seventeenth-century science from a 'utilitarian' point of view, parallel to that in which he related science and Puritanism. See *Social Theory*, chs. 18 and 19. It is tempting to speculate whether the 'norms' of science, particularly that of 'communism' resulted from an attempt to reconcile Weber and Marx.

[20] J. D. Bernal, *The Social Function of Science* (Routledge, London, 1939).

each of these approaches reminds us of a necessary component of a scientific ethics appropriate to the conditions of the present and the near future. The idealistic approach involved the assumption of the excellence of scientific inquiry as a permanent reality, and from this derived an enlightened ethical commitment of scientists as a logical consequence. My argument has used the same components, but has changed the mode of causality. I make the more realistic assumption that the health and vitality of scientific inquiry are not guaranteed, either by the objects of inquiry or by the social aspects of the work. Hence, unless there is an effective ethic, even more refined than a 'professional ethic', this very delicate and sensitive work will not long continue to be well governed or well performed. Yet it is unlikely that the ideological basis for such a scientific ethic can continue to be found in religion and philosophy, nor its practical roots in the refinement of the code of conduct of an élite class. Therefore the ethical basis of future excellence in science must lie in some other ideals and experience; perhaps in a humanitarian commitment, necessarily interpreted in a much more sophisticated fashion than ever before.

Part IV

SCIENCE IN THE MODERN WORLD

INTRODUCTION

Up to this point, our discussion has been restricted, for the sake of simplicity, to 'scientific' problems, where the goal of the work is the establishment of new properties of the objects of inquiry, and its ultimate function is the achievement of knowledge in its field. For an historical or philosophical study of science in the period preceding its industrialization, such a restriction is natural; then, it was only occasionally, and in certain specialized fields, that inquiry of this sort could be applied directly to the mastery of the natural world. But with the interpenetration of science and industry, this traditional separation has come to an end. Indeed, most of the tasks undertaken by the contemporary corps of 'scientists' have very different functions, and even different goals as well. The emergence of this new sort of work has led to a variety of new descriptive terms, which reflect the confusion over its nature and its relations with science of the traditional sort. The term 'science' itself has stretched to include technology, and even any sort of inquiry whose methods are modelled on those of the experimental and mathematical natural sciences. This multiplicity of meanings is all encased in the single plausible title of a projected new field of inquiry: 'the science of science'. Since these new sorts and extensions of 'science' are those which present the most severe problems in their relations with the social and natural environment, we must try to achieve some understanding of their distinctive character.[1]

From one point of view, all the new sorts of 'science' are entitled to that honorific appellation: their work consists of problem-solving on intellectually constructed objects; and there is no clear demarcation between the methods used in traditional pure science and those in the most speculative of social technologies. We can, however, make a useful distinction among the various sorts of problems by means of

[1] The 'philosophy of technology' is a subject which has attracted relatively little attention in recent years. The record of a symposium on the subject is published in *Technology and Culture*, 7 (1966), 301–40. In this collection of papers, that by Mario Bunge, 'Technology as Applied Science', bears the closest relation to the problems discussed here.

the classification of final causes involved in a task, that I developed in connection with the problem of quality control in science. We recall that the task itself has a goal, which is conditioned more or less strictly by the function which will be performed by the result of the accomplished task; and this in turn is governed by the ultimate human purposes which are expected to be served by the performance of that function. In the previous discussion, we were concerned with the need for a harmony between these 'objective' final causes and the 'subjective' ones brought to the task by the scientist: his own purposes, in what he is trying to achieve for himself in the work; and his motives, which bring him there and set his purposes within the constraints of the task. For the present problem, we shall concentrate mainly on the 'objective' final causes, and by their means distinguish between problems capable of description as 'scientific', 'technical', and 'practical'.

In each case, we consider the task itself as the investigation of the problem; the 'goal' of the task is the solution of the problem. We then ask, to what extent are 'functions' and 'purposes' involved in the specification of the goals of this sort of task. In the case of the 'scientific' problems that we have been discussing up to now, these higher final causes are not so influential. The function to be performed by the solved scientific problem, that is the contribution of new results for the advancement of the field, conditions the work only in a general way, through the controlling judgements of adequacy and value. The investigation of the problem may change course in mid-stream, if a more promising solution appears possible; and then the solved problem will perform a different function, to no one's regret. The ultimate purposes to be served by the scientific problem are quite remote, diffuse, and unpredictable in detail; we recall the words of Helmholtz on the radical separation between the applications of science and the goals of the scientific endeavour.

By contrast, we describe as 'technical' problems those where the function to be performed specifies the problem itself. The goal of the task is fulfilled, and the problem 'solved', if and only if that function can be adequately performed. A design team that produces a device that performs a function different from the one assigned, will not usually be congratulated on its initiative and independence. On the other hand, in a technical problem the work is conditioned only in a general way by the purposes to be served. Thus, in a commercial product, what counts is whether a sufficient number of people expect the product to serve their purposes, and so buy it; exactly why they like it does not matter except as a guide to the advertising.

Finally, there is a class of 'practical' problems in which the goal of the task, in principle, is the serving or achievement of some human purpose. The problem is brought into being by the recognition of a problem-situation, that some aspect of human welfare should be improved. Such problems involve intellectually constructed objects rather than ordinary 'common-sense', as we see from programmes and debates on such issues as 'education', 'poverty' and 'health'. We shall see that large-scale practical problems give rise to subsidiary technical problems, and perhaps to scientific problems as well. The fulfilment of its goals in a multitude of individual cases will usually require a large-scale project, with a hierarchy of decision and control and a division of labour, analogous to the projects required for the complete solution of a large-scale technical problem.[2]

By means of this distinction we can consider the similarities and differences among the three basic sorts of problem, as well as the various sorts of difficulties to which they are prone, individually and when they are related in practice. We shall see that technical and practical problems each have their cycles of investigation, and are subject to controlling judgements analogous to those of scientific problems. Even in the case of technical problems, there is no automatic mechanism enforcing quality control on solutions, and one can speak of a degeneration and even corruption in technology as in science. Practical problems, and the large-scale practical projects required for their execution, are by their nature subject to the most severe hazards and pitfalls. An understanding of these is most urgently necessary; for the increasing complexity and sophistication of our society throws up a rapidly growing set of practical problems. Even problems previously considered as purely 'technical' are

[2] My classification of scientific, technical and practical problems is closely parallel to Aristotle's classifications of 'theoretical', 'productive', and 'practical' knowledge. See the *Nicomachean Ethics*, vi, 3–5. 1139^{b}14–1141^{a}8, *Metaphysics*, E.1.1025b, and Sir David Ross's commentary on the latter text in his edition (Clarendon Press, Oxford, 1924). Aristotle distinguishes between 'making' and 'acting', and observes that the latter is not governed by any end outside itself. This corresponds to my distinction between 'performing a function' and 'achieving a purpose' for the goals of technical and practical problems respectively; the function to be performed by a device is always governed by some ultimate purpose or purposes.

increasingly being debated in their 'practical' aspects under the slogans of 'environment', 'pollution', and 'amenity'.

To achieve this understanding, we must appreciate the consequences of the ineffectiveness or immaturity of the disciplines (usually those concerning man and society) on which the investigation of practical problems depends. The analysis of the state of immaturity is not a pleasant task; the condition itself is generally regarded as at best a misfortune to be remedied quickly, or at worst as a scandal to be concealed. But without a knowledge of the deep and systematic differences in the practice of inquiry, and in the sort of results obtainable, between immature and mature disciplines, we will have no way of controlling the harmful effects (to themselves and to society) of ineffective fields grown beyond their appropriate size and prestige. The analysis of the credence given to the claims of patently ineffective disciplines leads us into a discussion of the category of 'folk-science', which applies equally well to the most sophisticated academic exercise as to peasant superstitions. With this battery of concepts we can return to a discussion of the hazards and pitfalls of practical projects; and see that the wreckage of so many of the great schemes for human betterment is, sadly, only the natural state of affairs, at least so long as our conception of 'science' is that inherited from Galileo and Descartes.

12

TECHNICAL PROBLEMS

As a result of the industrialization of science, the traditional lines of demarcation between 'science' and 'technology' are blurred and confused. One cannot distinguish the two sorts of problems by the formal qualifications or the formal title of employment of the research worker, nor even by the name and stated character of the larger project of which a particular investigation forms a part. Indeed, a single problem may have a mixed function, and so be both 'scientific' and 'technological' in varying degrees to the different parties involved in its planning, investigation, and application. If the real differences between 'science' and 'technology' are to be appreciated, and their interactions controlled for mutual benefit, there must be a clear understanding of the differences between the problems characteristic of each sort of work.

We can introduce the class of problems described here as 'technical' by recalling the cases where such problems have been mentioned in connexion with scientific inquiry earlier in this account. First, there are those problems whose goal is the establishment of new properties of the objects of inquiry, but where the function is assigned, externally and precisely. This is where certain information is required for the successful solution of a problem, and the provision of this information constitutes a subsidiary scientific problem in the work. The external, determining problem may be a scientific one, and the subsidiary problem may be investigated by the scientist himself, or by an assistant. But in this situation, the use to which the information will be put, its function, constitutes the major component of the value of the subsidiary problem, and it determines the standards of adequacy by which its solution will be assessed. The investigation of the subsidiary problem is then governed by criteria of quality which are not at all the same as those for a scientific problem; and even though the result has the same form as that of a scientific problem, we gave it a different name, technical problem.

A more pure case of a technical problem is also encountered in scientific research: the making of tools. Here the function determines the problem, but the solution of the problem, instead of being the conclusion to an argument, is the creation of a device, the means to the performance of the function. Sometimes such work involves only the application of simple craft skills, even if the objects themselves are defined in terms of scientific categories; this can be the case, for example, when one assembles standard electronic components to create a particular measuring instrument. Then the work is best described as a technical task, rather than as a technical problem. The latter title is better reserved for the work of those scientists who specialize in tool-making, where the sophistication of the tools is so great that the work involves drawing conclusions to arguments concerning the properties of intellectually constructed classes of things and events. Technical problems, in this sense, are not restricted to scientific tool-making; this is the sort of work required whenever there is a function to be performed, and its new means cannot be created by craft knowledge alone.

The penetration of industry by science in the present period is possible because the set of soluble technical problems is large, increasing, and continuously growing in importance. In early modern times, the only such problems were in the descriptive-mathematical sciences: navigation, surveying, and their allied fields, and arithmetic. A significant advance did not come until after the industrial revolution; although in the eighteenth century there had been a development of the application of an experimental approach to practical tasks (as in Smeaton's analysis of waterwheels),[1] and also some innovation which used established scientific results as information. But until the later nineteenth century, the intellectual objects of scientific inquiry were generally so removed from the physical objects of production, that attempts at analysis and innovation in their terms were generally fruitless.[2] The effective

[1] Dr. D. S. L. Cardwell and D. A. J. Pacey, of Manchester, have shown the significance for science as well as for engineering, of Smeaton's work; see D. S. L. Cardwell, 'The Academic Study of the History of Technology', *History of Science*, 7 (1968), 119–20.

[2] I am indebted to Professor A. Kauffeldt, of Magdeburg, for a systematic development of this point (private correspondence). He observes that the impotence of theoretical science, and its resulting irrelevance to the immediate needs of society, were

penetration of science into industry first came in those areas where the products themselves were pure and artificial, as in synthetic dyestuffs and electricity; and even to this day, it is more difficult to solve technical problems in industries whose materials and processes are largely traditional.[3] But the largest and most rapidly growing sections of industry in the advanced economies are those whose basis is scientific, and it is in their 'R. and D.' sections that the solution of technical problems has become a major industry in itself. We may consider 'technology' as the activity of investigating such technical problems, as 'science' is the activity of investigating scientific problems; and in this way identify the similarities and differences between the two sorts of work.

The Cycle of Investigation

By analogy with a scientific problem, we may define a technical problem as a statement of a function to be performed, whose means are to be established as the conclusion of an argument, with a plan for its accomplishment. The similarity to scientific problems is brought out by the mention of the conclusion of an argument, which implies a discussion of classes of intellectually constructed things and events. However, a technical problem will generally have less freedom to evolve than a scientific problem; to the extent that the function is externally assigned, and is not the speculation of an independent inventor, it cannot easily be altered if some unexpected possibility or difficulty appears in the course of the research. Also, the function will generally be specified in a more detailed fashion, each of its aspects being assigned some standards of adequacy of performance. This further restricts the freedom of the problem to evolve, and also makes its cycle of investigation significantly different from that of a scientific problem.[4]

useful in providing it with the autonomy in which to develop along lines which were ultimately very fruitful.

[3] 'No newcomer to textile research can help but be impressed with the degree of optimisation of the processes that has been achieved by trial and error methods extending over a long time. The same cannot be said for the conventional theories that have gone with these developments. One of the earliest needs of the textile engineer who wishes to modify existing processes is to disentangle the true purpose from the supposed purpose of many textile machines.' P. Grosberg, 'The Changing Role of the Textile Engineer', Inaugural Lecture (Leeds University Press, 1965), p. 11.

[4] This schematic account necessarily abstracts from many of the aspects of getting a technical problem under way, in particular, those of decision which we have discussed in connection with science. On this, see E. B. Roberts, 'Entrepreneurship and Technology', in *Factors in the Transfer of Technology*, eds. W. H. Gruber and D. G. Marquis

As an example, we might consider the creation of a new model of aeroplane.[5] The project is undertaken only when someone believes that a certain function could be better performed than at present. This is not merely 'carrying passengers by air'; it will concern a class of traffic defined by load in passengers and freight, distances travelled between stops, available landing and maintenance facilities, climatic conditions, and required safety and reliability of service, all of these with estimates of averages and extremes, and translated into a variety of capital and running costs and returns. Until someone conceives of a model of aeroplane that can adequately perform the particular function just described, we may say that a technical problem-situation exists, but not a technical problem. The problem comes to be when a possible model is tentatively specified. This specification will not be in terms of the function itself, but of the operating characteristics of the proposed model. This distinction is important, for it is the first of the conceptual shifts that occur when a technical problem is being investigated. For example, when 'function' is being discussed, landing facilities will be defined by length, orientation, and quality of runways, provision of guidance systems of different sorts, and relevant topographical and climatic features. The landing properties of the aeroplane, on the other hand, are defined by certain parameters of its own behaviour, including landing and stalling speed, angles of ascent and descent, taxiing distances, along with properties of its installed guidance system, undercarriage, and other relevant features. We should notice that both the function and the operating characteristics are defined in terms of scientific categories, rather than craft knowledge or informed common sense. The numbers by which they are described are based on tests using sophisticated equipment and a variety of theoretical assumptions, under specialized conditions.[6] Thus even

(M.I.T. Press, 1969), pp. 219–37. The same volume contains a review by D. J. de S. Price on the work on patterns of citation and communication in science and in technology: 'The Structures of Production in Science and Technology', pp. 91–104.

[5] A very clear account of the cycle of a technical problem, applied to a new model of an aeroplane is given by D. Norman and J. Britten in 'The Islander Project', *Advance* (University of Manchester Institute of Science and Technology), No. 6 (April 1969), pp. 44–9.

[6] Working engineers are necessarily familiar with the theory-laden and imprecise character of specified operating characteristics in their own field; the general public learns of them only incidentally. Thus, the 'True Boiling Point' of a petroleum distillate, the only way to 'establish its basic marketing specification', 'leaves a lot to be desired'. It requires a lengthy process for its determination, 'the results are useless in controlling a refinery', and 'the method is not very reproducible [*sic*] especially as it applies to the

in this early phase of the work, there is a transition from one set of intellectually constructed objects to another; and an argument is required to establish the conclusion that a device with these particular operating characteristics will perform the assigned function to the given standards of adequacy. In this example, the argument will be devoted to showing that an aeroplane with this particular landing behaviour will be able to make reliable and safe use of the anticipated landing facilities under an anticipated range of conditions. We should notice that such an argument involves approximations and invokes considerations of more or less, and invokes probabilities, as an essential part of its content.

We notice that the full specification of the 'function' includes rather more than the use to which the device will be put; we also find relevant aspects of the conditions under which it will operate, and of its maintenance in operation. Although this extension may appear to stretch the concept beyond its original meanings, it is easy to see that these additional components of the function must be specified in order that the device as designed will do its job. Also, each of these components of the function must be translated into appropriate operating characteristics. As in scientific inquiry, pitfalls can be encountered at every stage in the work; and in the process we have described so far, two sorts are possible. First, there may be faults in the translation from the specified function to the operating characteristics; so that even if the device is constructed to meet these, it may yet fail to perform its assigned function. This pitfall is most likely to be encountered when a concentration on the most important aspects of the function leads to the neglect of the implications of something apparently minor. This can happen most easily when the original set of operating characteristics proves to be unrealizable, and the task of re-designing is undertaken under pressure of time. The revised set of operating characteristics, tested against the main aspects of the function, may appear to be adequate, but it is possible either that some of the characteristics may be mutually incompatible, or that some aspect of the function, which was satisfied as a matter of routine in the first conception, is satisfied no longer. A more serious and fundamental pitfall is in the incorrect specification of the function. This occurs when there is insufficient craft knowledge of the actual conditions in which device will

initial and final boiling points.' Hewlett-Packard offers this information, as background to its improvement upon it. See advertisement in *Science*, 167 (1970), 1169–70.

operate, so that the categories in which the function is specified fail to include all the relevant aspects of the real situation.[7] This particular sort of pitfall is most likely when radical innovation is being attempted; it is impossible to be sure what the working conditions will be like until the device is actually working, and it is impractical to over-design in every conceivable aspect for safety. One mark of the great engineer is the ability to penetrate into the unknown in this way with success.

Once the operating characteristics are specified, the cycle of investigation of a technical problem proceeds through several further phases, each of them involving a translation between conceptual objects, an argument, criteria of adequacy, and characteristic pitfalls. The next phase is design, providing specifications of the structures, motions and actions by which the operating characteristics can be realized, as well as the materials in which the structures and processes are embodied. From the design, detailed plans are drawn up, providing the basis for the construction of the physical object itself. This end of the cycle can vary greatly in its complexity, depending on the sophistication of the device and of the process of production, and on the quantity to be produced. There may be a series of prototypes, produced under conditions increasingly

---

[7] The influence of the theory in whose terms a device is conceived on the specification of the function is shown well by an example of V. Ronchi, *Optics, the Science of Vision.* His experience of the design of range-finders is that they are designed to the highest possible accuracy, by the installation of extra lenses and plates. But this refinement is irrelevant when the instrument is used under conditions of low illumination, when all the extra optical parts cause a loss of energy to below the threshold of visibility. This neglect of the conditions of use stems from the neglect of the eye of the observer in physical optics, which he discusses at length (p. 283). The imposition of quite arbitrary criteria of adequacy in the performance of a function could be seen in certain characteristics of the Système International of measurements, as first announced in the United Kingdom. With a fine rationalist zeal, the designers of the system determined to reduce the number of official prefixes denoting magnitude, so that the memory of users would not be overburdened. One result of this was the banishment of the 'deci-' and 'centi-', denoting tenth and hundredth respectively. Then all new modules in industry were to be given in millimetres; this extended to builders' bricks. In this the designers of the Système neglected the well-known fact that the recording of measurements is onerous, liable to blunders, and also misleading in its implied precision, unless the common magnitudes are expressed as small multiples of submultiples of an appropriate unit. One group of scientists were able to force a violation of the 'thousands' rule for derived units, and achieved an official unit for pressure, very close to one atmosphere, which is 10^5 times the standard unit. On the other hand, the common run of humanity was expected to put up with measuring lengths either in decimal parts of a metre, or in large multiples of the millimetre. It appears from casual indications that recently this hyper-rigorous approach has quietly been relaxed.

resembling those of full-scale production; and the creation of the means for the production may itself involve technical problems of great difficulty. The full cycle, for a commercial product, is not complete without regular production, followed up by promotion, sales and service. As the cycle advances to its latter phases, the work thus changes from that of investigating a technical problem, to what we may call accomplishing a technical project. The scientific and technical component diminishes, and is replaced by routine and craft work, social organization, and (for most devices) commercial considerations. In one sense these are an integral part of the cycle, for it is only when the device brings in financial returns that the ultimate purposes for its creation can be achieved; and its success against competition is an important part of the test of its adequacy.

Technical and Scientific Problems: Differences and Similarities

The process of re-cycling during the investigation of a technical problem will have important differences from that of a scientific problem. Once a scientific problem has been set, the materials produced in an earlier phase serve as the basis for those of later phases; as 'causes' in the Aristotelian sense, they are material, and only when they are inadequate to their function is there a need for a revision of the plan of inquiry. Even then, the revision may be either in the direction of improving the materials, or in altering the goal of the task in the light of the experience with these materials. Of course, even in science such a re-cycling does not always succeed; the materials may turn out to be useless for the establishment of any worthwhile conclusion. In technical problems, the relation of the earlier phases to the later ones is more that of final causes; the assigned function delimits the possible operating characteristics, which in turn delimit the design, and so on. Re-cycling becomes necessary when it appears that the requirements set by the results of an earlier phase of the work cannot be met, either from technical incompatibility or (more commonly) from the preassigned limits on costs. Then every significant change in plan has effects running right back up to the definition of the problem itself, and hence the existence of the technical project. Thus a difficulty encountered at the prototype stage may require changes in design which alter the operating characteristics and so yield a modified device which does not perform the preassigned function as well as had been expected. Hence while the problem lacks the same freedom to evolve as a scientific problem has, the constraints imposed on the solution of problems at the later phases present challenges, and require control and co-ordination, to a degree which is not present in science.

In spite of this interrelation of phases, the difference between the earlier and later tasks enable them to be considered as distinct, in their organization as well as conceptually. Especially in the most sophisticated industries, the phase of 'research and development' can involve exploratory work where the preassigned function is not always narrowly defined.[8] In the case of such work, we can speak of the investigation of technical problems in the strict sense, independently of the technical projects of which they may form the earlier phase. Because of this natural division, it is possible for a technical problem to be successful in one sense while failing in another; the 'creation of a device' does not succeed or fail as an indivisible whole, as does the investigation of a scientific problem. This division is not a new one; the history of technology accords posthumous honour to men for the greatness of their conceptions, in spite of the occasional failure of their creations at the practical or even technical levels.

Although the ultimate purposes and even the immediate functions of technical problems are so different from those of scientific problems, the goals of individual tasks can be similar, and the methods of work nearly identical. In both cases, the solution of problems involves argument, based on controlled evidence, and cast in terms of conceptual objects. Because of this, someone with the training and outlook of a technician will be incapable of accomplishing the tasks; while a scientist, once adjusted to the new sorts of problems, will work as well as someone trained up to technology. In fact, the differences between scientific problems and technical problems become prominent only when the work is considered in the large. In routine, detailed work, the two sorts of problems overlap con-

---

[8] In his review of H. L. Nieburg, *In the Name of Science* (Quadrangle Books, Chicago, 1966), in the *Bulletin of the Atomic Scientists*, 22 (November 1966), 22–3, Bernard I. Spinrad distinguishes two stages of 'R. and D.': 'the production of a workable device illustrating a principle, and the production of a device using the principle for specific goals'. In the language of this analysis, one first sees whether a particular idea can be realized in a model whose operational characteristics are absolutely minimal; and one afterwards investigates the possibility of improving the design so that the device can perform some recognized function. He argues that some projects labelled as 'boon-doggles', as the 'Dynasaur' and the nuclear-powered aeroplanes, were actually serious projects; and for support he cites the nuclear submarine, which originated in just this way.

siderably, and a given problem may be capable of being considered in both classes simultaneously. The social distinction between science and technology, which is of great practical importance in England, derives not from any essential difference between the two sorts of work, but from a combination of inherited social attitudes (which are not so common on the Continent), together with academic prejudices imported from nineteenth-century Germany. But the idea that someone trained as a 'scientist' is thereby rendered less willing or less capable at working on technical problems has little relation to reality at the present time.

Yet there are certain deep differences between the two sorts of work; so that for the preservation of the health of both they must be kept distinct while yet in contact. It is well known that a scientific result may give rise to a technical problem, when it is realized that some function might be performed by means of its solution. Conversely, technical problems may give rise to subsidiary scientific problems which then become deep and fruitful investigations in their own right. Indeed, it can be argued that without this sort of 'hybrid vigour', as well as that resulting from an analogous interaction between different scientific disciplines, science would become sterile. But the creation of new scientific problems cannot be managed in an institutional manner as neatly as the exploitation of results; and so it would be dangerous to the integrity and survival of science if it were forced to seek its inspiration exclusively from the problems passed to it by technology. Also, in the long run the relations of science and technology to the external world will be very different. The process whereby a new field of science liberates itself from its technical origins and becomes governed by the advancement of knowledge for its own sake, is entirely legitimate, and necessary for the continued growth of science. A field of technology has no such freedom, of course; if it does not continue to contribute to the performance of functions in the real world, then it has lost its reason for existence and support.

We have seen that the classes of 'science' and 'technology' are continuous, and indeed overlapping. The continuum extends through technology, to more traditional sorts of work: invention and engineering. The defining characteristic of invention is its novelty: either the creation of a device to perform a new function, or one which performs an existing function in a new way. In engineering, the element of novelty is not so strong; here the task is the creation of a device to perform a known function in new conditions. Of course, a significant work of engineering will have a strong element of novelty, and may give rise to a host of inventions; here, as in any other sorts of work, the distinction is based on which problem is considered as fundamental, and which as subsidiary. Generally speaking, invention relates to the earlier phases of a technical problem, involving the conception of the device, while engineering is more concerned with making it work. Whether a particular problem should be classed as 'technology' rather than as invention or engineering depends on the demarcation adopted; that of Lord Bowden, that 'an art may become a science if it is concerned with less than about seven variables', is useful.[9] For it reflects the fact that in a real problem with many interrelated components, the abstract categories of scientific argument will not correspond sufficiently closely to the external world, to enable a technical task conceived in their terms to be accomplished. This is not to deny that on occasion a technologist may derive useful information from, say, a multivariate analysis on fifteen ill-defined variables; but unlike a scientist in an undemanding field, he does not enjoy the luxury of of being able to consider such materials as an adequate solution to his problem.

Because the solution of technical problems of any of these sorts is governed by functions and by criteria of quality assigned externally, those practitioners who rise above routine work do not have the same freedom as scientists to develop a specialized competence and style, ignoring every aspect of a problem that does not suit their taste. In this particular respect, good work in the solution of technical problems is more challenging to the intellect and to the imagination than such work in a self-contained field in a matured scientific discipline. Faced with a particularly refractory bit of the natural world in his experiments, the scientist can choose to move on to another problem; while the engineer in the same situation must fight it out right there, also knowing that the cost of defeat is far higher than a few months of his personal laboratory time. A good 'R. and D.' man in industry may find himself engaged in problems which run right across the spectrum from pure scientific inquiry to craft-based engineering; and in turning from one to the next, he must have flexibility of outlook and diversified skills, to

[9] B. V. Bowden, *Proposals for the Development of the Manchester College of Science and Technology* (The College, Manchester, 1956), p. 48.

apply technique and judgements appropriate to each case. The real intellectual challenge of good technological work is known to those who teach technology, and who try to recruit good students to it. It is then a paradox that English univerity students in technology are, on average, academically inferior to their colleagues in science; and so it would seem that technology can make do with inferior talent. The simple solution to this misleading paradox is that outside the traditional engineering tasks, recruitment to technology is independent of the title of the university degree, and there is then a sorting by ability, the more able becoming technologists, and the less able, technicians.[10]

Social Aspects of Technical Problem-Solving

The social aspects of work in technology have important differences from those in science; although the great diversity in the problems solved, and in the institutional setting, makes neat contrasts very difficult. In general, the basic form of property created in the work is a contribution to the 'art', and as such is registered and protected formally through a system of patents. But such property usually belongs to the employer of the man who actually created it, and for the establishment of his personal intellectual property in a solved problem, he must use social mechanisms more similar to those in science. These may be formal, as the publication of a paper in which he exhibits the solution of the technical problem or a related scientific problem, or purely informal, through an interpersonal channel of communication with colleagues. A great variety of strategies will be adopted, which depend not only on the character of the work done, but also on the regulations and traditions of the different institutions and occupational groupings.[11]

The mechanisms of quality control in technology will generally be different from those in science, and require a closer analysis. As we have seen, every solved problem is subjected to tests of adequacy; but although in the last resort these depend on craft judgements,

[10] N. D. Ellis, 'The Scientific Worker' (Ph.D. Thesis, University of Leeds, 1969), has shown that in his sample of several hundred industrial research personnel there is no significant difference in work tasks or even in attitudes between those with degrees in science and in technology. I am indebted to Dr. Ellis for many fruitful discussions on these and other matters dealt with in this book.

[11] N. D. Ellis, 'The Scientific Worker', discusses this and reviews the relevant literature, in the section on 'The Context of Non-Academic Research Q.S.E.s' Work Roles', pp. 123–30.

there is usually a more explicit framework, related to performance, in which these operate. Similarly, the value of a problem, when it is conceived purely in terms of its benefit to the person or organization sponsoring the work, is capable of assessment in largely quantitative terms. The assessment of the value of a new problem will naturally be speculative, and sometimes highly theoretical in itself; but it does not depend on such indefinable judgements as the contribution to the solution of future problems which do not yet exist.

In general, then, the criteria on which the judgements of quality are based are simpler in principle in technology than in science. The function to be performed provides an objective basis for criteria of adequacy; and the purposes to be achieved, in the way of financial return, do the same for value. Even those judgements which in science are so difficult and so sensitive to distortion, as the choice between investment in totally different fields, are here rendered straightforward by the common standard of commercial success. Hence these controlling judgements do not require so much in the way of accumulated wisdom and statesmanship among leaders to remain appropriate at least in the short run. Similarly, the degree of ethical commitment required for the maintenance of quality, will be less at each level of the structure of direction and control, than in science. This is not at all to say that scientists are ethically pure while technologists are impure; rather, that in technology the various components of the social mechanisms for the enforcement of quality control are in principle simpler.[12] The mechanism is then less dependent on ideological and ethical commitments, and can be managed adequately on a basis more similar to that by which society at large keeps going.

However, the solution of technical problems is not the same thing as business enterprise in a competitive market; there are important situations where the simple test of profit is either inappropriate or irrelevant to the assessment of the quality of a technical problem or its solution. Even in commercial products, competition by quality and price is distorted when one firm dominates the market or a handful share it out between them, selling indistinguishable objects and competing only in their advertising. Innovations which threaten such tidy arrangements can be killed at birth by a variety of means,

---

[12] N. D. Ellis, 'The Scientific Worker', shows the important respects in which industrial Q.S.E.'s do not form a 'profession' in any of the accepted senses of the term; see ch. 6, 'Professionalism in Practice: Its Meaning and Significance'.

given the superior resources of the established firms to those of any potential competitor. The suppression of inventions was a more popular theme before the Second World War, perhaps because the general depression in business affairs made speculative capital less available. But the belated emergence of alternatives to the petrol-engine in the later 1960s, under the political pressures resulting from recognition of the problem of atmospheric pollution, indicates that the process of entrenchment and stagnation of technology is still with us.

Moreover, there is a large and increasing number of situations in which there is no effective competitive market to make its contribution to the testing of adequacy of a device. This will include unique productions, such as roads, bridges, dams, and large buildings; once constructed, they occupy their niche for some considerable time at least, regardless of how well they perform their function. Similar to these, in principle, are products of which only one model can be in operation, as is the case with much military equipment. Those who use such devices may well have their opinions about their quality; but there is no automatic mechanism whereby the aggregate of these opinions can be translated into a recognized penalty or reward. In this way, important parts even of the 'market sector' of the capitalist economies, suffer from the defects now being recognized in the socialist economies. For the assessments of quality must be made formally, through institutional channels, and are thereby subject to certain characteristic weaknesses. First, the assessments will be derived as conclusions of arguments framed in scientific categories, in which the evidence is derived from tests. This work is carried out by experts acting on behalf of the purchaser; and if he is so minded the supplier can always find other experts to controvert their assessment. In all but extreme cases, there can be genuine differences of opinion; and when the experts disagree, there is little hope for the helpless layman, who lacks both the basis for comparison and the opportunity to switch to another brand. Also, the institutional framework in which the work of assessment is done may itself be subject to distorting pressures, acting through political channels or directly. The most direct method is through controlling the controllers: the purchasers' testing facilities are either weakened or subverted by the supplier, so that he can effectively control the assessments of quality at every stage of a contract. Such practices are best documented for the case of United States Government procurement. There, what is popularly called 'science' (mainly military and aerospace projects) is treated like any other commodity being sold to the government, and all the available profit-maximizing techniques are applied, vigorously and ruthlessly.[13]

In conditions such as these the profits earned by a particular device are by no means in any direct proportion to its quality; indeed, in some significant cases, the proportionality seems to be inverse.[14] These abuses are particularly likely to occur in the most sophisticated, speculative, and expensive sections of technology, as military and aerospace, where gigantic sums of government funds are involved, the competitive market is absent, and political considerations of prestige and 'defence' weigh heavily in the assessment of projects. In these conditions we can speak of a corruption of technology; and since it occurs in those sections recognized as leading, in finance, growth and prestige, its damaging effects can easily spread.[15] Moreover, there is no inherited ideological resistance to

[13] On this, see H. L. Nieberg, *In the Name of Science.*

[14] In '"Secret Study" of U.S. arms flops', *The Guardian* (27 January 1969), Richard Scott describes a *Washington Post* story of 26 January about an assessment of military aircraft missiles in a document then circulating in Government circles. It was alleged not only that costs were double or triple those estimated, and contracts two or more years delayed, but that most of the 'systems' studied were poor in performance: only four of thirteen worked to 75 per cent of specifications, and four others were discontinued or cancelled outright.

[15] One might raise an interesting philosophical question, as to whether shoddy weapons systems are morally superior to effective ones when they are intended for use in an unjust war. There is the precedent of the slave-labourers in the Nazi armaments factories who practised a quiet sabotage by producing substandard work whenever possible; but it is hard to cast the major American aviation companies in the role of secret agents for the Vietcong. Of less interest for moral philosophy, but of great importance for society, is the possible spread of corrupted technology into the ordinary market sector. It is one thing to recognize that mass advertising is generally meaningless pap; but it is far more serious if one cannot have a fair certainty that a device will function properly at all, or that the manufacturer or merchant will show any interest in repairing or servicing it. It is difficult to imagine the state of our civilization if 'shoddy technology' became ubiquitous; I think it would be demoralizing in every way. It is impossible to say whether the quality control on ordinary manufactured goods is less effective now than in earlier times; subjective impressions here are extremely unreliable. There is some evidence that American technology is becoming somewhat tattered at the fringes. A report from New York, after the minor fiascos with the Boeing 747, explained the lack of public concern by the fact that in New York at least, no one expects anything to work; and transport and communications systems least of all. See the article by M. Leapman, *The Times* (28 January 1970). A report of a decline in standards in a very sophisticated technology is given by I. Goldman in a letter to *Science*, 167 (1970), 237. Citing several instances of mislabelled and contaminated radioactive chemicals, he commented: 'Until recently, such slovenliness was confined mostly to companies that

the technological equivalents of entrepreneurial and shoddy science. The traditional engineering fields do have professional institutions which provide sanctions against dishonest and disastrously incompetent work. But these apply only in situations where the engineer is a true professional; in the large, institutionally ill-defined area of 'R. and D.', where the work is done by technological employees, there are no effective professional controls. And since commerce does not provide an effective 'hidden hand' for the maintenance of quality in these important sectors, the social problems of direction and quality control can become even more acute in technology than in science itself. These problems exist independently of those of the social and ecological effects of technological innovation, which I earlier described as 'runaway technology'; although they will naturally appear together in practice.

One can speak of the quality of the solution of a technical problem, quite independently of the commercial success of its result. We do this when rendering honour to great inventors and engineers of the past; and it is necessary when assessing the quality of any earlier phase of a full technical problem, as the specification of operating characteristics, or design. Here, the problem is set by the existence of a variety of requirements on the solution, which are independent and perhaps also incompatible. The specification of the function of the device will impose such constraints on its operating characteristics; and when these latter are given, there will be analogous constraints on the design. In one sense, every solution must involve a 'design compromise'; it is impossible to satisfy all the components of the solution to the maximum degree. A well-known example of this, is the array of 'ideal aeroplanes' seen from the different points of view of parties concerned; the operator will have seats crammed inside and perhaps tacked on outside too, the airframes man will have wings big enough for a glider, the maintenance man will have all the services running along the exterior, and so on.[16] And in a routine exercise, one will try to achieve a

enjoyed earned reputations for unreliability. The worrisome aspect of this problem is that the more reliable suppliers are now allowing this decline of standards.' Confirmation of this tendency was supplied by R. W. Fuller in a letter to *Science*, 167 (1970), 1562, 1564. He offered to provide the name of the supplier, and the lot numbers of the chemical, to anyone in doubt about his own sample. Neither letter mentioned the name of any offending firm; but the latter came from Lilly Research Laboratories, whose firm is thereby protected from suspicion.

[16] I am indebted to Mr. W. Houghton-Evans for this example, as well as for many stimulating discussions on the problems of design in technology.

solution which separately satisfies each of the aspects of the problem, while giving excellence to one or a few. Work of higher quality is done when the solution transcends a mere compromise; and by thinking the problem through as a whole, the technologist conceives new relations between the components of a solution, or new means of satisfying several of them simultaneously.

As in science, work of higher quality is necessary, although in itself not sufficient, for the solution to a technical problem to be more than ephemeral. Although the criteria of quality will generally be different from those of science, there are processes of selection and transformation of results closely analogous to those of the social phase of the development of scientific knowledge. We have already discussed the patterns of evolution of some devices, in connection with tools in science. The analogue to the stability and invariance of a scientific result is the successful performance of its assigned function, and the adaptibility to new functions, of a device. What can eventually survive through all changes of context and function, is the principle of a device, which becomes a basic, elementary piece of information, used in a variety of forms in the solutions of technical problems at all levels and all degrees of scientific sophistication. In this way, technical problems as well as scientific problems can yield a permanent contribution to our knowledge of the world around us.

From this analysis we can see why a 'classic' design, in which form and function are perfectly united, is as unique and unreproducible as a classic in aesthetic creation. For as the inevitable changes in function (including aspects of performance) demand change in design, the original unified conception must itself be modified. This may be done creatively, so that several 'marks' of a device maintain or even enhance its original excellence; but it is more common for changes to be made piecemeal, so that the later 'marks' become an incoherent pastiche of *ad hoc* solutions to new problems; eventually only the name and the memory of 'the way they used to build them' survive. The original conception is then essentially obsolete; and the craft of design (and with it the performance of the assigned functions) will decay until a new creative synthesis is achieved. This phenomenon can be seen in automobiles, and also in later 'stretched' versions of aeroplanes.[17]

[17] I am indebted to E. Paul Adler for this important point, which shows that technological progress is, even on the small scale, no more smooth and inevitable than scientific progress.

Varieties of Technical Problems

Although this discussion has concentrated on technical problems concerning physical objects and processes, the same analysis holds for a wider class of problems. For example, the largest and most important technical projects of the present include some concerned with 'information'; and in computers, the 'hardware' and the 'software' are correlative parts of the complete system. Thus, the performance of a function need not require a physical device; the organization of information, or of human activities, can constitute the solution of a genuine technical problem. Such solutions can vary in quality from the merely competent to the very deep. For instance, the systems of law and administration developed in the Roman Empire are a permanent contribution to human knowledge, as much as any principle of a physical device. There will, of course, be characteristic differences between technical problems of this extended sort, and those involving physical things; in particular, when the objects of inquiry and manipulation are themselves possessed of consciousness and purposes, they must be handled differently from inanimate things. But such differences exist within science itself, especially if (as in other European cultures) the concept extends across all fields of scholarship. The 'soft' technologies will have certain characteristic differences from the 'hard' ones; their associated sciences will usually be less matured, and in any event incapable of providing a doctrine which completely specifies all routine problems. But these differences, too, are of degree rather than of kind.

Thus, the distinction between technical and practical problems is not between non-human and human objects of inquiry and manipulation; but between the determining final causes of the task. In technical problems, the ultimate purposes to be served are reckoned with only as general constraints on the solution, commonly through the question of whether it will yield a profit; while in practical problems these purposes determine the functions to be performed, and from them the individual tasks to be accomplished. For example, the work of a surgeon is a practical problem; for unless he restores the health of the patient, his operation, however, brilliant, will have been a failure. Conversely, the work of administration, at any but the highest level, is of a technical character; for although people are involved as objects, the immediate problems are determined by the particular functions to be performed, while the ultimate purposes, even if they condition the work, are fixed and remote.

The correct identification of the distinction between technical and practical problems is of great importance for understanding the human aspects of modern physical science and technology. If one's language, in the term 'science', forces one to assume that work with physical things has an autonomous system of final causes, and that by contrast work with people is automatically concerned with welfare, then there is no framework in which these problems can be analysed, much less solved.

13

PRACTICAL PROBLEMS

WE now come to that class of problems, conceived in terms of intellectually constructed objects, where the ultimate purpose of the task directly determines the goal. We may define a practical problem as a statement of a purpose to be achieved, whose means are to be established as the conclusion of an argument, with a plan for its accomplishment. In some respects, such practical problems are the most difficult to solve, as well as including some of the most important tasks facing our society, those coming under the general category of 'welfare'. It is to these that the 'social sciences' are directed, as well as the embryonic sciences of 'the environment'. There is a class of practical problems which were the subject matter of the first learned disciplines to achieve maturity and social effectiveness: those which are the concern of law and theology. In the practice of the law for example, the particular tangle in which a client finds himself is immediately translated into the categories of legal theory; once an interesting case is underway he is reduced to the status of a name-tag; and yet the goal of the advocate's task is not to establish novel legal doctrine, but to win the case on behalf of his client. However, our main concern here will be with those large-scale practical problems whose solution involves the execution of practical projects, analogous to the technical projects necessary for the complete solution of a technical problem; for it is these that require understanding most urgently if their many pitfalls are to be avoided.

The Cycle of Investigation, and Characteristic Pitfalls

As in the case of the other sorts of problems, a practical problem has a phase of gestation, when it is a problem-situation: an awareness that things are not as they should be, but no clear conception of how they might be put right. Once it comes into being, its cycle of investigation may be described by five distinct phases: definition; information and argument; conclusion and decision; execution; and control.[1] Such a division of phases may be used to describe the accomplishment of any task; once the goal is set, one proceeds to consider the possible ways of fulfilling it; then decides between alternatives; proceeds to the operation; and during the actual work, supervises it to ensure that the goal is being satisfactorily fulfilled. The phases are quite sharply differentiated in practical problems. Thus the purpose to be achieved is necessarily stated in intellectually constructed categories, and so it is necessary to have the results from an inquiry of a scientific sort on which to decide the best means for its achievement. Also, the phase of execution of the work will usually be distinct from that of decision, in its agents as well as in its techniques; and finally, where there is any division of labour, the task of control cannot be neglected. Indeed, control may become so elaborated that it constitutes a technical problem, in the strict sense, with its own social mechanisms for each phase of its cycle, including that of controlling controllers in a variety of ways.

Comparing this cycle to that of scientific and technical problems, we notice certain systematic differences. Ultimate purposes, which are remote and diffuse in science, become a part of the criteria for the controlling judgements in technology, and here they determine the goal itself. Outside those limited fields where purposes are capable of being handled in terms of accepted explicit intellectual objects, the framing of a new practical problem is an essentially creative act. Whereas the setting of a problem in science involves the partial and tentative specification of a conclusion about artificial objects in a self-contained universe, and a technical problem involves imagining a device to perform a preassigned function, here the specification is of a state of affairs in human society which does not yet exist. Each of the controlling judgements of feasibility, cost, and value involves a multiplicity of factors, few of them reducible to quantitative or routine assessment. Moreover, the statement of the goal of the problem, as well as those of the controlling judgements, presupposes a social and moral philosophy. This may be

[1] This cycle might be compared with that given by J. Bray, *Decision in Government* (Gollancz, London, 1970), p. 269: 'objective – model – hypothesise control action – predict – vary hypothetical control action – revise prediction – repeat to select optimum – act – observe behaviour – refine objective – revise model – update prediction and optimisation – act again – continue to observe – all in the light of changing external circumstances'.

implicit and informal, and may seem obvious common sense to its proponents. But it is an ideology, a universe of reality and value, which itself is incapable of simple testing and scientific control. It will usually be articulated in a folk-science, whose peculiar features we will discuss later. The work of investigating and solving the practical problem, will necessarily be done within the framework of that ideology in whose objects the problem is conceived and first assessed. The categories in which the information is produced, and in which the argument about alternative means is conducted, will be determined by the ideology. The decision to be reached, depending as it does on the conclusion of that argument, is thereby limited in its range of possibilities. Indeed, once the problem has been framed in a certain way, the decision on its means of accomplishment may well be completely determined. In this way, the very first phases of the investigation of the problem have a closely determining influence over all those that follow. In technical problems, the relation is that of a sequence of final causes, each one delimiting the range of possibilities for the next phase: but here each phase also provides all the materials for the next, and so delimits it even more narrowly.

Although the categories in which the practical problem is solved are so narrowly restricted by its initial conception, this does not necessarily have the effect that the ultimate purpose to be served is dominant in all the tasks involved in the phase of execution. Several features of this work can cause a displacement of the effective goals of individual tasks into other hierarchies of final causes. The most universal cause is that all the agents will have private purposes of their own, which are not automatically in harmony with the goals determined by the ultimate purposes of the work; and the problems we discussed in connection with social behaviour in science, especially those of quality control and morale, have their analogue here.[2] Another cause, more noticeable in large practical projects, is that the work of execution gives rise to challenging technical

[2] Studies in the sociology of bureaucracies have analysed these tendencies to 'goal-displacement'. A very vivid account of a case-study of this process is given by Harry Cohen, *The Demonics of Bureaucracy* (Ames, Iowa, 1965), written in some ways as a sequel to the basic work by P. M. Blau, *The Dynamics of Bureaucracy* (University of Chicago Press, 1963). Cohen's analysis shows the pressures causing a deviation from the formal goals of the bureaucracy, and stresses the importance of the informal adjustments made by individuals to their tasks for their self-protection. He offers suggestions for counteracting these tendencies; but at the time of writing had not considered the usefulness of bureaucrats being answerable to their clients.

problems or projects; the performance of an assigned function becomes the focus of attention, while the purpose being served through that function is neglected. This is an entirely natural tendency; it is easier to specify a function than a purpose, just as it is easier to specify a goal than a function. Hence it is more convenient, and safer, for agents at any level and those who control their work, to restrict their consideration and responsibility to the most immediate final causes involved in their tasks. This tendency for immediate technical problems to displace the initiating practical problem in the execution of a project becomes very marked as soon as a stable organization has been formed, and 'welfare invention' has given way to 'welfare engineering'. This has its own effects, on the categories in which tasks are defined, and adequacy assessed. Like the scientist's conceptual objects, these cannot exist independently of their use. Although they may have been first conceived in a philosophical reflection on a social problem, when they come to be applied to the complex tasks of the project they will need additional specification for their new functions. This will be in the direction of appropriateness for the definition and control of the many routine tasks to be accomplished as conditioned by the individual and collective purposes of those involved at each level of a hierarchy. In time the original purposes which these categories expressed may well become inexpressible, and hence non-existent, in the framework of their elaborated, technical meanings.

The operation of all these tendencies, in a large and complex organization, can give rise to the characteristic features of a 'bureaucracy' (in the pejorative sense), in which the original defining purposes are forgotten and the agents concentrate on an internal political game, treating their nominal clients as irrelevant nuisances. Thus, although one can establish a clear distinction in principle between practical and technical problems, in terms of the importance of ultimate purposes in defining the goals of the task, in practice it is more difficult. There is no simple criterion in style of working to distinguish between those established organizations which were initially created to solve a practical problem and those whose only ultimate purpose was and is the personal benefit of their promoters and owners. This depressing conclusion accords with the experience of Socialist societies; and it raises deep problems for Socialist theory, whose tradition was generally negligent of the problems of bureaucracy. It would appear that the solution of such problems lies either in the reduction of the size of all political and economic units to the point where bureaucracies are unnecessary, or for the improvement of methods of accountability of large organizations to those who exercise ultimate control, and of answerability to those whom they should serve.[3]

The two features of ideological origin and bureaucratic execution, so different and indeed contradictory, combine to produce an extreme tendency to rigidity in the solution of a large practical problem, so that (unlike in science or technology) there is a minimum of re-cycling through the phases of the problem. In the earlier phases, any challenge to the conception of the problem is likely to meet with intense resistance from those whose personal ideologies, as well as careers, are involved in its planning and accomplishment. And once the routines have been set and organizations established, the whole enterprise has an inertia which is very difficult to overcome; for the costs, both political and administrative, of any major changes, will be seen by those involved as prohibitive.

This rigidity creates particular hazards. The pitfalls that are encountered whenever a system of intellectual objects makes contact with external reality, are reduced in incidence only by a tentative and exploratory approach to that contact, so that clues to their presence can be detected. Here, the close determination of the problem by its initial conception removes that warning system; and so practical problems are particularly prone to great and disastrous pitfalls. Similarly, the reduced re-cycling leads to the long-delayed recognition of pitfalls when they have been encountered, and a correspondingly greater difficulty in repairing their effects.[4] Hence

[3] This latter seems to be the content of the demand for 'Socialism with a human face', raised in Czechoslovakia in 1968; and it is also part of the movement for reform within the Roman Catholic Church. The concept of 'answerability' to those inferior in an hierarchy, was developed by myself in a paper, 'Power, Responsibility, Answerability', delivered to a teach-in on 'The Nature of the University' organized by the Leeds University Union on 12 December 1968.

[4] All these hazards are intensified by the natural tendency of men of vision to devote their years of power to the solution of the great problems of their youth, despite the gap of decades during which the problem may well have been completely transformed. For an example of this tendency, applied as an explanation of Cromwell's policies during the Protectorate, see H. Trevor-Roper, 'Three Foreigners', in *Religion, the Reformation and Social Change* (MacMillan, London, 1967), pp. 237–93. After describing the philosophy of the disastrous decade of the 1620s, and the changes wrought by intervening times, he comments: 'But Cromwell could not change his mind. It had been moulded, fixed and perhaps slightly cracked, in the grim and lurid furnace of the past. So now, as Lord Protector, he adopted a foreign policy that was twenty years out of date: ... Protestant reunion in Europe, Elizabethan war in the West Indies, and a

a common cycle for a new practical project is to proceed from a grand ideal conception, into a morass of difficulties of execution, and then either to lurch forwards into a final collapse or to survive, in a battered state, accomplishing tasks which have little relevance to the original purpose, or perhaps to any at all. This latter fate is a likely one even for projects that are well executed; for the time-lag between initial conception and large-scale execution may be so great that the real practical problem will have altered radically in the interval. And the worst feature of such a sequence is that those responsible for it are usually incapable of learning any lessons from the experience. The whole process is too complex to provide decisive tests of particular points of the ideology and folk-science in whose terms the project was first conceived, and any bureaucracy has built-in barriers to the recognition of failure in the performance of its functions.[5]

The work of solving practical problems also suffers from the inevitable absence of a consensus on the criteria by which quality or success is judged. In technical problems, these criteria are supplied, to some extent at least, by the objective tests of performance of function, and of commercial success; and in science it has been possible, in many important cases, for the members of the small community associated with a field to develop appropriate criteria for their work. But practical problems affect a variety of people in a variety of ways; there will be those who decide, those who execute the decision, those who are supposed to derive benefits, and those who incur costs. Such diverse groups will not only view the problem,

top-dressing of ideological mysticism which included the reception of the Jews' (p. 282). The essay is a magnificent study of the fortunes of Utopian schemes as they are tossed in the storms of politics.

[5] A particularly gloomy view of this process was taken by the social theorist Roberto Michels in his *Political Parties: a Sociological Study of Oligarchical Tendencies of Modern Democracy* (German original, 1911; translated E. and C. Paul, The Free Press, New York, 1949). He describes an 'iron law of oligarchy' whereby with the advent of a mass organization, bureaucracy kills the democracy within the movement. With the nineteenth-century Socialist movement in mind, he concluded his book with: 'When democracies have gained a certain state of development, they undergo a gradual transformation, adopting the aristocratic spirit, and in many cases the aristocratic forms, against which at the outset they struggled so fiercely. Now new accusers arise to denounce the traitors; after an era of glorious combats and inglorious power, they end by fusing with the old dominant class; whereupon once more they are in their turn attacked by fresh opponents who appeal to the name of democracy. It is probable that this cruel game will continue without end' (p. 408). See R. Nisbett, *The Sociological Tradition* (Basic Books, New York, 1966; Heinemann, London, 1967), pp. 148–50 for a discussion of Michels.

and its effects, in ways depending on what they consider their self-
interests; they will also have different universes of reality and value,
with their corresponding folk-sciences, in which the words used for
the description of the problem relate to radically different things.
In the debates over particular practical projects, the various sides
will inevitably be talking at cross purposes about different things.[6]
Appeals to 'the 'facts' are of little use; for 'facts' are even more rare
and evanescent in collections of such informal systems than they
are in science. The situation is analogous to that in an immature
science, but even worse in that the most fundamental objects of
argument, values and purposes, are inherently incapable of being
the subjects of universally accepted demonstrated conclusions.[7]

Yet with all these inherent weaknesses, it is necessary to solve
practical problems as I have defined them, if certain sorts of

[6] In a study of the historical background to the Quarry Hill Flats in Leeds, Dr. A.
Ravetz has identified the following series of practical problems associated with housing
and the poor: 'insanitary areas' from the 1830s; 'bad housing conditions for the working
classes' by the end of the century; 'a housing shortage' at the end of the First War;
'slums requiring clearance' in the 1930s; 'housing as a social service' around the same
time; and 'overcrowding' in the mid-1930s. The earliest problem received attention as
a menace to general health conditions, and the relevant function to be performed was to
provide water, sewage, and fresh air. For this last, houses would be destroyed, the fate
of the inhabitants being of secondary concern or none at all. After the First World War,
the first Council estates were built under the slogan of 'homes for heroes' and their being
too expensive for the slum-dwelling poor was an irrelevance. In spite of this, some leading Socialists interpreted the measure as a step towards 'housing as a social service' and,
on these grounds, bitterly opposed the later policy of giving priority to re-housing the
desperate slum-dwellers. In Leeds, the Labour movement was itself split on this
disagreement about the nature of 'the housing problem' and the split was one of the
contributory factors in the defeat of the Revd. Charles Jenkinson in the local elections of
1935 after the brief period of two heroic years in power.

[7] The common feeling of inferiority of cloistered academics with respect to successful
practical men of affairs may be assuaged by the experience of Descartes when he left
the world of books and entered that of real action. As he tells it, he went out into the
world seeking the way to Truth, 'For it seemed to me that I should find more truth in
the reasonings which a man makes with regard to matters which touch him closely, of
which the outcome must be to his detriment, if his judgement has been at fault, than in
the reasonings of a man of learning in his study, whose speculations remain without
effect, and are of no further consequence to him than that he may derive all the more
vanity from them the further removed they are from good sense, because of the greater
skill and ingenuity he has to employ to make them plausible.' Unfortunately, 'It is true
that so long as I did nothing but reflect upon the behaviour of other men, I found no
grounds in it for assurance; indeed I perceived in it as many divergences as I had formerly found in the opinions of philosophers.' *Discourse on Method*, translated by A.
Wollaston (Penguin Books, 1960), last page of 1st Part, p. 43. Descartes did draw lessons
from this experience of the real world, ones which could have been (and perhaps were)
derived from a reading of Montaigne; and he could then have recourse to the '*lumière
naturelle*' of his mind, to achieve his distinctive philosophy.

purposes are to be achieved. It is naïve in the extreme to believe that the common sense and good will of all right-thinking people could suffice for the accomplishment of important social tasks. For when there are situations where there is a consensus on ends and spontaneous co-operation on means, then the practical problem does not arise in the first place, and the social task is accomplished without attracting any notice. But where these favourable circumstances do not exist, any co-operative attempt to fulfil collective goals while also serving the personal purposes of the agents without any definition of the tasks can yield nothing but a shambles. Nor can one rely on the presence of a 'hidden hand' which will ensure that an aggregate of purely private purposes will somehow be equivalent to a beneficial public goal. An equal naïvété at the other extreme is found in the belief that 'scientific method' can be applied in a simple and straightforward fashion to this class of problems.[8] This claim was part of the propaganda for science in earlier times; and the history of social reform is littered with the failed schemes of socially conscious intellectuals and tidy-minded reformers who blundered into the first pitfalls in their way. To explain such failures exclusively by the stupidity and selfishness of politicians and of other groups of people is merely to ignore the difficulties inherent in practical problems and to be just as incompetent on the next attempt.

'Scientific' Aspects of Practical Problems

The sphere of activity comprised in the solution of practical problems is so vast and complex, that its proper study requires, and has already created, an extensive research effort. Here we will restrict ourselves to considering those aspects of practical problems which relate closely to science. The relation is twofold: the influence of the specifically 'scientific' character of practical problems as distinct from tasks; and 'practical' aspects of technical problems.

One may say that a set of tasks constitutes a field of practical problems, when it is recognized that mastery of some demonstrative discipline is necessary for their accomplishment. Medicine and law

---

[8] Karl Pearson, in *The Grammar of Science*, strongly advocates the application of the methods and the results of science to social problems. His chosen example was the problem of the poor; and he used Waissman's theory of the germ-plasm (with reservations in his account of the theory but not in his practical conclusions) to condemn that ill-conceived philanthropy which encourages such 'inferior stocks' either to increase their numbers or to mix with those superior to them. (Everyman ed., introductory, paragraph 9, pp. 27–30.)

have been so recognized for many centuries, and they have provided the model for the 'learned professions'. This recognition is not a simple thing; the group of practitioners must convince both the State and the public that a particular set of purposes can be achieved only by their efforts, and that competence for attempting these tasks must be assessed and certified by themselves. The problems of the would-be professions in the modern world were anticipated by medicine over the course of centuries; since the training for 'physic' was largely vacuous in content, and practical competence was not tested to any degree of rigour, the physicians had to contend on all fronts with a variety of competitors.[9] Indeed, the learned professions did not have a very strong claim to their monopoly of practice until quite recently.

It was during the last century that the welfare of the population came to be seen as involving practical problems as I have defined them; and with the movement for the creation of corps of skilled agents of various sorts, and their proper training and certification, the professions themselves, old and new, gradually put their houses in order. Many of the projects that were established required the work of new sorts of technicians, in particular those involved in the detailed work of inspection and regulation. Although their tasks are sometimes defined in principle by certain ultimate purposes (such as the protection of some or all of the population from bad food or bad working conditions), their working contact is with physical things (or symbols) rather than with individuals. But there is another class of skilled agents, who (in principle) serve individuals to improve their welfare; in this respect they have an affinity with professionals, and we can describe them as 'practitioners'.

Both technicians and practitioners are restricted in their work to routine tasks; difficult cases are referred to superiors, and genuine innovation is reserved to the highest level in a bureaucratic project. Their training will be correspondingly straightforward and detailed, and less theoretical and 'liberal' than that of those destined for higher positions. Since they are put into contact with a reality

---

[9] See R. H. Shryock, *The Development of Modern Medicine*, for a general history of this problem. The struggles between the College of Physicians and its rivals (apothecaries and 'chemical physicians') in the seventeenth century is particularly well documented; see Sir George Clark, *History of the Royal College of Physicians*, vol. i (Clarendon Press, Oxford, 1964); and C. Webster, 'English Medical Reformers of the Puritan Revolution: a Background to the Society of Chymical Physitians', *Ambix*, 14 (1967), 16–41.

external to that of the artificial categories of the project, they are exposed to the divergence between theory and the external world. This can happen even to technicians, who may find themselves deprived of the resources with which to do a proper job of inspection or regulation on behalf of the public.[10] Practitioners may find themselves in an even more contradictory position; for the administrative case-law which they are required to enforce may be designed more for convenience and parsimony than for the welfare of their clients. By the nature of their work, they are accountable to their superiors, but they are liable to be answerable to their clients; and the two sorts of goals imposed on their tasks may be in direct opposition. However, we shall see that their training for their work can provide a protection against such discomforts.   There is a very important class of skilled agents involved in practical problems, who fall between technicians and practitioners on the one hand, and true professionals on the other. The 'experts' are not restricted to routine tasks, but will study genuine problems, involving the exercise of judgement and the deriving of a conclusion from an argument. If they work directly for a client, they may merely suggest a decision and leave the execution to him; or they may undertake that work as well, depending on conditions. The objects of their work may be human, non-human, or a mixture. The expert is distinguished from the true professional by his position, entailing the controls exercised on his work. For he is an employee of a firm, accountable to his superiors there who have the powers of penalty and reward over him. His answerability to his clients is very slight, and to his colleagues in his specialism, not much greater. This difference is of course crucial; and is recognized in our language. Given two men, of identical skills, and ostensibly offering the same service, the employee is an 'expert', and the independent agent, the true professional, is a 'consultant'.[11]

[10] An example of this phenomenon is provided by A. D. Woolf, 'Industrial Accidents: Can we Stop this Suffering?', letter to *The Times* (Business News Section), 24 March, 1968, p. 28. He quotes Mr Plumbe, the Chief Inspector of Factories: 'The Inspectorate has never aimed at, and certainly has never achieved, a rigorous enforcement of the (Factories) Act such as a Teutonic country might attempt.' He comments, 'The pathetic number of prosecutions set against the annual toll of accidents amply confirms his statement.'

[11] The difference between 'expert' and 'consultant' becomes quite clear from examples of situations where they work in competition. Thus: 'Recently, in California, a small group of independent entomologists has found that advising cotton growers can be a profitable business. Growers contract to have their fields checked once a week by these

As we have seen, an important class of technical problems are incapable of being subjected to quality control through the automatic mechanisms of the market; and when disputes arise because some person or group considers their welfare to be neglected or injured, experts are called in by both sides. In some ways their role is analogous to that of an advocate; but they do not operate within the sophisticated etiquette and ethic of that profession whereby a man can argue for his client without losing his integrity. Rather, the expert tries to argue (perhaps with sincerity) as if he were a scientist, establishing his conclusions on supposedly known and irrefutable facts. The absolute loyalty of experts to their organization can sometimes be quite touching; even when a project is revealed to have stumbled into the most ghastly pitfalls in technical execution, the firm can usually find an expert to assure the public that the matter is really in competent hands. As a result of this aspect of the experts' task, their assertions come in that class to which is applied the maxim, 'Never believe anything until it's been officially denied.'[12] It is perhaps less serious that the general public is ignorant of these limitations on the reliability of expert pronouncements, than that their superiors in positions of decision-making must, by bureaucratic etiquette, place some reliance on their conclusions and recommendations.[13]

Even in the case of technicians and practitioners, the tasks of

men for a per acre fee that is usually less than the cost of one insecticide treatment. They furnish the same service for a fee as the industry salesmen furnish free. The difference is that they are selling only the service and not the chemicals.' Kevin P. Shea, 'Cotton and Chemicals', *Scientist and Citizen*, 10 (1968), 209–20. (This journal was subsequently re-titled *Environment*.)

[12] A classic case of 'official denial' occurred after one corner of a point-block of London council flats (Ronan Point) had gone down like dominoes following a gas explosion in one upper room. See *The Times*, 17 May 1968.

[13] 'There are three roads to ruin: women, gambling and technicians. The most pleasant is with women, the quickest is with gambling, the surest is with technicians.' M. Georges Pompidou (quoted in *The Sunday Telegraph*, 26 May 1968). Thus the Pacific Gas and Electric Company fought a long battle against conservationists and geologists to establish a nuclear power station at Bodega Head, north of San Francisco. Only after five years of argument was a proper seismic survey done on the site on which a large hole had already been bored, and this showed the bedrock to be disastrously faulted. It would appear that the directors of this utility, so shrewd in other respects, must have believed what their 'experts' were telling the public about the suitability of the site. See Joel W. Hedgpath, 'Bodega Head—a Partisan View', *Bulletin of the Atomic Scientists*, 21 (March 1965) 2–7, and Sheldon Novick, *The Careless Atom* (Houghton Mifflin, New York, 1969). Examples of the way that 'experts' discharge their duties, in England no less than in America, are not hard to come by; but under present law one hesitates to mention these in print.

training and certification require a formal body of doctrine presented as a 'science'. Without this, the work would be recognized as merely a craft; and with skills transmitted purely interpersonally, by example and precept, it would be impossible to enforce any uniformity of standards for certification, or of practice on the job. A stable bureaucracy, with its complex internal technical problems of organization and control, could not operate under such conditions. Hence if an appropriate science does not exist for a particular practical problem, it is necessary to invent one. Once it is available, its further development can perform several valuable functions for its field. On the social side, the claim to possess a body of theoretical doctrine, necessary for competent practice, helps the occupational grouping of practitioners and experts to consider itself as equivalent to a profession; by this technique, it can demand status for its present members, and try to influence the flow of certified recruits in the direction of smaller numbers and superior social origins.[14] Also, the theoretical discussion of the body of doctrine, and the activity of research in its terms, enhances the claim to truly professional status, and also provides the basis for a demand that the subject be recognized as properly belonging to the world of higher education. Although the body of doctrine in question may be nearly vacuous, an incompetent rationalization of a craft practice, it can still perform several functions in the internal workings of the field of practice. As it becomes elaborated, it can become the dominant or sole framework for the definition of the tasks accomplished by practitioners and experts, excluding common sense altogether from

[14] For the penetrating comments of Max Weber on the role of education in the maintenance of a bureaucratic élite, see R. Bendix, *Max Weber, an Intellectual Portrait* (Doubleday Anchor, New York, 1962), pp. 229–30, 461. Robert M. Hutchins tells the following story: 'I shall never forget—though I have often tried—the last meeting of the Big Ten [Midwestern Universities] presidents I attended, fifteen years ago. The President of the University of Michigan said "Say, I want to ask you fellows, what are we going to do about embalming?" He went on to report that the embalmers in his state wanted to become a profession for the double purpose of limiting competition and raising their social standing. This they proposed to accomplish by establishing a school of embalming at the University of Michigan and requiring all practitioners to have a degree from the school before being permitted to embalm any resident of the state. He was consoled by the president of the University of Minnesota, who assured him that "the embalming program at that great university had not interfered with its smooth operation".' See Clerk Kerr and others, *The University in America* (Center for the Study of Democratic Institutions, Santa Barbara, California, 1967), p. 5. The Michigan embalmers were aiming for truly professional status by requiring their recruits to train at the prestigious arts and sciences University of Michigan rather than at the land-grants Michigan State University.

their conception of their work and from the criteria for its quality. This is convenient both for the agents and for those who exercise decision and control over them; the raw world of untidy and contradictory values and purposes of real people can be henceforth ignored in favour of the neat conceptual world of the field. In this way, practitioners at every level are at minimum risk; and the ultimate purposes of the work, to the extent that they are reckoned, are so defined in the terms of the theoretical structure that they are automatically served every time a unit task is accomplished.[15]

The situation of the true professional presents certain barriers to this invasion of his field by science, or pseudo-science, which are conveniently absent in the case of the bureaucratic practitioner and expert. For the professional is approached by a client, who can choose the best means of having his purposes achieved. If the client goes away dissatisfied, he can try a different man the next time; and the profession as a whole always faces competition on the fringes of its field, from unqualified practitioners who advertise themselves as achieving the same purposes by different means. But the bureaucratic practitioner almost always enjoys a monopoly; if he distributes welfare on behalf of the State, his clients, through ignorance or poverty, cannot turn elsewhere; if the task is regulatory in any way, his monopoly will be enforced by the State; and if his service is provided by a private firm, it is likely that it will be one of a few which together dominate the market, and have identical styles of operation. Hence the practitioner need know nothing at all, from

---

[15] Tendencies in this direction are described by Barbara Wootton, *Social Science and Social Pathology* (Allen & Unwin, London, 1959). 'In some cases this emphasis on the hidden psychological issues that are supposed to be uncovered by what has come to be known as the "casework process" has gone so far as to lead to almost deliberate disregard of the practical problems which were the immediate occasion of the relationship being established' (p. 278).

Another very important function of the theoretical doctrine in whose terms practitioners conceive their problems is as a folk-science for themselves, enabling them to cope with the intolerable situations they encounter. These may arise in interactions with people whose values and way of life are abhorrent and incomprehensible, or whose suffering is so severe that the practitioner cannot identify with it. The natural defence is to dehumanize such 'patients', in the former case by applying moral judgements to their situation; such indeed was the attitude of much 'charity' in former times. The theoretical doctrine offers a means to a more sophisticated and depersonalized defence, for in its terms the moral and emotional threat is explained away. The practitioner can then allow his natural sympathies to operate without danger to his own person. Needless to say, the formation and adoption of a doctrine is governed by many other functions as well, and so it is not always well suited for this one. I am indebted to Dr. H. W. S. Francis for this important point.

direct human contact, about the purposes of the people he is claimed to be serving. To him they can be merely the sum of the theoretical attributes which he memorized in the course of his training; and the assessment of his fulfilment of his goals will be based on these categories, as modified by an administrative case-law.

The new technology of information has recently provided tools which supplement the work of the bureaucratic practitioner, and in some cases even make him redundant. The routinization of tasks is pushed to its logical extreme, when every particular case is completely characterized by an array of holes on a punch-card. This technique not only saves manpower and money; it also provides materials convenient for analysis by that most fashionable of sciences, mathematics. With such data, research can become a routine, and the members of the 'scientific' field associated with an area of practical problems can generate the pseudo-property of research titles with no more labour than the hiring of a programmer. The human beings whose purposes are involved in the field, are now at one remove further from the perceptions of those who control their welfare. Even if their particular situation is capable of translation into the categories in which the practitioner or expert operates, they are at risk of being overlooked by the systems analyst who constructs the computer program, or even of being the victim of an error by the key-punch operator. Should that occur, they cease to exist as anything remotely resembling their true selves, and either vanish as unclassifiable or re-appear in foreign garb. There are few defences available to the individual against such a destruction of his identity; as a mute protest against the machine, one can fold, mutilate, *and* spindle one's punch-card if one is fortunate enough to handle it;[16] but otherwise one is truly up against a System.

The natural propensity of bureaucratic practitioners and experts to reject and condemn any particular case which falls outside their routine categories is exaggerated and freed from constraints when the 'case' is not even a file of papers but only a punch-card. The computer is, after all, only a very rapid clerk which strictly 'works to rule', and which makes many of the rules itself by its physical limitations. More sophisticated categories of its objects require an

[16] I am indebted to Dr. Robert S. Ravetz, of Philadelphia, for this technique. It serves to bring one's grievance to the attention of those who can deal with it, unlike a mere letter which is automatically shunted to a 'complaints' bureaucracy.

elaboration of procedures which is costly to achieve, and whose operation will waste the computer's precious time; and non-standard cases, spat out by the computer, are a serious nuisance. Hence the natural tendency is to replace any judgements of attributes by simple sets of parameters, and in case of doubt to apply 'fail-safe' to the benefit of the organization. It is for this reason that the proposed 'data banks' are a threat to civil liberties. Their materials will be raw data from a variety of disparate sources; it will be converted into information by automatic procedures; and then interpreted as evidence for practical conclusions in the context of a computerized bureaucracy. Any citizen will then be at risk of being victimized, with no possibility of redress, because of some deviance (real or erroneously imputed) from some expert's imagined 'norm'.[17]

It is always important to remember that such developments as these arise from natural causes and also bring great benefits along with their dangers. Practical projects come to be when large-scale purposes cannot be served in any other way. The practitioners who accomplish the routine tasks may see reality only through the lenses of their official doctrine; but unless this is completely corrupted it is likely to be more appropriate to their genuine tasks than the mixture of misinformation and unconscious prejudice which is provided by their untutored 'common sense'.[18] And the punch-cards and computers, if used rather than abused, can clear away the mountains of

[17] A recent example of this approach, applied in England, was revealed by *The Sunday Telegraph* (14 December 1969). For some years the South Eastern Electricity Board has secretly operated a 'points' system on consumers, an account losing 'points' every time a payment is, for whatever reason, delayed. The intended function of the system is to identify likely defaulters and accelerate the sending of reminders and (finally) the twenty-four-hour notice of cutting off of supplies to them. Naturally, the twenty-four-hour notice has gone out to people who never consider themselves as slow payers; and (because of delays within the postal and accounting systems) some time after the bill has been paid.

[18] I am indebted to Dr. Leslie Walsh for this observation. A confirming instance is provided by Bob Bailey and Lou Smith, 'Operation Bootstrap', *Center Diary*, 16 (1967) (The Fund for the Republic, Santa Barbara, California), 37–44. At their unorthodox training school in a Los Angeles black ghetto, they had visits from students at the University of Southern California. 'These are people majoring in race relations who will probably become the heads of Human Relations Commissions or who will be getting human relations jobs in industry. They had never, up until the time they came to us, talked to the very people they're going to be making decisions for. In one of our sessions we had a white girl who was just terrified. She finally admitted it. She said "I'm scared to death of Negroes." Somebody asked her why and she said, "Well, they go around raping people." This is a girl majoring in race relations' (p. 43). One can hope that by the time she took her degree her formal instruction would have imparted a more sophisticated understanding of the problems she would handle in a professional capacity.

routine paper so that the genuinely difficult cases can be quickly recognized and dealt with. But there are real dangers in an extension of this pseudo-scientific style of handling practical problems. The artificial world of the experts and systems analysts, the construction of whose objects may not have involved much intellect at all, may be given effective reality in virtue of their institutional and political power. The solution of a practical problem is, as I pointed out, determined by the categories in which it is conceived; and if these determining categories become those which are convenient for a computer program, we will in that measure live or die by the computers.[19] It is less likely that man will be made over into the image of the punch-card, than that when reality does obtrude itself, it will do so in an opposition to the imposed systems which is revolutionary and destructive of them, and of much else besides.

Practical and Technical Problems Compared

The difference between technical and practical problems has been widely observed in recent years, in the contrast between the great successes of America and the Soviet Union in the one sort, and their abysmal failures in the other. The exploration of space is certainly an astounding technical exploit, which requires inventiveness, ingenuity, administrative skills, and quality control to an extremely high degree. Yet these same qualities were insufficient for America to impose its solution on the conflict in Vietnam, or to resolve the crisis of its black cities; and the Soviet Union could neither solve its own problems of management and production, nor permit an associated state to solve its problems in its own way. The differences are clear. In the space race, the ultimate purposes were simple, and established by decree of national leaders: the enhancement of national prestige. Within this framework, the choice of goals was not complex, and work could then proceed on purely technical

The artificiality of 'common sense' is well expressed in an aphorism by John Maynard Keynes in the concluding notes to *The General Theory of Employment Interest and Money* (Macmillan, London, 1936). 'Practical men who believe themselves to be quite exempt from any intellectual influences are usually the slaves of some defunct economist. Madmen in authority who hear voices in the air are distilling their frenzy from some academic scribbler of a few years back' (p. 383).

[19] The possibility of civilization dying quite suddenly by the computers was a very real one in the early 1960s, when American 'nuclear strategy' relied heavily on computer simulations of real crises. Sir Solly Zuckerman was the first eminent military scientist to expose this lunacy, in 1961; his essay is reprinted as ch. 5, 'Judgment and Control in Modern Warfare', of *Scientists and War* (Hamish Hamilton, London, 1966).

problems; and the harmonizing of their goals with private purposes of those on the work, was not too difficult. But practical problems cannot be reduced to a matter of technique, however great the desire or the financial investment.[20]

The deepest and most urgent practical problem-situations are not discovered or invented; they are presented to us, frequently against our desires, by the processes of human history acting through time up to the present. The framing and solution of practical problems is at risk of encountering a multiplicity of pitfalls, so that the purposes served can turn out to be quite different from those intended. Some at least of these can be avoided by an awareness of the different phases of the cycle of a practical problem, and its relation to the associated scientific and technical problems. The earlier phases involve inquiries of a scientific sort, to determine what the problem is, and how it might be solved. The first common pitfall is that the objects in whose terms the inquiry is conducted are so tightly bound to a particular ideology that the conclusion is determined before the work begins. But if the inquiry avoids 'theory' and becomes 'empirical', it can encounter the pitfall of simplifying its objects of inquiry to homogeneous populations defined by classes of simple data; then the complexity and contrariness of the situation, which created the problem situation in the first place, is lost from view. Thus, to the extent that the conclusions of the inquiry are simple, they are likely to be over-simple; but a conclusion full of nuance, subtlety, and personal wisdom is inappropriate for its function as the basis for a decision. Another pitfall of the phase of 'information and argument' is the imposition of excessively high standards of adequacy on the work, so that a lengthy and expensive research programme is undertaken as a preliminary to any decisions. This may well serve the private purposes of the scientists concerned,[21] but the resulting delay may involve costs of its own, not

[20] In his article 'Can Technology Replace Social Engineering?', *Bulletin of the Atomic Scientists*, 22 (December 1966), 4–8, Alvin Weinberg argues cogently that 'social problems can be circumvented or at least reduced to less formidable proportions by the application of the Technological Fix'; but he is fully aware that 'social engineering' cannot be dispensed with.

[21] In 'The New Estate', *Bulletin of the Atomic Scientists*, 20 (February 1964), 16–19, Alvin Weinberg gives a good description of this tendency. 'With respect to the style of intellectuality, I refer to the tendency to deal with difficult techno-social problems—such as the population problem, or aid to the underdeveloped countries, or the lag in our civilian technology—by deifying research and denying engineering. I have read

least those of being interpreted by aggrieved parties as a deliberate attempt to defer any decision indefinitely.

The function of the conclusions of the phase of enquiry is to provide a basis for a decision; and this must be capable of execution as a technical project, by whose means the desired general purposes may be achieved. The technical project itself must be capable of being analysed into a multitude of routine tasks, governed by an hierarchy of decision and control. Further pitfalls are encountered here; even if the intended functions are appropriate to their ends, the aggregate of unit tasks, as they are articulated and then controlled by a bureaucracy, may come to be governed by goals which are contrary to this original function. The operation of overall control over the phase of execution then constitutes a new practical problem in itself, with its distinct phases of investigation; and the neglect of this will put the whole practical project in jeopardy. Thus the inherent complexity of a practical problem, both in its objects and in the cycle of its solution calls for a diversity of operations, skills, and approaches even greater than in the case of technical projects. The application of 'scientific method', not merely as a crude imitation of the mathematical-experimental sciences but even in the sense of scholarly inquiry, is insufficient by itself. What is required is an appreciation of the variety in the nature of the problems encountered, the difference in their methods, the criteria of adequacy of their solution, and in their characteristic pitfalls; and an awareness that there exist some important problems, of any of these sorts, which are incapable of solution under any possible circumstances.

'Practical' Aspects of Technical Problems

We saw earlier how a practical problem can tend, in its phase of execution, to lose its original defining purposes, and to become just

many reports by presidential panels and panels of our National Academy of Sciences on a variety of socio-scientific problems. The reports surprisingly often have the same, basic structure: first, a statement that the problem is difficult and that not enough is known about it; second, that more research and therefore more students and more fellowships are needed; and finally, that much more government money should be directed toward research on these matters. All these things are good; but I submit that in dealing with socio-technical problems the university community (that tends to dominate the panels that prepare such reports) is influenced too much by its tradition of scholarship and research and too little by the engineering notion of doing the best one can with the available knowledge. It is not that more knowledge is not needed; it is

another technical project, serving some other external purposes or perhaps none at all. But there has recently developed a converse tendency, in the politically effective awareness of the effects of technical projects on the welfare of others, independently of their relation to the organization as customers or users. Thus in America there have been successful campaigns against the sale of the insecticide D.D.T., and in England, an equally successful campaign against the first proposed site for the third London airport. In each case, the argument was based on injury to 'third parties', whose welfare should be taken into account as a final cause in the decision on the project. Regulation of certain industries on behalf of a 'public welfare' of one sort or another has being going on for some time; what is new in the present situation is the growing strength of the assumption that every enterprise must be assessed in these more inclusive terms. This is not due to a sudden increase in the ecological and social effects of industrial activity, but to the increasing sophistication of politics now that 'welfare' means more than a subsistence wage for the great majority.

As yet, the tendency to assess technical projects as practical ones lacks both a coherent ideology and a stable political base. The traditions on which it draws are scattered and thin, mainly those of 'conservation' of unspoiled areas, and 'preservation' of the countryside. In the capitalist societies, the dominant tradition of thought about the accountability of industrial enterprise has naturally tended to advocate a minimum of public control; and the opposition traditions of socialism, based on the struggle for the redivision of wealth, concentrated on social ownership as an automatic guarantee of democratic control Each of these traditions neglects, in its own way, the problems of controlling technical projects by considerations of social welfare. Until very recently, the training of engineers and scientists, and the rules of their professional institutions, gave no indication that such practical problems are relevant to their work. Quite suddenly within recent years an awareness has grown, partly because the limitations of a technology of affluence are beginning to be felt, in immediate small problems as well as in large philosophical ones.[22] The exhaust-fumes of the automobiles of Los Angeles may

simply that in complicated technico-social situations complete knowledge is never at hand, and the engineering approach is as appropriate as is the research approach.'

[22] An example of this new awareness is the symposium on 'Technology for Man', published in *Technology and Culture*, 10 (1969), 1-19.

turn out to be a more genuine harbinger of the leading technological challenges of the future than the magnificent freeways on which they are generated.

It is important to realize why such problems cannot be either wished out of existence, or solved either by generalized good will or general rules. Our analysis of the withering away of ultimate purposes in a bureaucratic practical project is relevant here, but it must be supplemented by certain features of technical projects. We have seen that any successful device must be capable of performing several functions; and its design must take these into account. In the case of a mass-produced commodity, these various functions will be in a rough correspondence to the purposes of different groups of users; and the device will be commercially successful if it achieves superiority in a range of functions which appeals to a large group of prospective purchasers. Thus, the private automobile can function as commuter transport, suburban transport, long-range transport, all for different numbers of passengers, as well as status symbol, fantasy-object, and love-nest; and each particular model will embody a design compromise, within the constraints of cost, among such possible functions. However, there are other devices in which the success of a particular design compromise cannot be tested retrospectively by the market; as an example, let us consider a dam.

Among the functions of a dam, there can be hydro-electric power, water storage, flood control, irrigation, and recreation. Each of these will serve the purposes of some particular group, through some benefit conferred by its actions. But the dam will also have 'costs' or dysfunctions, in the sense that some purposes will be injured by its effects; thus, it will inevitably be dysfunctional in respect of habitation and farming of the area to be drowned, as well as of other sorts of recreation; in its destruction of a natural habitat it may be dysfunctional for the ecology of the region and for the sciences dependent on those materials for data; and if material monuments are also drowned, it will be dysfunctional for the national culture. Each function or dysfunction is relevant to the interests of a particular group of individuals; and the groups may have little or no purposes in common which could serve as the basis for an acceptable compromise. The dam might even become the focus of political protest, as when the rural areas to be drowned in the interests of the city, belong to a different nation; thus the Welsh nationalists see the reservoirs on their land as an instrument of English imperialism.

Special social mechanisms are necessary for the achievement of a decision in such cases. After public hearings in which injured citizens vent their aggravation and tame experts display their erudition, the actual deliberation is done in camera; and the decision, like all such difficult ones, is probably agreed informally between three men going down in the lift, or encountering each other in some other place conducive to relaxed and confidential discussion. In earlier times, the decision process did not involve so much agony; but this was because one interest could frequently ride roughshod over all others. Things usually got done more quickly, but at the price of producing ill-designed monstrosities, which ravaged the landscape and blighted human lives. In our more enlightened age, each special interest may, if it can organize itself, put its point of view to those responsible for controlling such projects on behalf of the public. But the control on these controllers is extremely indirect; although their decisions are eminently political, the traditional structures of political power are not adapted to the exercise of influence on such decisions. For even if the project is executed by the State, it is the province of the executive rather than the legislature; and the relevant branch of the executive can always muster enough loyal experts to keep the odd inquisitive legislator at bay. Hence to the extent that influence is exercised, it will tend to be done informally, to the benefit of those who command the channels of informal influence.[23]

---

[23] In the United States there is a clear 'life cycle' for regulatory commissions, which starts with public outcry, and symbolic reassurance, but then is followed by political quiescence and the conversion of the commission to the furthering of the purposes of the corporate bodies supposedly being regulated. See Beryl L. Crowe, 'The Tragedy of the Commons Revisited', *Science*, 166 (1969), 1103–9.

Hints of a deeper understanding of the problem of the control of regulatory commissions are provided by Lincoln Steffens, in his chapter 'How Hard it is to Keep Things Wrong', *Autobiography*, pp. 561–9. 'The responsible attorney for a railroad and conscientious railroad men have told me—and convinced me, too—that you cannot run a railroad without corrupting and controlling government. All discussion of public ownership is foolish; either the State will own and operate the railroads and other public utilities or these public corporations will "own" and govern the State' (p. 565). But this form of unofficial government has its own severe problems. For as his informant explained to him, 'The Southern Pacific Railroad and all the companies and interests associated with us are not rich enough to pay all that politics [i.e. corruption] costs.' Hence, to keep the machinery of government complaisant, 'We have to let these little skates [i.e. the crude, petty, "dirty" grafters] get theirs; we have to sit by and see them run riot and take risks that risk our interests, too. We can't help it' (p. 567; the order of quotations is inverted here).

Political activity on such issues cannot be along traditional lines, since there is rarely a mass consensus whereby pressure can be exerted on a national party. It is for this reason, among others, that non-violent direct action, conducted as a symbolic gesture bringing attention to the problem and bringing embarrassment to the bureaucrats, is becoming popular as a form of practical political activity.

In any debate over a proposed innovation, the most loud and consistent voice will be that of the group promoting the new device. At best, they will be honest practitioners of traditional myopic engineering; their problems are defined in hard, quantitative terms, and their projected solution may well be the only one possible on the assumption that the technical and social context of the problem remains unchanged.[24] But the engineers have recently been joined by the apostles of runaway technology, who when all other arguments fail have the last refuge of 'progress'. In practice this means that if an existing device can be 'improved', usually by being made larger, faster and more expensive, then it is violating a law of nature to abstain from doing so. The ideology of such 'progress' is given support by the history of technology, both folk-history and scholarly. Whereas the history of science has tended to be Whiggish, the history of technology is quite Hegelian, seeing the efforts of the past as the unfolding of the Idea of the Perfect Device of the present or of the near future.[25] If one restricts 'technology' to the devices

[24] In his criticism of a proposal for a 'master drain' to remove brackish ground-water from the San Joaquin Valley of California, Frank M. Stead, 'Desalting California', *Environment*, 2, No. 5 (June 1969), 2–10, says 'But let us not jump to the easy conclusion that because the drain plan is environmentally unsound in the long haul, the engineers who propose this drain are incompetent or venal. I know this is not so. The problem stems from the shortsighted basic precepts that guide our efforts in the management of environmental resources in both the public and private sectors. If we want something better than the San Joaquin Master Drain, you and I are going to have to change the environmental management ground rules' (p. 2).

[25] The strength of this conception of 'technology' is well illustrated by a story of Lewis Mumford: 'Unfortunately, so firmly were the nineteenth-century conceptions committed to the notion of man as primarily *homo faber*, the tool-maker, rather than *homo sapiens*, the mind-maker, that, as you know, the first discovery of the art of the Altamira caves was dismissed as a hoax because the leading paleo-ethnologists would not admit that the Ice Age hunters whose weapons and tools they had recently discovered could have had either the leisure or the mental inclination to produce art—not crude forms but images that showed powers of observation and abstraction of a high order.' See 'Technics and the Nature of Man', *Technology and Culture*, 7 (1966), p. 309. For an example of a recognition of the limits of this historiographical tradition, we may cite T. K. Derry and T. I. Williams, *A Short History of Technology* (Clarendon Press, Oxford, 1960), Epilogue, p. 710: 'From this standpoint (the happiness of the individual) the greatest indisputable benefits of modern technology are perhaps those conferred by

created by the mechanical and civil engineering of the last few centuries, and concentrates on a simple sequence of advances, one can render a plausible story along these lines. The time-lag between advances can be explained by unfavourable technical or commercial circumstances, and the delays in diffusion to other places can be similarly explained in terms of their 'backwardness' in some respect. The moral of 'inevitable progress' always emerges from such an history, partly because of its inherent plausibility and partly because it was built into the study in the first place.

We have already seen that a new device is successful only if it can perform some functions whereby certain purposes are served. In the case of a device put on the market, both consumers and producers must derive benefit from its production and sale; and together they must be able to override the objections of those who are injured by the innovation. These latter can include those whose interests are threatened by the displacement of an existing system of devices (either capitalists or workmen), and those who will suffer incidental harm from the adoption of the new device. Over the last couple of centuries, the capitalist system has been well suited to the fostering of innovations and the overcoming of resistance, in manufacturing and agriculture. With the passing of the guilds of Medieval times, and of the Royal patents for monopolies of a later period, the protection of capital against innovation was lost. The artisans and workmen whose work would be degraded or displaced by innovation could be suppressed by economic and political techniques. And those who were injured by an innovation could either move away if they had the money, or remain and suffer if they didn't. Outside the manufacturing industry and agriculture of the nineteenth and early twentieth centuries, the constraints on innovation were not so easily swept away. House-construction, for example, is only now moving forward from the materials and organization developed millennia ago; and the 'technology of the home' can be seen as far more complex cultural phenomenon than might appear from a casual glance at the latest refrigerators and televisions.[26] Also, as it becomes increasingly difficult to escape from the degradation and

branches which the present *Short History* has lacked space to emphasise as they deserve, namely the revolutionary changes in medicine and surgery.'

[26] A case study in the complexities of the diffusion of innovations is given by A. Ravetz, 'The Victorian Coal Kitchen and its Reformers', *Victorian Studies*, 11 (1968), 435–60.

squalor of mindless industrialism, sections of the articulate and educated classes have joined the protests.

The inadequacies of the simplistic 'progressive' view of technological development can be seen most clearly in the case of the automobile. On the one hand, it is possible to trace a development from the first primitive machines, through a sequence of models of increased speed, comfort, reliability and safety. Yet even this internal history is not one of steady, autonomous improvement. The safety of automobiles has received serious attention only recently, in response to campaigns waged by independent critics. And the speed of automobiles has stopped increasing; although they can be made to ride comfortably at some ninety miles an hour, in all the countries where the automobile is a standard piece of domestic equipment it is recognized that at more than seventy the weapon becomes too lethal. Also, the external constraints on the diffusion of the automobile are now becoming recognized. For years, highway engineers attacked the problem of traffic congestion by building bigger roads; but recently they have become aware of the principle that any convenient automobile channel soon becomes clogged beyond capacity. Finally, the damage done to the atmosphere, and to the social environment by the present stage of progress in private transport, is recognized as serious and urgent. Thus, in this most important sector of the modern economy, the belief in the necessity of simple 'progress', and the duty to participate in it, has collapsed in the face of the realities of technology in its social and ecological context.[27]

The problem of achieving effective democratic control on the decisions of technological innovation is beset by the many difficulties we have seen: the combination of contrary purposes, with bureaucratic operation, and the different ideologies of the various groups involved. Yet if this problem is not solved, our social lives will inevitably come to be ruled to an increasing extent, by blundering bureaucrats and experts. Worse yet, this will be recognized as a new form of oppression, which can secure acquiescence neither by the possession of direct police powers by the decision-makers, nor by

[27] An illuminating survey of the intricacy of the 'welfare' aspects of technical projects is given by *Transportation and Community Values* (Special Report 105, Highways Research Board; National Academy of Sciences, Washington, 1969). The projects discussed there are the urban motorways, designed for the convenience of the suburban whites, and causing distress and disruption to the urban poor, mostly black. Only after the great urban riots did the planning engineers discover that there is a genuine conflict of values, not easily reducible to quantitative terms for a systems analysis.

the legitimacy derived from a semblance of answerability to the population. Such problems have already appeared explicitly in the Socialist countries, where the bureaucracy is quite obtrusive; and they can be interpreted as the cause of much of the current 'student' unrest in the capitalist societies as well.[28] The first ideas towards a solution, which involve drastic changes in our inherited conceptions of democratic politics, have already been put forward. To his great credit, Mr. A. Wedgwood Benn made a positive contribution to this long-term problem even when he was, as Minister of Technology, enmeshed in the destructive practicalities of the present.[29] But the full solution of this very deep political problem is a task for the decades to come; and on it may well depend the survival of our civilization.

[28] See Hannah Arendt, *On Violence* (Harcourt, Brace & World, New York, 1970). The dominion over man is today exercised by a 'bureaucracy or the rule of an intricate system of bureaus in which no man, neither one nor the best, neither the few nor the many, can be held responsible, and which properly could be called rule by Nobody. If we identify tyranny as government that is not held to give an account of itself, rule by Nobody is clearly the most tyrannical of all, since there is no one left who could even be asked to answer for what is being done. It is this state of affairs, making it impossible to localise responsibility and to identify the enemy, that is among the most potent causes of the current world-wide rebellious unrest, its chaotic nature, and its dangerous tendency to get out of control and to run amuck.' (Quoted from a review by Fred J. Cook, *The Nation* (6 April 1970), 606.)

[29] Nigel Calder, *Technopolis* (MacGibbon & Kee, London, 1969) discusses the Czech investigations of this problem before 1968, along with the proposals of Mr. Wedgwood Benn and various Americans, in ch. 17, 'Democracy of the Second Kind'.

14

IMMATURE AND INEFFECTIVE FIELDS OF INQUIRY

OVER the centuries, philosophical inquiry into the nature of scientific knowledge has concentrated its attention on fields which were already matured and effective. These provided an example of what could be achieved through the study of the natural world; and it was plausible to assume that they could also serve as a model for the methods of other, less fortunate fields. But such fields are a minority, even within the study of the natural world. To ignore the others is to present a distorted view of scientific inquiry, and to mislead scientists who are trying to develop their fields to the same state of effectiveness as those whose success is universally recognized. For when the fully matured fields are taken as the normal, the others are implicitly classed as abnormal, and their condition is a matter for concern and even shame. Attempts to improve them by a mechanical imitation of the methods of the paradigm successful disciplines, in our times the matured mathematical-experimental fields of natural science, can do little good and much harm.

In order to analyse this particular pathology of scientific inquiry, we will need to use a more sophisticated battery of concepts than has been developed hitherto. The dominant traditions in the philosophy of science have assumed a simple dichotomy between genuine science, either already matured or capable of maturation by the application of some straightforward techniques, and pseudo-science, doomed to ineffectiveness by a misconception of its objects of inquiry and its methods. The list of modern pseudo-sciences was fixed during the seventeenth century; it included astrology, alchemy, geomancy (divination by local features of the earth's surface), and phytognomy (ascription of medicinal powers to plants by their suggestive shapes, as developed in folk traditions). Its characteristic was the assumption of sense and meaning in the ordinary objects of the external world, and its methods involved a measure of dialogue with them. By contrast, genuine science assumed no human properties in its objects, and proceeded by some mixture of observation and reason.

This simple distinction had a very important ideological function during several centuries, while 'science' was still involved in a struggle against 'dogma and superstition' for authority in pronouncing on the natural world. The problems of the philosophy of science were organized around the specification of those methods that would yield genuine scientific knowledge, and then the character of that knowledge as a system yielding truth or its acceptable substitute. The appellation 'pseudo-science' conveyed the deliberate connotation that such fields were not merely ineffective, but also deceitful, and pernicious, standing for darkness against enlightenment.

Because the historical associations of the term 'pseudo-science' are still so strong, its mention inevitably suggests a deep and blanket condemnation of a field so described. I shall therefore avoid it and instead employ terms focussing on one or another aspect of the state of ineffectiveness. We recall that we could define this state, in terms of the absence of 'facts', a condition caused by the absence of criteria of adequacy appropriate for the detection and avoidance of pitfalls in research. A more commonly used term for this condition is 'immaturity', and I shall occasionally use it as well. But this has the connotation of denying the complete ineffectiveness of the field, and also promising a development towards matured and effective state; and as such may well be misleading. Indeed, one of the most delicate problems of the history of science is the assessment of a field now successful when it was in an earlier, ineffective state. The historian must decide whether it was then 'immature', containing within itself the needs of later successes; or whether the subsequent innovations were so deep as to involve the transformation and rejection of the essential features of the inquiry as then practised. On this assessment will depend the evaluation of the field as it was then, and the consequent practical decision on its inclusion in programmes of teaching or research. The most famous instance of such a problem is the field of mechanics in the later Middle Ages, where some formal 'anticipations' of Galileo's results are to be found; and the significance of these for the development of classical mechanics from Galileo onwards is still under debate.

The problem of assessment is even more acute in the case of a field which is currently ineffective, but whose leaders claim it to be on the point of emerging from immaturity. This amounts to a claim that the objects of inquiry and their associated methods are appropriate for the development of the field; while those who disagree would consider further investment in the field as it stands to be futile. The correct assessment can only be made retrospectively, and even then involves deep difficulties. Because 'immaturity' and 'ineffectiveness', although quite distinct in their implications, are connected by so many nuances as well as being difficult to distinguish in practice, I shall use both terms in any discussion nearly interchangeably.

At the present time, the disciplines that present the most obvious evidence of ineffectiveness or at least immaturity, are those which attempt to study human behaviour in the style of the mathematical-experimental natural sciences. But it is important to keep in mind that each of the natural sciences in the past, and special fields within them at the present, have had and do have this character. Earlier generations of historians of science tended to make a sharp distinction between the darkness of the 'prehistory' of a subject (perhaps as lightened by one or a few great men speaking truth before their time), suddenly transformed into maturity; hence synthetic accounts of the evolution of scientific disciplines in these terms are rare. For medicine, it is possible to interpret its history from its beginnings to the end of the nineteenth century, as reactions to the insoluble and yet pressing problems posed by its being in a state of ineffectiveness.[1]

As we shall see, the difficulties of working in an immature or ineffective field are serious and manifold. Added to the basic difficulties of trying to do research in a field where the pitfalls are still unidentified, there are the social constraints forced by the pretence of maturity. The situation becomes worse when an immature or ineffective field is enlisted in the work of resolution of some practical problem. In such an uncontrolled and perhaps uncontrollable context, where facts are few and political passions many, the relevant immature field functions to a great extent as a 'folk-science'. This is a body of accepted knowledge whose function is not to provide the basis for further advance, but to offer comfort and reassurance to some body of believers. Our discussion of the applications of immature or ineffective fields depends heavily on this idea of 'folk-science', and so two sections on it are inserted in the chapter. We

[1] For such an interpretation, see R. H. Shryock, *The Development of Modern Medicine.*

should observe that not every folk-science is associated with an immature academic discipline; some have none, while others are associated with fully matured disciplines. Conversely, not every immature or ineffective field is associated with a popular folk-science; but the presence of such an association can explain an otherwise incomprehensible popular trust in a discipline of this sort when it occurs.

Using these materials, we can study the state and causes of the condition of ineffectiveness, and then see in what ways knowledge can come to be achieved in spite of it. We then move on to the social aspects of inquiry in such fields, discussing the reasons for the pretence of maturity and its effects on research and teaching. Passing to the relations of such disciplines with the outer world, we consider the really complicated situations that arise when they are developed in connection with technical and practical problems, and also entangled with ideologically-sensitive folk-sciences. Since the most important applications of 'science' to the problems of society have just these features, an appreciation of these difficulties and pitfalls is important for the proper use of science in the modern world.

Ineffectiveness—the Absence of Facts

In a matured field, the processes of selecting and transforming the results of research for the achievement of facts can proceed in such an orderly manner that the nature of the 'facts', as stable products of these operations, can pass unnoticed. But here as elsewhere, the pathological state can provide clues for understanding the normal. The indubitable and public symptom of ineffectiveness of a field is the absence of facts, in the sense in which I have defined them. Moreover, in such fields the deficiency is most obvious at the level of elementary teaching. There, unlike in a matured field, the students do not encounter a collection of standardized materials, presented in a digestible form, and utterly reliable and incontrovertible in themselves in spite of the slightly different guises in which they are presented in different books or lecture-courses. By contrast, in the ineffective or immature field, the student is presented with one out of several sets of supposedly basic materials, and can discover other sets by reading textbooks not on the recommended list. These materials themselves consist of intuitive generalities dressed up as empirical laws, and insecure theoretical speculations masquerading as fundamental explanations.[2] The lay public sees the same phenomenon from a slightly different aspect. Watching the activity of a field over a period of years, one does not witness the steady cumulation of new facts, perhaps superseding but never completely destroying the old. Instead, there is a succession of leading schools, each with a manifesto which is more impressive than its achievements, and each passing into obscurity when its turn on the stage is over.[3]

The ineffectiveness of a field, as revealed by the absence of facts, is a trying state of affairs in the best of circumstances. There are many natural reasons why the leaders of such a field should try to pretend that it is otherwise, or to claim and hope that the condition is easily remedied. Most such claims, and the research strategies organized around them, involve a concentration on one aspect of the work which is seen to be deficient, and whose improvement should produce full maturity. Heroic attempts have been made to amass empirical data; to apply mathematical and computational tools for the production of information; to construct elaborate systems representing objects by symbols and manipulating them in a formal argument; and finally to develop methods and a methodology appropriate for the discipline. In each case, the attempt is to reproduce what is believed to be the crucial feature of an established science, where this is learned more from philosophers of science than from the successful practitioners themselves. But almost all these one-sided efforts fail utterly. Those which survive over time, through the strength of the founder of a school, and institutional stability, are likely to produce results which become elaborate and refined to the point of being grotesque. For the condition of ineffectiveness is not

[2] This instabililty in the ineffective sciences was of crucial importance in the development of the ideas of Thomas Kuhn. In the *Structure of Scientific Revolutions*, he reports on his experience at the Center for Advanced Studies in the Behavioural Sciences, '. . . I was struck by the number and extent of the overt disagreements between social scientists about the nature of legitimate scientific problems and methods. Both history and acquaintance made me doubt that practitioners of the natural sciences possess firmer or more permanent answers to such questions than their colleagues in social science. Yet, somehow, the practice of astronomy, physics, chemistry or biology, normally fails to evoke the controversies over fundamentals that often seem endemic among, say, psychologists or sociologists' (p. x).

[3] In his Editor's Introduction to *Handbook of Modern Sociology* (Rand McNally, Chicago, 1966), R. E. L. Farris discusses the early history of American sociology which proceeded from manifestoes, through the establishment of formal institutions, to methodological debates. The process started in 1893; he says 'In a sense, sociology became organized before it began to exist.' Only some thirty years later did viable results begin to appear, in the work of Thomas and Znaniecki, and the latter could hardly be counted as 'American' at that time at least (pp. 26–7).

an accidental deficiency in some component of the materials of a field, but is a systematic weakness in those materials and in the social activity whereby they are produced. Projects for immediate and radical improvement in a field must necessarily assume a simplicity of the means to their end; but they then become similar to those for radical reform or revolution in society, requiring the same qualities of dedication or fanaticism in their leaders, and incurring the same risks of patent failure and illusory success.[4]

In analysing the absence of facts, we naturally consider first those materials which should, but do not, achieve this status: the conclusions of arguments, and the evidence, information, and data on which they are based. The failure of nearly all of these to survive even in the short run, is an indication that most of the work of investigating problems is vitiated by pitfalls, encountered sooner or later in the work. The results of research are generally weak, or even vacuous. This condition prevails even in fields where the leaders and their associates spare nothing in their endeavours; but the absence of a body of appropriate methods of inquiry nullifies their efforts. For it is through such methods, ranging from the techniques of production of data, to the judgements of adequacy on an argument, that pitfalls are identified, and ways around them are charted. Because of the subtlety and sophistication of scientific inquiry, these methods are a craft knowledge, built up by successful experience. But an ineffective or immature field has no such experience; and so the improvement of its methods is not a straightforward operation.[5]

[4] As an influential English example of the many attempts in this direction, we may cite Baroness Barbara Wootton, *Testament for the Social Sciences* (Allen & Unwin, London, 1950). Learning from her sources in the philosophy of science that science consists of observation, hypothesis-testing, and measurement, and thus being spared any intimation of the artificiality of the objects of scientific inquiry, she could conclude that simple diligence and care could bring the social sciences to the same state as physics. She was troubled by one sign of immaturity: that the classics of political and social theory do not become obsolete in teaching. Her explanations are not entirely consistent: that these practices are retained through a prejudice against the scientific status of these disciplines; and that 'perhaps the need adequately to fill a curriculum has something to do with this' (pp. 29–30). In fairness to her, we should record that in her *Social Science and Social Pathology* (Allen & Unwin, London, 1959), derived from her practical experience, she recognizes the pitfalls to be encountered in the simplistic application of a science (in this case psychiatry) to social problems.

[5] The absence of appropriate criteria of adequacy in the rapidly-growing field of 'technological forecasting' is shown clearly in Martin Shubik, 'Processing the Future', *Science*, 166 (1969), 1257–8, a review of R. U. Ayres, *Technological Forecasting and Long-Range Planning* (McGraw-Hill, 1969). Commenting on the author's enthusiasm for computer analysis of 'hundreds of thousands' of scenarios, he says, 'Tons of pages of

The weaknesses in the social aspects of inquiry also contribute to the self-perpetuating condition of ineffectiveness. The mechanisms for the processing of results, and for the exercise of quality control, cannot be stronger than the materials on which they operate. For social reasons it is necessary to give the formal authenticity of publication to masses of results which are very weak; and so the effective standards of quality cannot meet those of a matured field. Because of the rapid succession of separate schools, each with its own objects of inquiry and principles of method, there is little opportunity for results of potentially high quality to survive and become established as facts. And in this unstable and frequently false social situation, the mechanisms for the control of quality and the maintenance of scientific integrity at the highest levels do not exist.[6]

To bring a field out of an immature state requires first of all a strengthening of its objects and methods; and this requires the endeavours of a man of exceptional talent and dedication. Such a man will have a strongly marked personal style, both in the scientific and social aspects of his work, and probably a personal commitment to a goal which is deeper than the mere establishment of positive knowledge in the field. Hence, as we have discussed in connection with style, a field in a phase of maturation will still bear certain marks of its earlier history, both in its social organization and doctrines. Also, the existence of a body of established facts is not in itself sufficient to ensure the proper operation of the social mechanisms of direction and quality control. A field can enter a degenerate and stagnant state, where its members and leaders are incapable or unwilling to devote their efforts to its further advancement, but are content to produce mediocre or shoddy work while basking in the reflected glory of their predecessors. But this problem has already

computer output are in general a sign of a badly understood job. Most simulations have a value inversely related to the fourth power of the quantity of computer output.' For corrective reading, he prescribes G. A. Miller, 'The Magical Number Seven, Plus or Minus Two—Some Limits on our Capacity for Processing Information', *Journal of the Acoustical Society of America*, 22 (1950), 725–30. Of course, in this case as in others, the Americans compensate for their invention of technological monstrosities by their ready wit in describing them; thus Shubik's point is expressed by the acronym GIGO—Garbage In, Garbage Out—for vacuous computing.

[6] Paul E. Meehl, 'Theory Testing in Psychology and Physics: a Methodological Para dox', *Philosophy of Science*, 34 (1967), 103–15, shows how sophisticated significance tests give a spurious 'confirmation' of hypotheses; even random data would pass the tests h alf the time. In spite of his distinguished position in American psychology, he has been unable to prevent the widespread adoption of such lax standards of adequacy.

been discussed in connection with ethics and quality control, and we can concentrate here on genuinely immature fields.

Nor should it be imagined that the problems discussed here are restricted in their relevance to entire disciplines which are recognizably ineffective or immature. We have already seen in connection with methods of research, that any major advance in science involves an innovation, which temporarily at least is not capable of assessment by the existing criteria for work in its field. Hence whenever a discipline extends a deep salient into the unknown, that point will have the characteristic features of immaturity to a greater or less extent.[7] In a mature and healthy field, the social processes of selection and transformation of results will bring back from the research front, for more general use, materials that are 'matured' in the passage, and add them to its stock of facts. By contrast, in an ineffective or immature field, there is nothing there behind the front lines; and finally, a sign of stagnation and senescence in a field is when it is all completely matured and safe, hence completely safe and boring.

Knowledge outside the Matured Mathematical-Experimental Sciences

One reason for the difficulties of immature and ineffective fields is that their model of genuine science is a very specialized one, which

[7] A most illuminating description of the immature state of a new field in physics is given by D. C. Montgomery and D. A. Tidman, *Plasma Kinetic Theory* (McGraw-Hill, 1964), preface, p. vi. On theory and experiment: 'there is so far very little overlap in plasma physics between accurately measurable and accurately calculable quantities. The experimentalist becomes entangled with situation-dependent parameters—percentage of impurities, details of vacuum systems—while the theoretician concentrates on one-dimensional, unbounded, and otherwise highly idealized situations, so that opportunities for comparing results have been regrettably few. The indications of the last two years have been hopeful, however, and this disjointedness may well be on its way out.' On the theory itself, we read: 'The tendency in kinetic theory, since long before the days of plasma physics, has been to confuse what is assumed with what is proved. This has not been disastrous in the area of classical molecular gas dynamics, since there was always a steadily accumulating wealth of experimental data to provide ballast for the theoretician. In plasma physics, however, most of even the simplest theoretical predictions go quantitatively unverified experimentally, and there has been a profound and unfortunate separation of what has been calculable and what has been measurable. It is a truism of modern science that the rewards for writing papers are greater than those for reading them. Plasma physics, perhaps more than any other branch, has suffered from undue attention to this dictum and has generated, in so doing, a literature so large, unwieldy, and hastily put together that the newcomer to the field finds that it is nearly impossible to verify the truth or falsity of any given theory to his own satisfaction.' I am indebted to one of the authors, D. C. Montgomery, for telling me of this source, and for fruitful discussions of these problems, when we were both in Utrecht in 1964–5.

may be quite inappropriate to their own tasks. To understand the special problems of immature fields, and the possibilities for their development, we can derive guidance from the history of scientific inquiry in the period before the rise to dominance of 'positive science'. Then, there was a clear distinction between two sorts of inquiry, 'history' and 'philosophy'; and they were in turn distinguished from 'art'. The former meant description, not merely of past times, but of any class of objects; and 'natural history' still survives as the title of some long-established museums. 'Philosophy', on the other hand, meant reflection and explanation, as applied to any problem. Thus Dalton's atomic theory was announced in his 'New System of Chemical Philosophy', and what we now call 'physics' was known as 'natural philosophy' in England through the nineteenth century. Each type of inquiry had its own criteria of adequacy, the one emphasizing faithful and comprehensive accounts, and the other coherence of argument. Of course, there was plenty of variation in these, over time and place, and many were the debates in which each side accused the other of being 'unphilosophical'.

Between the two sorts of inquiry, 'history' and 'philosophy', they omitted something distinctive of positive science, and also maintained something that has since been lost. For the positivist ideal of a system of advancing knowledge, in which bold speculation proceeds in a tight connection with particular facts from experience, was not the ruling one. Newton's work is a partial exception to this generalization, but even he had recourse to 'Queries' when he published his views on the deeper problems of what we now call physics. On the other hand, the philosophy of nature was, in subject matter and in spirit, continuous with philosophy in general; the difference between them in the nature of the evidence and the conclusiveness of the arguments was one of degree only. Although it would not be true to say that all the productions in natural history and natural philosophy were good of their kind, these conceptions of the nature of the inquiry were appropriate both to the state of knowledge, and to the conditions of social organization of the fields. The activity of 'research' requires a certain degree of maturity of the field, and also elaborated social institutions, if it is to be effective.[8] When the study

[8] It is only very recently that the term 'natural philosopher' was driven out by 'scientist'; in England the conquest was not complete until the present century. See S. Ross, 'Scientist, the story of a Word', *Annals of Science*, 18 (1962; published 1964), 65–86.

of nature was pursued largely by isolated individuals, in fields far from maturity, it was as well that each one could pursue the sort of problem that has a lengthy cycle of development and can be worked in an individual style.

The recognition of 'art' as a category distinct from inquiry could also contribute to a better self-understanding of immature fields. The traditional term was not restricted to subliterate handicrafts; for Aristotle, it extended to the set of principles defining the methods of any class of tasks. Arts could be 'liberal' as well as 'mechanical'; and could involve a sophisticated (and genuine) scientific component, as the Renaissance 'art of navigation'. Now, a literate art would naturally be based partly on the 'history' of its objects, and be informed to some extent by a 'philosophy' of its principles. On occasion, a rising art would develop a 'philosophy', with the function of enhancing its prestige; such was the case with architecture during the Italian Renaissance.[9] But these related inquiries would be ancillary and incidental to its real work, whose strength and success was independent of theirs.[10]

In the modern situation, we have many examples of successful special 'arts' arising out of inquiries in immature or even nascent fields. These will usually be methods for solving technical problems which, although very restricted in comparison with the goals of the field as a whole, are quite useful. It is known from experience that the methods work, and a body of genuine craft skills can be developed for the tasks; even though the theory on which the method was first based may have been discredited, and no scientifically satisfactory explanation has been produced. It is a natural error to cite such arts as evidence for the maturity of the field, as if they arose as applications of a solidly established body of fact.[11] But it would be more honest and more fruitful to accept such arts as those points where the

[9] This is not to say that those who proposed the 'philosophy' were necessarily conscious of this social function. In the case of the Renaissance theory of harmonious proportion, the motive could well have been the extension of the Platonic ideal of knowledge to all areas of experience. See R. Wittkower, *Architectural Principles in the Age of Humanism*, 2nd ed. (Alec Tiranti, London, 1967).

[10] Even in the case of Renaissance architecture, the greatest architects were not the greatest theoreticians; and greatness was achieved by creation within a style rather than by working out exercises in the application of its theory.

[11] Thus, E. G. Boring, *A History of Experimental Psychology*, 2nd ed., reviews his assessment of 1929, that American psychology was still immature; and cites its quantitative growth and its applications as evidence for a satisfactory state of maturity (pp. 742–743).

discipline has somehow made an effective contact with the external world, and with some humility to consider how this might have occurred.

If the leaders of ineffective and immature fields at the present time were able to recognize their condition publicly, and to institute appropriate methods, then this distinction between 'history', 'philosophy', and 'art' might be conducive to a more healthy atmosphere. Where the objects of inquiry have but a tenuous relation to the real things and events they purport to describe, and are themselves ill-formed and unstable, an isolated investigation devoted to a supposedly 'empirical' test of some hypothesis about their relations, is highly unlikely to yield worthwhile results. As an alternative to such 'research', we can imagine a sort of 'history', conducted in a disciplined fashion and using all appropriate tools, whose objects of inquiry are those of a trained common sense, and which has a less formalized, and correspondingly more extensive and perhaps deeper, contact with its sources. The discipline which we now call history, works in this fashion on the traces of human activity in the past; and it can offer tested methods, including criteria of adequacy, to this sort of work. At the other extreme, 'theory' could be recognized as a sort of 'philosophy', and could be governed to some extent by the methods of that discipline. It would then not be a cause of surprise or shame that effective new insights come only very rarely, and that they are only slightly developed by the subsequent efforts of lesser men; and it could be recognized that an esssential part of a genuine education in the discipline is a dialogue with its great masters. Finally, the successful 'arts' could be recognized as the most genuine tested experience of the field, and studied and developed as such. Each of these three sorts of inquiry could eventually yield knowledge of its characteristic sort and, in their interaction, offer mutual criticism and support.

It might be asked, what would then happen to the middle ground between 'history' and 'philosophy', which constitutes the distinctive achievement of positive science: the framing of general laws, universal in application, which are tested and confirmed by controlled experience? The question is not relevant to this discussion, for here we are considering just those fields where this does not exist; such laws are one variety of facts. It is even doubtful that this should be the goal of the development of every field of disciplined inquiry; but in any event the simplistic pursuit of that goal has clearly failed to bring results in many cases. Even with the enriched conception of the work in an immature field suggested here, the 'quality cut-off point' would still be higher than in those where routine technician's work is sufficient for the production of competent results. For a good history requires personal insight into the phenomena, and philosophy is a very demanding inquiry. But the resulting reorganization of the materials of the field and of the conception of the work could save many a talented recruit from years of futility.

Moreover, it is not necessary for a discipline to be fully 'positive' in the sense of imitating physics for it to make a contribution to the advancement of human knowledge. Facts of the ordinary sort can be achieved in history by processes which are the same in principle as those we have described for experimental natural science, although not so closely controlled in all their operations. In philosophy too, there are certain problems and solutions which are widely accepted as fundamental, and this 'classic' doctrine can be and is successfully taught to students. However, the eventual products of the processes of selecting and transforming 'facts' of these sorts have a different form from that of knowledge in the matured mathematical and experimental sciences. A very deep result in these other disciplines will generally not survive as a simple, impersonal, and apparently elementary and indubitable assertion. It is more likely to appear as an aphorism: an expression of a deep personal understanding of its objects, in a condensed and communicable form. When the ideal of knowledge is that of an impersonal and rigid system, aphorisms are scarcely recognized as bearers of knowledge; and this has been the situation in the philosophy of science, and in the dominant currents of epistemology, since the seventeenth century. But for Francis Bacon, one of the pioneers of the 'new philosophy', who had deep roots in an older, humanistic tradition, knowledge was distilled rather than deduced, and aphorisms were correspondingly important. He believed that the deepest and most general principles of nature would be expressed as aphorisms; and his *Novum Organum* was itself organized as a series of aphorisms. He contrasted the fruitful and living character of aphoristic knowledge, to the rigid and frequently sterile character of deductive systems. In this, he doubtless had in mind the standardised versions of Aristotelian doctrine that were taught in the universities. But of equal importance to him was theology, of which he said that a collection of aphorisms from English sermons of the preceding half-century would be 'the best work in divinity which had been written since the apostles' times'.[12]

Thus, to the extent that a discipline differs from the state of being matured mathematical or experimental science, its permanent knowledge, in whose terms its facts are organized, will be embodied in aphorisms rather than in impersonal universal laws. This applies both to immature fields and to those whose materials do not enable the establishment of evidence from rigorously controlled experience, or of arguments with a complex and rigid structure. And at the far extreme, where a craft knowledge finds verbal expression, all its principles will be expressed as aphorisms. When compared with 'scientific' knowledge, aphorisms suffer from several defects. Since they are not the conclusions of a tightly-structured argument, they are incapable of modification through criticisms of particular

[12] It seems likely that in his earliest years, Bacon hoped that personal wisdom could be obtained directly from a collection of aphorisms; a record of this endeavour is preserved in his *Colours of Good and Evil* (*Works*, vii, 75–92), and in the *De Augmentis Scientiarum*, Book VI, ch. 3 (translation in *Works*, iv, 473–91). Although he apologized for these in the earlier piece, and relegated them to the section on transmission of knowledge in the later book, there is evidence that he considered aphorisms to carry the deepest truths about God's creation. For, in ch. 1 of Book III of the *De Augmentis*, he gave a tentative list of 'axioms' which would constitute *Philosophia Prima* or *Sapientia*; and showed that they hold true in all disciplines of nature and man. Indeed, he considered them as 'plainly the same footsteps of nature treading or printing upon different subjects and matters'. An example of this class of axioms is : 'Putrefaction is more contagious before than after Maturity', true for 'physics' (? medicine) and for morality (translation pp. 337–9).

Bacon's remarks on aphorisms as a source of knowledge are reiterated through his writings, starting with the *Advancement of Learning* (*Works*, iii, 292). There he condemned the 'over-early and peremptory reduction of knowledge into arts and methods', using the argument that things which come to a perfect shape too soon seldom grow further. In the *Cogitata et Visa* of a few years later (*Works*, iii, 593–4; translated in Farrington, *The Philosophy of Francis Bacon*, Liverpool, 1964, p. 75), he used the precedent of the ancient Greek philosophers (now called the 'presocratics'). He praised the use of aphorisms for giving a plain account of things discovered, and being silent where there was nothing discovered; and also for stimulating the mind to judgement and discovery. Part of this passage was taken over into Aphorism 86 of Book 1 of the *Novum Organum* (translated in *Works*, iv, 85); there he added the comment that the pretence of systematic knowledge in incomplete sciences, results from the 'craft and artifices' of those who handle and transmit them; with a resulting sterility in those sciences.

The remarks on divinity are near the end of the *Advancement of Learning* (*Works*, iii, 487–8). It must not be thought that Bacon was opposed to all systematizations. The consolidation and reform of all the branches of English law was of great concern to him through his entire public career. His short paper of 1616, defining a project to that end, was read out in Parliament by Sir Robert Peel in 1826, when he was attempting the first phase of the achievement of Bacon's programme. See *Works*, xiii, 57–71.

details. Nor do they go through the same complex processes of evolution and standardization as scientific results, and so the social phase of their development does not involve such rigorous tests. Also, they express a private understanding as much as a public knowledge, and so their very statement will include terms with important connotations of meaning which are lost when they are removed from the context of their first announcement. Because of this, they can be seriously misleading, when taken over by later generations in ignorance of their history.

The other sorts of disciplines cannot hope to achieve knowledge of the same sort as the matured mathematical and experimental sciences; an aphorism will never have the permanence and objectivity of a law. Whether they are thereby rendered inferior is another matter; if their problems are important, and can be effectively solved to yield some sort of genuine knowledge, then they are worth pursuing. The same considerations hold for immature disciplines; although the results will be weak, and it is impossible to predict when and how they will reach maturity, if the problems are real then one continues the work as an investment for the future. The question, then, is how to make this investment most effectively. I have argued that it is unrealistic to proceed as if a discipline is mature when in fact it is not, and that it is ill-advised to force it along a path of mechanical imitation of matured disciplines. Whatever model is adopted, the forcing of inquiry into the mould of 'research', investigating small-scale problems totally within the world of intellectually constructed classes of things and events, is unlikely to bear scientific fruit, either in the present or for the future. By pretending to be what it is not, the immature or ineffective discipline condemns itself to remain in that state, rather than engaging in the sort of work that would open paths to its achieving maturity of a character appropriate to itself.

The Pretence of Maturity

If scientific inquiry could be conducted in a social vacuum, the failure to recognize immaturity, and to adopt methods of inquiry appropriate to the condition, would be a sign of incompetence or dishonesty in the leaders of an immature field. But the present social institutions of science, and of learning in general, impose such constraints that the growth and even the survival of an immature field would be endangered by the simple honesty of public announcement of its condition.[13] For these institutions were developed around mature or rapidly maturing fields in the nineteenth century. Their operation presupposes that good and first-class research is possible, that it is governed by a set of stable and appropriate methods, and that it has already produced a body of facts and techniques in which students can be trained and examined. If the representatives of a discipline announce that they do not fit in with such a system, they can be simply excluded from it, to the benefit of their competitors for the perennially limited resources. The field would be relegated to amateur status, and thereby pushed over to the very margin of the world of learning; it would be deprived of funds and prestige, its members would need to pursue their inquiries in the time left over from an unrelated job, and recruits would come in only through self-education. It is true that the study of the natural world as a whole was in just such a social condition in most places up to the present century; but it could rely on men of independent means as recruits, and there was no established institutional structure devoted to the advancement of learning, from which it was being excluded; the Academies included natural science in their concerns, and the universities were at best teaching establishments. Hence under present conditions it is natural, and tactically appropriate, for the leaders of immature fields to conceal their condition, from themselves as well as from their audiences. We will later see that the simple pretence of maturity is not the key to the institutional success of an immature field; for this it must be believed in as a folk-science by some educated public. But to secure such a belief, it must affiliate to the general folk-science of 'Science', and this is equivalent to the concealment of immaturity.[14]

[13] In the academic hierarchy, unpopular immature disciplines are in a social position comparable to that of the category 'nigger': they must be gold to pass for silver. This became clear to me when I learned of the deliberations of a Committee of Professors on the filling of a vacant Chair in the History of Science at a distinguished foreign university. An acceptable candidate was required to have been an accomplished scientist *and* a classical scholar; to be at home with historians of all descriptions *and* also to be offering a message relevant to modern problems; as well as having eminence in a special field of research in the subject. One wise old man commented, 'They want a sheep with five legs.' At the same time, powerful Professors on the committee were filling old and new Chairs in their departments with men who made no pretence of being anything but productive research workers.

[14] Some leading men in the field of psychology in Britain have taken the plunge into public honesty. See G. Westby, *Behaviour Theories and the Status of Psychology*, Inaugural Lecture (University of Wales Press, Cardiff, 1963). He concludes with a recom-

The pretence of maturity throws up a host of practical problems, which aggravate the already severe difficulties of scientific inquiry in such conditions. Recruits are generally given no warning that their research work is likely to be very hazardous, and after some years of producing results which inexplicably fail to consolidate into facts, they can become demoralized; the real state of their field becomes a shameful secret. Nor do they have the security of knowing that their years of specialized training in the subject gives them a monopoly of practice in it, as in a learned profession or a matured science. For the founders of the field, on whose insights all subsequent work is directly based (without the enrichment that occurs in matured fields), are likely to have been philosophers, polymaths, or amateurs, rather than certified experts; and even at the present it is possible for an amateur to crash into the field, or to analyse the practical problems of its concern with more success than its practitioners.[15] Moreover, since the social mechanisms for quality control and direction in the field cannot function properly, the safeguards against the abuses of prestige are weakened, and the assessments of a lay audience, based on popularizations, can be of more practical importance in the politics of the field, than those of the community of experts. Under such conditions, the pretence of maintaining the

mendation for 'patience and boldness' for this 'infant science', together with a link with common sense experience. (I am indebted to Miss Helen Gower for this reference.) For a critical discussion of American sociology, in which many of the points made here about the character and problems of immature disciplines are argued at length and with examples, see *Sociology on Trial*, eds. M. Stein and A. Viddach (Prentice-Hall, 1963). (I am indebted to Mrs. J. Wootton for this reference.)

It is noteworthy that economics, by contrast, does not have any significant debates over its methods and their appropriateness. The *Festschrift* to Oskar Morgenstern, *Essays in Mathematical Economics*, ed. M. Shubik (Princeton University Press, 1967) contains a variety of papers on theoretical economics and games theory, but not one discussing the critical analysis of *On the Accuracy of Economic Observations*. The reason may be that if Morgenstern's recommendations were to be accepted, and economists did somehow learn that a number is an estimate and not a fact, then both research and teaching would need to be drastically overhauled, and many flourishing lines of research postponed for at least a decade until worthwhile data could be obtained. But it cannot be said to be a healthy situation when the published analysis and recommendations of an eminent member of a discipline are recognized only informally, and ignored in reasearch and in teaching.

[15] In his *History of Experimental Psychology*, G. Boring identifies four very great men in psychology's history: Darwin, Helmholtz, James, and Freud (p. 743). Only the last two had any concern to be considered as 'psychologists', and Freud was, of course, excluded from academic psychology all his life. Comparing Francis Galton to Wundt, Boring said of the former 'He had the advantage of competence without the limitation of being an expert' (p. 462).

social mechanisms appropriate to a matured discipline, and even more the task of improving their real condition, are rendered yet more difficult.[16]

Teaching an Immature Discipline

It is possible to mask the effects of immaturity on the research activities of a field; but in the work of teaching, which is necessary both for the training of recruits and for a vocational base, there are problems which are both acute and insoluble. In some respects, the tasks of teaching and of learning in an immature field can present more challenge, excitement, and true education than in one where there is an enormous body of standard information and tools to be mastered before the student is accepted as competent to think for himself. For in the absence of a great mass of established doctrine to be imparted, teacher and student can participate in a common search as near-equals, and the teacher's particular role can be Socratic rather than magisterial. The student gains the experience of inquiry into unsolved problems, and the teacher is continually refreshed and rejuvenated by his dialogue with new students on new problems.

Unfortunately, there are so many obstacles to the realization of this ideal state, that it can be achieved only in very exceptional circumstances. Even for the attempt to be made, there must be a prior public recognition that the field is so immature that there is no large body of facts to be taught; and, as we have seen, such a recognition involves such heavy costs for the field that only leaders who are both courageous and powerful can afford to make it. Also, such a style of teaching makes very heavy demands on the teachers; they must have exceptional intelligence, commitment, and confidence for the dialogue to be genuine and constructive. Discussions on the materials of an immature science will, if they are free, range over all levels, from data to methodology. Without skilful control, they can easily degenerate into intellectual chaos at the one extreme, or be structured into a set routine on the other. The teacher must be able to steer this delicate path, year after year, avoiding the temptation to use his superior knowledge of the teaching situation to guide the discussion along familiar and relatively safe channels, and still pre-

[16] In connection with his discussion of academic cliques in the social sciences, C. Wright Mills quotes an anonymous comment about a nameless statesman of social science: 'As long as he lives, he'll be the most eminent man in his field; two weeks after he dies, no one will remember him' (*The Sociological Imagination*, p. 109).

serving his personal stability. Under the conditions of our present culture, it is even more difficult to find students who can participate effectively and benefit from such an experience. They must be highly intelligent, with minds well trained for abstract reasoning; otherwise the teaching situation degenerates into an exchange of personal feelings and impressions. But their previous training at secondary school has been in the manipulative mastery of established doctrine; an entirely new set of skills and attitudes are required here, and the transition can be unsettling both intellectually and emotionally. It requires unusual strength and understanding to master difficult technical material which is patently unsound, for the sake of the training it gives in abstract thought, and as a corrective to naïve common sense.[17] Finally, even when the right teachers and the right students come together, this is an intensely personal situation, charged with excitement and with danger. It may succeed brilliantly for one year or for five, but it is a most difficult thing to institutionalize so that it can carry on for decades without becoming stale or corrupted.

Even when it succeeds, this approach to teaching has its costs. As an anomaly within an educational system conceived and organized around the transmission of doctrine and the objective testing of the mastery of information and skills, it is subject to a variety of external pressures and internal strains. The problem of examination (an essential part of the university's function as a certifying body) is severe; for as the doctrinal component of the student's training diminishes, the difficulties of objective and fair examination increase. Also, the students are deprived of the security which comes from mastering information and skills at a rate which can be tested against a public standard. Their situation subjects them to all the anxieties which afflict research students, and for which the institutional setting of their undergraduate work provides no protection. A student for

[17] See Ely Devons, 'Applied Economics—the Application of What?', in *The Logic of Personal Knowledge* (Michael Polanyi *Festschrift*) (Routledge, London, 1961), pp.155–69. Devons argues that 'what we are applying in theory are the elementary propositions and common-sense maxims'; and finds these extremely difficult to teach. 'It is only when the student has been through a theoretical drill well beyond the elementary level that the elementary notions really sink in deep.' And on the theory itself, he says earlier, '. . . but the position is more pathetic when economists by using the language merely deceive themselves. For the use of the language may give the illusion of a great understanding, whereas in fact it often merely conceals ignorance in a mass of esoteric jargon. And it is easy and dangerous to mistake description and classification of situations in a special economic language as answers to problems' (pp. 164, 165, 161).

whom such a course is inappropriate cannot settle down to the achievement of a modest competence as in an orthodox undergraduate subject; success is an all-or-nothing affair. Finally, until the peculiar skills and attitudes gained by a successful student of such a course are recognized as being genuine and important, the graduates will find themselves without a market for their labour. The strains of preparing oneself for nothing are severe on all except those who are confident of passing on to research, and those others who are not concerned to fit in to society. In fact, undergraduate study in an immature discipline is successful only at two extremes. For those who merely want intellectual fun, without the grinding discipline of an established subject, an easy course can be very enjoyable. And for those who can quickly learn to grapple with unsolved and ill-defined problems, a demanding course can be the best possible education. The latter group form a natural élite, and are equipped to function in modern society analogously to the social élite of former generations whose education mainly consisted of training in the personal and social characteristics of leadership. But we are as yet far from recognizing this quality; and in any event the difficulties of distinguishing objectively between weak and demanding courses in an immature discipline further delays such recognition.

In the great majority of cases, the teaching of an immature or ineffective discipline will be attempted imitation of that of matured disciplines. In the interests both of academic respectability, and of the needs of the less gifted students, instruction will be of straightforward doctrine, of which students will be expected to achieve manipulative mastery before voicing their opinions on open questions. But in immature fields, the straightforward doctrine does not come to be as the outcome of a natural evolution. There is no corpus of basic standardized facts, utterly reliable in themselves, which can be presented in a digestible form. Nor can the obscurities at the foundations of the discipline be buried out of everyone's sight by a mass of technical material. Those students who need to believe what is taught to them in order to understand it are thereby indoctrinated, perhaps for life. Those others whose critical sense impels them to question the doctrine, find themselves trying to play a different game from the one accepted by their colleagues and demanded by their teachers. Should they persist, then crises, intellectual, emotional, and social, are bound to result. Nor is it easy for a teacher, at least for one who is not a naïve believer. He must purvey the doctrine as if it is physics, and attempt to answer the unanswerable questions of the critical handful of students; and to the extent that he is aware of the true state of affairs, the contradictions in his position will be severe.

Technical and Practical Problems and their Hazards

It is when immature sciences are enlisted for the solution of practical or technical problems that the most severe strains arise. For this engagement inevitably leads to deceptions, of self and others, compared to which those on the academic scene are but minor slips. A discipline which is unable to establish facts even within its closed world of controlled experience is much less capable of genuinely drawing conclusions about the problems of a raw and unstable reality. Of course, a master of the field may possess a personal wisdom of its phenomena which enables him to perceive the real situation and its problems more deeply than a person with no special experience; but his conclusions are derived more from his intuitive knowledge, built up informally over a long period, than from any programme of piecemeal research.[18] Yet there are many temptations to pretend, and to believe, that a large-scale research programme is necessary before adequate decisions can be taken on an urgent practical or technical problem. Both scientists and decision-makers have such temptations; and these, with their effects, are worth reviewing.

First, a scientific field, like any other institution, has an innate tendency to expansion, if nothing else than as a means of survival. To refuse support for an expanded research effort, on the grounds that the intended function of the research might not be performed, is to possess such an otherworldly purity of character as would prevent a man from reaching the point of choice in the first place. Mixed with ambition is the ever-present hope that the field is really on the point of maturation, and that engagement on such a problem might bring it over the top. Even if the leaders of the field have private doubts about its effectiveness in this respect, they can legitimately reassure themselves that in the absence of its intervention

[18] A striking example of the eruption of common-sense is cited by Wright Mills, in *The Sociological Imagination*, using comments by A. W. Gouldner on the work of the 'Grand Theorist' Talcott Parsons. It appears that when Parsons analysed post-war Germany with a view to policy recommendations for its social reconstruction, he used a straight Marxist analysis of class structure, with no evidence of the 'normative structure' so strongly developed in his purely theoretical writings (pp. 43–4).

ignorance, prejudice, and self-interest would dominate any decision to be taken. Moreover, the problem-situation will frequently be one whose solution will involve a large project, with the creation of a corps of technicians, practitioners and experts. To have a share in their training will provide the field with influence and a firm economic base; and it will also protect the clients from the disastrous effects of the agents' untutored common sense in the relevant respects. And even if some of the answers are known already, they will legitimately carry more weight if they are derived as conclusions from a mass of rigorously controlled data, than if they are offered as aphorisms distilled from personal experience.

Those who must take the decisions on practical problems will welcome research, for their own reasons. If nothing else, it imposes a delay on the decision; and in the meantime the problem may dry up and blow away, or they themselves may shift to a less exposed position. Also, the conclusions of a research project, in which the categories are tidily defined and recommendations neatly listed, seem more amenable to translation into the terms of a technical project, and the articulation of routine tasks, than the aphoristic generalizations of an informal study. And, to be sure, if other things were equal (although in the case of immature disciplines they are not) a scientific argument provides a more secure foundation for its conclusions than an impressionistic essay.

Engagement on such technical and practical problems is certainly necessary for the rapid institutional growth of a field, and under favourable circumstances may contribute to its maturing. But it also presents dangers, for which an appropriate term is 'hypertrophy':[19] a rate of growth so rapid that the existing social mechanisms of direction and quality control cannot perform their functions. In response to the urgent calls for helpful research, a clever mediocrity can build an empire and attain power and prestige at the expense of those with more caution or scruples. In the absence of effective controls from within the discipline, or from neighbouring subjects, the worse excesses of entrepreneurial and shoddy science can occur; and if the problems are partly military in significance, some of this can be dirty as well.[20] The growth in volume of research provides an

[19] I am indebted to Mr. M. J. Wilson, of Imperial College, London, for this term.
[20] During the period of intense examination by students of Government-sponsored research at the Massachusetts Institute of Technology (1968–9), certain projects in 'political science' were judged by students to belong to all these categories simultaneously.

opportunity for an expansion of the institutional apparatus, including an academic base. There soon appears a structure of postgraduate and then undergraduate courses, mainly vacuous in content and largely taught by a mixture of mediocrities, philosophers *manqués*, and entrepreneurs. All the contradictions inherent in the teaching of immature sciences are made more acute, as the small core of successful craft techniques and aphoristic wisdom is embedded in a doctrine imitating a matured science, theoretical and applied. 'Pseudo-science' is not the most appropriate term for such fields, for they are not essentially misconceived in the problems and objects of inquiry. But if we describe them as 'cliché-sciences' we will characterize their distinguishing feature: the genuine insights at their base, which may well be valuable in the education of students whose previous experience is utterly foreign to the area of inquiry, become reduced to clichés as teachers and researchers in the field rub them together in an attempt to produce a plausible fascimile of scientific arguments. And their conclusions too, to the extent that they are not vacuous academic jargon, will be nothing but rearrangements of the clichés that constitute the materials of the field, organized for the best performance of the political functions of a result.

The graduates of courses in such cliché-sciences then emerge as manpower-units with spurious qualifications for taking their places as technicians, practitioners, or experts on the growing industry of vacuous research or misconceived technical problems.[21] In such circumstances we can speak of corruption; for there is a sufficient penumbra of uncertainty about the nature of the enterprise, that while it is not universally recognized as a straightforward racket, there is an awareness of something false about it which is best not discussed too openly.

The hazards of hypertrophy of an immature field are thus very real, especially in the present period when new practical problems,

[21] An indication of these dangers has been given by Paul Lazarsfeld: 'But sociology is not yet in the stage where it can provide a safe basis for social engineering.... It took the natural sciences about 350 years between Galileo and the beginning of the industrial revolution before they had a major effect upon the history of the world. Empirical social research has a history of three or four decades. If we expect from it quick solutions to the world's greatest problems, if we demand of it nothing but immediately practical results, we will just corrupt its natural course.' The quotation is taken from C. Wright Mills, *The Sociological Imagination*, p. 100, with a reference back to 'What is Sociology', a duplicated paper at Oslo University, 1948. Mills italicises 'a safe basis for social engineering', as the relevant passage in the context of his discussion.

and new technical problems concerned with human behaviour, are recognized at an increasing rate. The dangers extend to the whole world of science, for a rise to dominance of immature disciplines could, regardless of the intentions of their leaders, wreck the subtle system of direction and quality control at the highest level, whereby the health and vitality of the entire scientific community are maintained. And like any genuine practical problem, this one admits of no easy solution. It is not merely that any new practical or technical problem requires some new scientific materials to aid in its solutions, and that these will tend to come from new and hence immature fields. Another factor in the situation is the intimate relation of practical problems and immature disciplines with folk-sciences; and we must discuss these in some detail in order to analyse the problem.

Folk-Science

The category of 'folk-science' has long been recognized; but it has always been believed to apply only to preliterate or subliterate cultures. It is a part of a general world-view, or ideology, which is given special articulation so that it may provide comfort and reassurance in the face of the crucial uncertainties of the world of experience. In earlier ages, some of the leading sciences and educated arts grew out of popular folk-sciences; among these were the ancient art of prediction, as well as other forms of magic, along with the very sophisticated arts of alchemy and astrology. After the relegation of such fields to the status of pseudo-sciences in the seventeenth century, it appeared to scholars that there was an absolute distinction in character between rational, progressive 'science', and the superstitious and retrograde folk-sciences. It could not be denied that in the historical order genuine science emerged out of folk-science; but the causes of this deep transformation remained a mystery, and survivals of the old in the new presented a serious embarrassment. But the functions performed by folk-sciences are necessary so long as the human condition exists; and it can be argued that the 'new philosophy' of the seventeenth century, with its disenchanted and dehumanized world of nature and its appreciation of closely controlled experience, itself functioned as a folk-science for its audience at the time. For, as it appeared then, it promised a solution to all problems, metaphysical and theological as well as natural; and it gained a stable social base long before its achievements were commensurate with any of its claims.[22]

Indeed, we may say that the basic folk-science of the educated sections of the advanced societies is 'Science' itself, in various senses derived from the seventeenth-century revolution in philosophy. This is quite explicit in figures of the Enlightenment such as Condorcet;[23] and a basic faith in the methods and results of the successful natural sciences, as the means to the solution of the deepest practical problems, can explain the absence of genuinely critical studies of the natural sciences as cultural and social phenomena until very recently.[24] Experiences which are inexplicable within the categories of this 'Science' (as those indicating powers of mind independent of matter) are denied significance or even existence, for they constitute a challenge to an entire world-view, and hence threaten chaos and disintegration. A more visible and vulgar effect of the role of 'Science' as the dominant folk-science of our time is the status that the title itself confers on any field of inquiry or practice. Grabbing for such status, with its attendant social and material benefits, is most widespread in America, where the folk-science of 'Science' seems strongest and where control on titles is weakest; 'mortuary science' is a well-known case of a craft dubbed science, with its attendant academic apparatus.[25]

[22] J. Ravetz, 'What was "The Scientific Revolution"?', *Indian Journal of History of Science*, 1 (1966), 15–21.

[23] See Condorcet, *Esquisse d'un Tableau historique des progrès de l'esprit humain* (1795), ed. O. H. Prior (Boivin, Paris, 1938); 'Neuvième époque', pp. 145–202. 'Toutes les erreurs en politique, en morale, ont pour base des erreurs philosophiques, qui ellesmêmes sont liées à des erreurs physiques . . .' (p. 191).

[24] It is significant that through the entire nineteenth century none of the great sociologists analysed science as a social phenomenon. Comte took his idea of 'positive' from the successes of the natural sciences; and for Max Weber natural science was the purest realization of 'rationality'. The earliest attempt I know at analysing the style and metaphysics of natural science as the product of a particular culture was that of Max Scheler, *Die Wissensformen und die Gesellschaft*, 1926. See J. R. Staude, *Max Scheler, an Intellectual Portrait* (The Free Press, New York; Collier MacMillan, London, 1967), pp. 180–91. Scheler's insights seem to have derived from his long-standing contempt for the bourgeois and his belated conversion to a mystical philosophy. There may have been a current of Marxist analyses already; shortly afterwards quite similar ideas were expounded by Marxist writers.

[25] A list of American 'sciences' has been provided by C. Trusedell: 'A short random search of the catalogues of Graduate Schools delivered, in addition to the ubiquitous "Social Science", "Political Science," and "Computer Science," some special examples: "Meat and Animal Science" (Wisconsin), "Administrative Sciences" (Yale), "Speech Sciences" (Purdue), "Library Sciences" (Indiana), "Forest Science" (Harvard), "Dairy

The relations between disciplined inquiry and the folk-science aspects of a subject are naturally complex and variable.[26] To make the distinction clear, we can consider the case where the same field could, in different aspects provide both the technical setting for a scientist's research, and also a folk-science for his personal belief. This was quite common in the period when science was close to 'the philosophy of nature' in spirit and name. The quotations from Helmholtz and Einstein given earlier (pp. 39 and 66) each show how reflection on some more or less extended part of science, either in its methods or results or both, could provide comfort and reassurance. Indeed, the Helmholtz quotation, and our discussion of ethics in science, show that without some personal adherence to an ideal of science as providing some very important knowledge, this very demanding work could not be sustained at the highest level for more than a few generations.

The external groups for which a subject functions as folk-science can vary enormously in their size, sophistication and influence; the particular folk-science itself can be more or less central to their ideology, and more or less durable. The style of work in the folk-science will depend on such circumstances; for a sub-literate audience it needs fewer trappings of academic jargon and titles, while a sophisticated audience requires a reasonable facsimile of a leading branch of 'Science', such as physics. But in any case, the clue to a subject's being a folk-science is that belief in it is quite independent of its achievements in producing facts or knowledge, or in accomplishing genuine solutions of technical or practical problems. Since the functions performed by a folk-science are so different from those of the solutions of problems investigated in a disciplined style, the relevant criteria of adequacy and value of its results will be correspondingly different. Value is determined by the degree to which a problem-situation is central to the experience of the audience; and adequacy by the success in offering reassurance and the promise of understanding. For an educated audience, the mere existence of its folk-science as a recognized academic discipline may be sufficient to these ends.

Science" (Illinois), "Mortuary Science" (Minnesota). *Essays in the History of Mechanics* (Springer Verlag, New York, 1968), p. 75.

[26] A brief but illuminating analysis of the character of the interaction between the 'scholarly' and 'folk' aspects of a discipline, from a Marxist point of view, applied to Marxism itself, is given by A. Gramsci, 'Marxism and Modern Culture', in his *The Modern Prince and other writings* (Lawrence & Wishart, London, 1957).

On occasion, adoption as a folk-science by some influential group can provide a maturing field with social support (in the way of prestige, jobs, finance, and recruits) sufficient to enable its firm establishment; such seems to have been the case with history in the nineteenth century. Conversely, some particular results of a matured field may (with suitable interpretation and vulgarization) suddenly become a component of a leading folk-science, and embark on a new career independently of the thoughts and desires of their creator; such was the case with 'Darwinism'.[27] But there can also be conflict, especially when a challenge to a scientific orthodoxy is made from a base in a popular folk-science, and uses a style of inquiry and tactics of struggle appropriate to that base. The ensuing struggles can be very bitter, for the threat is not merely to a school, but to the autonomy of the world of established science in its goals and methods. Even when a challenge to scientific orthodoxy is made by a man thought to be a scholarly crank attacking a strong field from a shallow popular base, the shadow of a threat can be sufficient to cause a violent reaction; this was seen in the extreme measures taken by the American community of physical scientists against Emmanuel Velikovsky and his theories of recent cosmological catastrophes.[28] Immature sciences are even more vulnerable, of course; and (as we shall see) more closely related to folk-sciences. Hence in them the struggles between competing schools can involve not only doctrines and personalities, but rival ideologies, their related folk-sciences, and perhaps their political associations as well.

Folk-Science and Ideological Conflicts

The conflicts between academic science and strongly based folk-sciences provide many insights into the social situation of science,

[27] The interpretation of a chapter in intellectual history in terms of the folk-science functions of an idea has been done by J. W. Burrow, *Evolution and Society: a Study of Victorian Social Theory* (Cambridge University Press, 1966). He states, in the Preface, 'It will be argued in this book that the seeds of modern sociological theory were to a considerable extent implicit in the doctrines which were becoming current in the 1860s, but they were stifled by the overriding needs of an evolutionism which provided what the Victorians sought in theories of society' (p. xiii).

On 'Social Darwinism' itself, the standard work is R. Hofstadter, *Social Darwinism in American Thought* (Harvard University Press, 1944): for the English scene, see G. Himmelfarb, *Darwin and the Darwinian Revolution* (Chatto & Windus, London, 1957), especially chs. 14 and 19.

[28] For a well-argued presentation of Velikovsky's side of the story, see *The Velikovsky Affair*, ed. Alfred de Grazia (University Books Inc., New York; Sedgwick & Jackson Ltd., London, 1966).

and into some insoluble problems inherent in that situation. Such conflicts have been little studied by historians from this point of view, and so are worth some extended discussion here. These conflicts take two very different forms. The first occurs when the results of disciplined scientific inquiry contradict the beliefs of a folk-science, usually a popular one which is also adopted by the established cultural organs of society. The second sort occurs when a popular folk-science, perhaps with a radical social and political message, attacks academic science for its style and connections as much as for its content. The former sort of conflict is better known; the history of such conflicts can be read as a series of triumphs for intellectual integrity and freedom of thought, against popular superstition and imposed dogma. Indeed, there was a strong tradition of such histories, starting in the Enlightenment and continuing through the nineteenth century; for the struggles were real and bitter, and the camp of 'science' stood for the highest values of civilized life.[29] However, the work of science, when it is so engaged, is something more and something less than that of a self-contained 'positive' discipline. There is an added dimension of social and philosophical involvement; and those who inherit the successes of that struggle, both in the knowledge achieved and the secure social position won, cannot honestly wear the mantle of the heroic reformers of the past. On the other hand, 'positive' science can justly claim autonomy from society and the state, on the grounds that its particular results are neutral in their social and ideological effects, while the maintenance of the activity of research is beneficial in a general and diffuse way.

If we are to understand the moral problems involved in the struggle for 'freedom in science', we must recognize that those who attack established or official folk-sciences, are advancing doctrines that are 'ideologically sensitive': their effects may be disruptive of the stability of society. They may claim the freedom to make their attacks, but they must then base their claim on some higher principle. One such principle is the absolute right of anyone to proclaim Truth; another is the ultimate value to society resulting from the free clash of conflicting doctrines. But to claim the right to utter unsettling doctrines, while denying any responsiblilty for their effects, comes perilously close to the arrogation of power without responsibility, a

[29] The classic in this type of historical literature is A. D. White, *A History of the Warfare of Science with Theology in Christendom*.

traditional variety of immorality.[30] For some generations up to the present time, most academic natural science has been 'ideologically neutral'; and indeed this neutrality has been invoked by propagandists for science, as a reason for its complete autonomy of goals from outside interference. The exceptions to this neutrality have been little studied; one which may be historically significant is the close association, in the early twentieth century, of the science of genetics with a generally reactionary and élitist folk-science of 'eugenics'.[31]

[30] Galileo early received clear warning of the ideological sensitivity of his Copernican doctrines from his friend and protector, Maffeo Barberini, later Pope Urban VIII. In March 1615, Galileo received a letter from Ciampoli reporting a conversation with Barberini, advising him not to 'exceed the limits of physics and mathematics', for an author's assertions on matters in the Scriptures can, on publication, be transformed out of recognition. He cites the example of Galileo's analogy between the earth and the moon in the casting of shadows on the surface: 'somebody expands on this, and says that you place human inhabitants on the moon; the next fellow starts to dispute how these can be descended from Adam, or how they can have come off Noah's ark, and many other extravagances you never dreamed of.' See S. Drake, *Discoveries and Opinions of Galileo* (Doubleday, New York, 1957), p. 158. There is no evidence that Galileo recognized the force of this problem up to the time of publication of the *Dialogue on the Two World Systems*. There is, however, an odd passage in the later *Discourses on Two New Sciences*, which may indicate that after the struggle was over, Galileo finally saw what it was all about. It comes when one of the interlocutors praises Salviati (Galileo's mouthpiece) for a particularly elegant demonstration, and remarks how little such achievements are appreciated. Salviati replies with some reflections on those who claim to be competent in a field of study, and yet still publicly deny new truths in spite of believing them at heart, 'merely for the purpose of lowering the esteem in which certain others are held by the unthinking crowd'. (Edizione Nazionale, p. 204; Dover ed., p. 169.) This particular audience is hardly ever mentioned in Galileo's published works; and so one is tempted to suppose that at that very late stage he had finally discovered its relevance to the struggle which he had lost. One may put it that he had recognized the category of 'dirty truth'. Modern students come to a study of Galileo affair with healthy libertarian prejudices; for the education of those in my tutorials, I put the question, 'Would you try to prevent someone from giving away coloured comics depicting scenes of sexual violence against children, at a junior school?' Those who would are then in a position to appreciate that there was a genuine tragedy in Galileo's conflict with the Church.

By contrast, Descartes saw the point of ideological sensitivity as soon as he heard the news of the condemnation of Galileo. He promptly suppressed his completed work *Le Monde*; and three years later, when explaining this, mentioned 'certain persons to whom I defer, and whose authority governs my action no less than my reason governs my thoughts . . .'. It seems clear that he had never previously imagined that there could be a contradiction between these two governing principles, especially for someone so shrewd, cautious and honourable as himself. The crucial text here is the opening section of the sixth part of the *Discours de la Méthode*, in which Descartes finds himself arguing both sides of a question, without a clear answer coming out of the application of his Method.

[31] Of this particular association between science and social affairs, which was conveniently forgotten during the denunciations of Lysenko's campaign against orthodox genetics, some indications are given by Paul Gary Wersky, '*Nature* and Politics between

Even to this day, the social sciences are ideologically sensitive, and affixing the label 'positive' to any one of them does not alter this reality. Also, the problems of 'reckless science' and of the sciences associated with runaway technology have resurrected old moral problems, and seem likely to raise ideological issues analogous to those involved in earlier conflicts with popular or official folk-sciences.

When academic science finds itself on the defensive against challenges made by popular folk-sciences (as distinct from counter-attacks by the servants of an official orthodoxy), it becomes involved in the deepest problems of politics; and these show how precarious is the neutrality and universality of scientific inquiry and scientific knowledge. For academic science has hitherto inevitably been restricted to an élite leisured class and its occasional recruits; and its conventions and etiquette are meaningless to the less privileged sections of society. The institutions of science have usually been so far removed from the perceptions of the lower orders that both sides are happily ignorant of its position in a class-divided society. But there have been occasions when a section of the prevailing academic science has been attacked, for its content, its style, and its social position, by representatives of a radical movement demanding its replacement by a populist folk-science. The Paracelsian tradition in early modern Europe provided a basis for such challenges, of which the best known is that of the 'sectarians' during the English Civil War.[32] A similar pattern can be discerned in the attack on the *Académie des Sciences* during the French Revolution;[33] and again in the career of the Soviet agronomist Lysenko. In each case the challenging folk-science related to a 'romantic' philosophy of nature, combining a stress on craftman's manipulation, a personal involvement in the work, a democracy of participation, and a distrust of

the Wars', *Nature*, 224 (1969), 462–72. I am indebted to him for this reference; he has work in preparation which will show the connection fully.

[32] For the English Civil War, the seminal paper is P. M. Rattansi, 'Paracelsus and the Puritan Revolution', *Ambix*, 11 (1963), 24–32. The theme is developed further by the same author in 'Politics and Natural Philosophy in Civil-War England', *Actes du XIe Congrès International d'Histoire des Sciences*, ii (Ossolineum, Warsaw, 1967), 162–6. The basic printed documents for this controversy are the polemical works on the universities, published in the early 1650s; see A. G. Debus, *Science and Education in the Seventeenth Century: the Webster-Ward Debate* (Oldbourne, London, 1970).

[33] For the ideological conflicts over science in the radical phase of the French Revolution, see C. C. Gillispie, 'The *Encyclopèdie* and the Jacobin Philosophy of Science: a Study in Ideas and Consequences', in *Critical Problems in the History of Science*, ed.

abstract or mathematical reasoning.[34] Viewed from the point of view of academic science, such movements are all doomed to end in tragic failure, since they are incapable of producing worthwhile scientific knowledge.[35] But it may be that their leaders have been less concerned to achieve the same sort of specialized knowledge in a different style than to bring about a total revolution in human life, and make their contribution as Utopians rather than as reformers.

In the folk-history of academic science (that is, the folk-science of its own past), the defence of the autonomy of science has been seen in terms of attacks by simple obscurantists, or by the State. This other, more troubling sort of conflict, in which the enemies of science rise from a popular base and seek to bring back the philanthropic ideal of knowledge, has been little explored. But the involvement of contemporary science, not merely with élite culture, but also with the technical and administrative government of a nation, is bound to

M. Clagett (University of Wisconsin Press, 1959), pp. 255–89. Gillispie shows that Lavoisier's reform of chemical nomenclature was seen as a move involving politics and ideology: denying that chemistry is a branch of 'natural history', accessible to every man; and converting it into a branch of abstract, quantitative, élitist physics. Independent confirmation of this effect of the reformed nomenclature is provided by a letter of Thomas Henry, the Manchester manufacturer, to James Watt, of 1790: 'Chemistry being a science so intimately connected with many of the Arts, and consequently studied by many illiterate men, should have its language plain and intelligible and not made up by words compounded from a dead language which none but men of learning can understand.' Quoted from A. E. Musson, and E. Robinson, *Science and Technology in the Industrial Revolution* (Manchester University Press), p. 242.

[34] For Lysenko, see Z. A. Medvedev, *The Rise and Fall of T. D. Lysenko* (Columbia University Press, 1969). The 'romantic' aspect of Lysenko's style does not appear in his main polemical writings, but it was detected by C. A. Waddington. In a discussion of the cluster-planting of trees, Lysenko is quoted by him as having made the point, '. . . that it would be against the great laws of organic life if members of the same species competed with one another, and did not help one another, and he went on that nothing is worthy of being called true science unless it exhibits the great underlying order of the cosmos in general'. Waddington felt that Lysenko's underlying philosophy was not so much dialectical materialism as having 'a pretty strong flavour derived from orthodox Russian theology, with God left out'. See C. H. Waddington, 'Talking to Russian Biologists', *The Listener*, 69 (1963), 119–21. The French biologist Marcel Prenant also discussed this topic with Lysenko, and came away with the same impression, comparing Lysenko with Bernardin de Saint Pierre (see Medvedev, op. cit., p. 168). It is possible that German *Naturphilosophie*, rather than religion, was the prime source of Lysenko's biological philosophy; it came into nineteenth-century Russia with political radicalism, and if a link could be established through agronomy, a strong circumstantial case could be argued. The matter is a delicate one, for Marxism, like other nineteenth-century German rationalist philosophies, concealed its connections with *Naturphilosophie* and officially despised it.

[35] The futility of the 'romantic' style of science is the theme of a lengthy survey of the history of scientific ideas, by C. C. Gillispie, *The Edge of Objectivity* (Princeton University Press, 1960).

lead to political challenges to its position. One of the main targets of
the Cultural Revolution in China was the prevailing style of academic
teaching and research; it was argued that these were perpetuating
the rule of a bureaucratic élite, who would inevitably lead the nation
along the 'capitalist road'.[36] In the event, no rival folk-sciences made
a strong appearance, and some branches of Western-type technology,
as the nuclear weapons programme, were protected by the State; and
so the Cultural Revolution did not proceed through its full course
in this field.

The Chinese experience offers some clues to the patterns of future
political challenges to academic science. As the traditional handicraft
industries are destroyed by technological change, the folk-sciences
based on them, which run continuously back to the magic of earlier
times will inevitably vanish. Also, since a modern society, unlike most
earlier ones, is so crucially dependent on a sophisticated physical and
social technology, an attack on its institutional and cultural basis can
succeed only at the price of destroying the fabric of social life. But
these conditions in themselves do not prevent the appearance of
revolutionaries who do not know of, or care about, such consequences. With the rapid growth of an educated and relatively leisured
class in the advanced societies, radical movements can fasten on to
folk-sciences which relate to man's spiritual existence in quite
sophisticated terms. The first example of this is the succession of
movements and groupings, strongest in America, which has produced 'beats' and 'hippies', not all of whose members can still be

---

[36] It is insufficiently realized, by Marxist as well as by liberal scientists, that the very
real 'democracy' of science is strictly bourgeois democracy: an abstract equality for those
who have had the opportunities enabling them to use it. I can argue this with a paradox:
a boy now growing up in an urban black ghetto has a greater chance of becoming
President of the United States of America than of becoming a distinguished physicist.
For it is easy to imagine a brilliant youth teaching himself to read during adolescence and
then embarking on a successful political career; but for physics he would be well past
his most creative years by the time he was in a position to do original research. It may be
that this contradiction in the social position of science has prevented historians sympathetic to 'populist' tendencies in the study of nature from achieving a fully synthetic
account of them. This can be seen in J. Needham, *Science and Civilisation in China*,
vol. ii (Cambridge University Press, 1956). He gives a masterly account of the different
aspects of Taoism, including European parallels to 'empiricism and mysticism' and the
ideals of social benefit; but since their political attacks were on the Confucian *political*
philosophy, he can assimilate the democracy of modern science to that conceived by them
(pp. 86–132).

For an account of the Chinese Cultural Revolution in relation to science, see C. H. G.
Oldham, 'China Today: Science' (Cantor Lecture), *Journal of the Royal Society of Arts*
116 (1968), 666–82.

youths. Although their message is one of gentleness and love, they are far more deeply radical than the hyper-militant activists, who complain about the system of intellectual culture only because it is not tailored to their desires.[37]

Up to the present, however, the important folk-sciences of the educated classes have not been involved in such fundamental struggles. Sometimes a fully matured academic science could also function as a folk-science; thus there was a great stream of 'popular science' of the eighteenth and nineteenth centuries, in which the wonders of physical science were cited either to show the cleverness and beneficence of the Creator, or alternatively His nonexistence, depending on the audience. And when sections of the educated classes (perhaps in association with some of the self-educated) took up an 'unrecognized' science as a folk-science, as phrenology, it was not used as a lever for the displacement of the whole system of official culture.

An intermediate situation, of great importance, is the adoption of an immature discipline as a folk-science. This is indeed necessary, if it is to survive beyond the personal endeavours of its founder and a small band of enthusiasts. Once the discipline is established, its 'disciplined' and 'folk' aspects maintain a subtle but deep interaction. The adherents of the folk-science will describe and hence perceive the world in terms conditioned by the conclusions of the science; their common sense is thus partly a product (at some removes) of disciplined inquiry, if not scientific then at least philosophical. On the other hand, the objects of inquiry of the discipline were themselves likely to have been distilled from the common sense of the founders of the field, who could well have had an ideological motivation for their work; and so the folk-science as it existed at the inception of the field can have a continuing influence. Moreover, in an immature field the objects of inquiry are not modified and enriched solely by the internal processes of the solution of problems and the creation of facts; the internal weakness of the field makes it more subject to external influences, even in the modification of its objects of inquiry by the changing currents of common sense in its lay audience, as conditioned by their personal experience and the teachings of popular expositors. Thus the 'folk' aspect of a science can react back on its 'disciplined' aspect; and although individual

[37] For a penetrating and sympathetic account of this movement, see T. Roszak, *The Making of a Counter Culture* (Doubleday, New York, 1969; Faber, London, 1970).

productions in the separate aspects will superficially have a very different style and content, they are to a great extent merely performing different functions within a single enterprise.

The human and social sciences show this admixture of 'folk' and 'disciplined' aspects to a strong degree; and this is significant in view of their being basic folk-sciences for large and important sections of the educated population. The question of the nature of the groups adhering to them is a complex one; but the 'swing' from (natural) 'science' among adolescents, and the disproportionate support for student revolts among those in the social sciences, are clear indications of a strong nucleus. Their influence is far more pervasive than is indicated by the university scene alone. For some decades, psychology has been an important folk-science in America, generating an enormous handbook literature on all aspects of the art of living. Economics is doubtless the folk-science of all those committed to an economy planned to any degree; in spite of the vacuity, or irrelevance of most of its theory, and the patent unreliability of its statistical information, it ranks as the queen of the sciences in the formation of national policy.[38] Debates over economics as a discipline are almost all debates over political economy; and while the critics, of the far Right and libertarian Left, ruthlessly expose its fallacies and inadequacies, academic students are carefully ushered through the accepted doctrines in blissful ignorance of these difficulties.

The ideological function of a subject as folk-science can vary enormously; in the nineteenth century economics was the 'gloomy science' which demonstrated the necessity of the harsh conditions of

[38] See Ely Devons, 'Statistics as a Basis for Policy', *Lloyds Bank Review* (July 1954); reprinted in *Essays in Economics* (Allen & Unwin, London, 1961), pp. 122–37. We read: 'Considered in this light there seem to be striking similarities between the role of economic statistics in our society and some of the functions which magic and divination play in primitive society'; and Devons argues the analogy at some length, along the lines which I describe as the functions of folk-science. His criticism was of economic statistics as they are actually used; in principle they might be used more 'scientifically'.

One of the leading and constant dogmas of applied economics in Britain since the war is that wages should be kept down, because of the obvious and necessary relation between wages, internal prices, and export prices. Any rise in unit wage costs in manufacturing will, we have been told incessantly, lead to such increases in export prices as to cause loss of markets, national bankruptcy, etc. A set of charts of these three variables in the *Lloyds Bank Review* of April 1964, p. 54, showed that over the previous decade they had correlated reasonably well for the United Kingdom; but for Italy, France, and Japan they went their own ways independently. As befits practitioners of an immature folk-science, the economic advisers to the State took not the slightest notice of this inconvenient information.

unrestricted free enterprise, while now it is the folk-science of planning (although, to be sure, many distinguished economists are quite conservative). Similarly, although the great tradition of nineteenth-century social philosophy was generally conservative in cast, in the present century sociology has become the folk-science of liberal reformers.[39] Thus when the American Supreme Court needed a doctrine on which to base its decision that racially segregated schools cannot be 'equal', and had available no inherited common sense capable of being translated into constitutional precedent, it boldly produced an explicitly sociological argument. Although such a procedure is more honest than the traditional one of twisting precedents to suit one's prejudices, it also carries with it the danger of endangering the autonomy of the law, and making it subservient, in explicit detail as well as in principle, to the conclusions of a folk-science, which, as a scientific discipline, is still immature. Such support at the very highest levels of society gives a practical influence to the favoured disciplines, out of all proportion to their scientific strength. Yet the situation is not one for which blame can or should be assigned; for the close relation of practical problems, immature disciplines and folk-sciences make such practices inevitable.

Immature Fields and Practical Problems : Some Hazards

We can now return to our discussion of the difficulties that arise in the attempt to 'apply' sciences that are ineffective or immature, in the solution of practical problems. This application takes place first through programmes of research when the problems are first publicly recognized, and then in the training of experts for the technical projects through which it is hoped that the problem will be resolved. It is only natural that the disciplines involved in new practical problems should be immature; for the novelty of the problem takes it outside the domain of competence of established fields of inquiry.

The difficulties that ensue are only partly the result of the condition of immaturity or ineffectiveness of the field that assumes the load of research and indoctrination. For the recognition of a new practical problem is a political phenomenon, involving a change in the real world as perceived by an important section of the public.

[39] See R. A. Nisbett, *The Sociological Tradition* (Basic Books, New York, 1966; Heinemann, London, 1967).

Indeed, it frequently occurs that a problem-situation is recognized and analysed by isolated scholars for years before it suddenly comes to exist as a political event. The eventual popular recognition may be a result of serious disturbances of the accepted order of things (such as mass action in the form of riots, an economic crisis, or a war); or a particularly eloquent account of the problem may reach the public just as its previous cares are losing their dominance on its perceptions. However, even for that new perception to take place, it must be coherent with the informal categories of the ideology of that public; and it is perceived in the terms of its accompanying explanation, cast in the relevant folk-science. The function of such explanations is to minimize the threat displayed by the problem-situation; this is done either by explaining it away altogether, or, if this is impossible, to give assurance of the possibility of its comfortable resolution. If these tasks cannot be accomplished, then the ideology of that public faces collapse, and the people themselves are threatened with the disintegration of their world.

Thus the emergence of a new problem-situation is not merely a political event; it also presents new problems for the folk-science of those concerned. Similarly, the solution of the practical problem is twofold: both to identify and remove the causes of the distressing situation, and to restore the threatened sense of security. One aspect is handled by the scientific inquiry leading to the establishment of a practical project; and the other by the folk-science. But these two aspects intermingle; the announcement of the investigation of the practical problem is itself reassuring; and since the objects of inquiry of the immature field are already so closely related to those of the folk-science, there is a tendency for any inquiry to be conducted along those lines which support the underlying ideology of the threatened group. Also, there are many situations where the beliefs and expectations of the individuals of the relevant public influence the reality they collectively experience; and so the reassurance offered by the folk-science may have a chance of being a 'self-fulfilling prophecy'. In cases where this applies, the function of the relevant scientific disciplines in the solution of practical problems cannot be entirely different from those of the ancient 'pseudo-sciences' where the client's belief (or credulity) was essential to success. Should the academic study of the discipline ignore these similarities, it can encounter pitfalls whereby a new 'science' of supposedly rational man becomes nearly isomorphic in its structure to an ancient pseudo-science of man involved in a humanized and enchanted world.[40]

When an immature field takes on the task of expanding its research effort for the solution of some urgent practical problem, there will be a tendency for the outcome of its labours to be a weighty argument establishing the conclusions that its sponsors and its public wanted all along.[41] This result seems to be independent of the style of the inquiry, whether it is large-scale sponsored research by experts on the American pattern, or the smaller, largely amateur inquiry by committees of mandarins on the British pattern. The British style does at least avoid the worst dangers of hypertrophy; and it has the additional benefit that the leaders of the inquiry, free of any professional involvement in its conclusions, can continue to learn about the problem from the debate that follows the publication of their report.[42] However, the tendency to confirmation of the folk-

[40] Some years ago I argued that nuclear strategy is isomorphic in its structure to classical astrology; see J. R. Ravetz, 'The New Astrology', *SSRS Newsletter* (Society for Social Responsibility in Science), No. 140 (April 1964). I now believe that this identification is mistaken, and unfair to astrology. Ptolemy's defence of the science in his Introduction to the *Tetrabiblios* (translated by F. E. Robbins; Loeb Classical Library, Harvard University Press; Heinemann, London, 1940), pp. 3-35, can apply to any human science whose predictive power is as yet limited; and in the mathematical elaboration on common sense foundations, classical astrology provides a model for any more recent mathematical social science. The ancient science to which nuclear strategy is truly the successor is gematria, as applied to conflict situations. There, a number could be assigned, by a standard coding procedure, to the name of each contestant; and there was a prediction algorithm based on comparison of the two numbers. If both sides used the same code and predictive system, the science would tend to confirm its predictions, and also avoid unnecessary bloodshed. See the *Secretum Secretorum* of Roger Bacon, ed. R. Steele (Oxford, 1920), p. 251. At the time of this writing, nuclear strategy has lost all credibility and respect; and the nuclear arms race is recognized as depending only on its own insane momentum. Those who did not live through the later 1950s may find it hard to believe that this patently absurd pseudo-science was taken seriously in councils of state; but it was, and its subsequent discrediting is a useful precedent for those who wish to criticize such new pseudo-sciences and cliché-sciences as may arise.

[41] In his review of Ayres, *Technological Forecasting and Long-Range Planning*, M. Shubik (op. cit.) puts 'Project Hindsight' (a study of the origins of technology in 'pure' research) into perspective. 'In general all one needs to know about a review project is who is the sponsor, who is inventing the criteria for judgement, and who is running the project; at that point the odds are overwhelming that the conclusions can be guessed before the study had been done. It is my belief that Project Hindsight has many of the signs of a hatchet job and Ayres is being naïve in his treatment of it.'

[42] Of course, only a minority will take this opportunity. Among these is Dr. F. S. Dainton, the chairman of the committee which produced the report on the 'swing from science' in England, *Enquiry into the Flow of Candidates in Science and Technology into Higher Education*, Cmnd. 3541 (H.M.S.O., 1968). The report itself made a number of recommendations for changing the secondary-school syllabus, with the implication that these would be sufficient to arrest the 'swing'. After an intensive public discussion of the

science is not absolute, except of course where the folk-science itself is tied to a rigidly enforced political orthodoxy. Independent inquiries can reach contrary conclusions, and offer them to the public. But these are relatively easy to ignore, since their categories and language will be foreign to the original audience. Only if these inquiries relate to the folk-science of some other group will they force themselves into recognition. But their challenge will be perceived mainly as a political one; and in the ensuing debate, the different levels of argument, from the scholarly to the popular and political, will be inextricably entangled and confused.

It is only to be expected that the application of scientific inquiry to new practical problems should be even more hazardous than the management of deeply novel results within science itself. To the extent that the investigation of problems loses its protective framework of accepted and successful methods, it becomes exposed to pitfalls of every sort. On being associated with an influential folk-science, an immature field, in chaos internally, experiences the additional strains of hypertrophy, and its leaders and practitioners are exposed to the temptations of being accepted as consultants and experts for the rapid solution of urgent practical problems. The field can soon become identical in outward appearance to an established physical technology, but in reality be a gigantic confidence-game, combining the worst features of entrepreneurial and shoddy science. The dangers of such corruption are at present more acute for some of the social sciences and technologies (especially those using mathematical and computational tools) than for the natural sciences, since they are related to the most urgent practical problems and they lack a base in fully matured disciplines. Protection against these dangers is not to be found in mechanical imitation of the methods of the matured disciplines; pseudo-research is one of the symptoms of the diseased state, and sophisticated criteria of adequacy of results (as 'falsifiability of theories') are irrelevant to situations where conflicting purposes and ideologies are central to the problems, and the discipline's function as folk-science cannot be eliminated. To thread one's way through these pitfalls, making a genuine contribution both to scientific knowledge and to the welfare of society, requires a combination of knowledge and understanding in

report, Dr. Dainton came to see that the problem is much deeper, and ascribes the 'swing' to the new mood of the young, who see science both as irrelevant to human concerns and also as hard work. See *Science Journal* (October 1969), p. 13.

so many different areas of experience, that its only correct title is wisdom.

Conclusion

From our analysis, it appears that the condition of ineffectiveness or immaturity in a field is not a simple absence of some desired characteristic of a matured science. Indeed, a field can be kept in such a state by inappropriate methods, adopted partly to achieve a state imagined to be that of a mathematical-experimental natural science, and partly to pretend that it is already there. Under such circumstances the very real difficulties of achieving worthwhile results, of teaching, and of operating the social mechanisms for the guidance and control of research, are made still more severe. However, the external constraints on such a field, in the present social arrangements of science, would make it very difficult for the leaders of an immature field to adopt methods appropriate to its condition, recalling the old distinction between 'philosophy', 'history', and 'art'. When immature and ineffective sciences are enlisted in the effort to resolve practical problems, their difficulties are still further compounded. In addition to the hazards of hypertrophy and corruption, an immature academic field is particularly liable to the complex influences from its associated folk-science; and to become submerged in cross-currents of political and ideological conflicts. The picture is a gloomy one, but since immature and ineffective fields are due to be involved in public affairs to an increasing extent as our social problems become ever more complex, an awareness of their limitations is necessary if their application is to systematically produce more good than harm.

Also, with an appreciation of the naturalness of the state of immaturity, we can see it as one phase of a full cycle of development, rather than imagining that every science emerges from a futile pre-history, onto a permanent plateau of the 'positive' state. With the state of immaturity corresponding to the infancy of a field, from which only the hardy survive, we can see the phase of maturity giving way, in its turn, to one of senescence. The lasting achievements of the phase of immaturity will be a few aphoristic insights of a philosophical character, and perhaps some tools and techniques; those of the phase of maturity will be a mass of positive knowledge. But when the basic insights are exhausted and the leading problems shift elsewhere, the field enters senescence, useful only for its standardized information and tools purveyed through teachers.[43] This cycle is easily observed on the small scale, in descendant-lattices of problems and their associated schools; but it may also apply, on the large scale to whole sections of the world of scientific and scholarly enquiry. To be involved in a field just entering maturity is the most rewarding career for a scientist; for then one can make great achievements at relatively little risk. But estimating the points of transition between phases is a very delicate task; a field or area of science which is approaching senescence is a dreary place; and immature fields with the hope of imminent maturation are, with all their attendant hazards, the place where the greatest challenge is found.

[43] The senescence of academic physics has been indicated by the distinguished physicist, A. B. Pippard: 'But we should not forget that the great era of academic physics may well be drawing to its close, and that we are very likely entering a new era to be dominated by fundamental advances in molecular biology, biochemistry and pharmacology, all of enormous industrial potentiality.' The last phrase indicates that the senescence may extend to all of academic science, although the context of the remarks is the problem of scientists in industry. See A. B. Pippard, 'Innovation in Physics-based Industry', Annex D of 'The Swann Report', pp. 100–3. I am grateful to Dr. J. Wootton for this reference.

Part V

CONCLUSION: THE FUTURE OF SCIENCE

CONCLUSION: THE FUTURE OF SCIENCE

For several centuries, the understanding of science has been conditioned by a belief in the separateness of knowledge and society. The faith in the attainment of human knowledge which is absolute and unconditioned inspired the pioneers of the new philosophy out of which natural science grew; and it has been the easiest line of defence of the autonomy of science against its many enemies. That simple faith is no longer adequate for its function of maintaining the integrity and vitality of science. Attempts to refine it through purely epistemological analyses do not provide a basis for defending science against the dangers and abuses arising from its new social conditions; but to consider science as merely a special branch of industrial production would lead to its speedy degeneration. The argument of this book has attempted to exhibit the ways in which genuine scientific knowledge can be a product of a social endeavour, and yet embody truth, at least within the fundamental metaphysical framework of the civilization in which it is achieved. From this analysis we have been able to study the conditions under which science can advance towards knowledge, to identify diseases and abuses to which science is subject, and to examine the special features of the application of science to the solution of technical and practical problems.

Recapitulation

Our analysis necessarily started with an abstraction from a complex reality: considering the investigation of a problem as the unit-task of scientific inquiry. We saw that this work comprises several distinct phases, each involving sophisticated craft skills. Contact with the external world is made in the production of data; but this must be converted into information, and then used as evidence in an argument. The argument concerns artificial objects, intellectually constructed classes of things and events; and it is about these that the conclusion is drawn. Since no argument in science can be formally valid, and hence no conclusion necessarily true, the acceptance of conclusions must be governed by criteria of adequacy. These impose a complex and subtle structure on the argument, and ensure the avoidance of the known pitfalls which can be encountered in manipulation or in inference. They belong to the body of craft knowledge of the methods of the field, along with particular techniques of using tools and with other controlling judgements such as those of value. The methods are informal and even tacit, and are transmitted interpersonally rather than publicly; they are incapable of being tested scientifically themselves, but arise out of the collective craft experience of the field.

The conclusion of an adequately solved problem is still far from being knowledge or even a fact. The research report on a solved problem must be assessed by a referee before it is certified through publication in a recognized journal. Even then, it must prove its significance (by being put to use), its stability under testing and repetition, and its invariance under the changes in conceptual objects which inevitably occur as new problems are investigated. Of all the facts which are so established, the great majority remain within the descendant-lattice of problems deriving from their original, and sink into oblivion when that field is exhausted and forgotten. Those facts which survive to become scientific knowledge have a different path of evolution. Rather like successful tools, they are also extended to other fields, in standardized versions performing a variety of functions and taking diverse forms. When such facts have survived the demise of their original problem and its descendants, and remain alive through their many uses, they are recognizable as knowledge. It is paradoxical that the different extant versions will be incapable of being reduced to a single, standard statement; and that the obscurities latent in the original formulation will frequently remain unresolved. It is also paradoxical that the whole process of evolution and selection is accomplished by fallible individuals, and governed by a craft knowledge of methods. Such a conception of the nature and origins of genuine knowledge runs counter to the hitherto dominant traditions of the philosophy of science and epistemology. But in them, the basic problem was how an individual could quickly achieve truth or the best substitute. Here, the guiding principle is 'veritas temporis filia'; and as the daughter of time, transformed and tested by a great variety of contacts with the external world, recognizable scientific knowledge emerges from a complex and lengthy social endeavour.

For simplicity this first analysis was restricted to matured fields of 'pure' scientific inquiry, and it presupposed the presence of social mechanisms whereby the private purposes of individuals would be harmonized with the collective goals of the endeavour. An examination of these mechanisms was necessary as a preliminary to any analysis of the conditions under which the health and vitality of science can be maintained. We saw that in the protection of the intellectual property embodied in an authenticated research report, the inherited formal system of journals and citations must be operated by an informal etiquette if it is not to be abused and destroyed; and the introduction of new forms of property lacking the controls of the journals requires a most refined etiquette if they are not to lead to a degeneration of the work. The management of novelty has in the past presented some of the most severe practical problems for science. We saw that neither the assimilation of old materials to new, nor the choice between competing research strategies, can be accomplished on the basis of general rules; but that destruction and conflict are inevitable in science, and the social organization and style of work in each particular area will determine whether the outcome of a 'revolution' will be renewed vitality or stagnation. The social task of quality control in science bears most directly on ethics, for the immediate private purposes of most individual agents are served by skimping, however slightly, on the quality of their accomplished tasks. No formal system of imposed penalties and rewards will guarantee the maintenance of quality, for the tasks of scientific inquiry are generally too subtle to be so crudely assessed; nor will the advantages to an individual of a good reputation of his group be sufficient to induce a self-interested individual to make sacrifices to maintain it. Only the identification with his colleagues, and the pride in his work, both requiring good morale, will ensure good work. Science possesses a hierarchy of quality control, informal except at the lowest level where research reports are assessed for publication; and the controllers are controlled by rewards of prestige in various ways. At the top of the pyramid of control are the leading scientists at the leading universities, who control each other and their fields by the most informal of techniques. They are neither answerable to their inferiors, nor strictly accountable to those who provide their support; and their work of direction and control requires both wisdom and the highest ethical commitment if it is not to degenerate. The ideology and social context of this ethical commitment are inherited from an earlier age; and the conditions of industrialized science present them with problems and temptations for which their inherited 'scientific ethic' is totally inadequate.

Science becomes directly involved with society at large when it is applied to the solution of technical problems, involving the production of the means for the performance of a function, or practical problems, involving the achievement of the purposes of individuals or groups of people. Each of these other sorts of problems have their characteristic cycles of investigation, their appropriate criteria of quality, and their particular pitfalls. The most sophisticated technical problems, in which the scientific component is strongest, are encountered in those industries where the automatic mechanisms of quality control through a competitive market are weakest; in them there is the danger, not merely of runaway technology advanced without regard for human welfare, but the corruption of the activity of technical problem-solving itself. The investigation of practical problems, and their solution through large-scale practical projects, encounters every pitfall of scientific and technical problems, and then some peculiar to itself. Conflicting ideologies and purposes are at the heart of every urgent practical problem; they lack the accepted criteria of quality for their solution; the sciences involved in them are usually immature; and in their execution they are prone to distortion by the natural tendencies of bureaucratic operation. Because of the increasing recognition of new practical problems, immature sciences are assigned tasks which they are not strong enough to accomplish properly; to their internal difficulties (aggravated by the necessary pretence of maturity) are then added those of hypertrophy. When involved in the solution of practical problems, they also function as folk-sciences; and the resulting confusion of the different sorts of problems and their appropriate styles of work can result in total demoralization and corruption. These difficulties and dangers are directly relevant to the future health of the traditional, established, matured natural sciences. For their internal difficulties of recruitment and morale, as well as those of their relations with society at large, are practical problems of the sort analysed here. Their resolution is urgent, because of the delicacy and vulnerability of scientific inquiry; but the pitfalls to be encountered here are no less dangerous than on any other practical problem.

Science in History

Our analysis of genuine scientific knowledge showed that it is the outcome of a lengthy process operating through history; and we have also indicated various ways in which scientific inquiry as a whole is an historical phenomenon: conditioned by its social and cultural environment, and subject to cycles of growth and decline. The very long period of the flourishing of the matured natural sciences was plausible evidence for the comforting belief that science, in its academic and positive period, had truly reached evolution's end, and would thenceforth experience a simple progress onwards and upwards indefinitely. The 'idea of progress' with which the rise of modern science was intimately associated received its mortal blow in the First World War and its aftermath; but in science itself the assumption survived for nearly another half-century. Appreciating that that long 'golden age' of science is now definitely ended, we can see it as one temporary phase in the history of man's attempts at understanding and control of the perceptible world around himself. Extending back to remote antiquity, this history has the common pattern of continuity and change, and gains and losses, as its successive phases appear. Seeing ourselves in a new phase of this history, we can face its problems as inherent in its conditions, and not as merely accidental difficulties to be removed by exhortations or by administrative devices.

While recognizing the novelty of the problems of industrialized science, we would be quite mistaken to think that the whole social activity of science has been completely transformed in the last two decades. A large part of scientific research proceeds as before, in a social context which is still mainly 'academic' rather than industrialized. The radical difference is that certain new tendencies resulting from industrialization are developing rapidly, and that the self-consciousness of science, as reflected in the pronouncements of its leaders, has changed from a simple optimism to a troubled uncertainty. Even if these developments continue and intensify, one cannot predict with any assurance just how serious their effects will be within any given time. The work of scientific inquiry is now embedded in a very large social institution which performs other essential functions in an advanced society, including higher education and the investigation of technical and practical problems. Even if all these sectors encounter increasingly serious problems they will continue to receive support from society so long as they are considered as performing these functions better than any feasible alternatives. Historical change can take a long time to work itself out; and we must avoid the naïve rationalism characteristic of radical reformers, who believe that once the insoluble problems of an institution have been exposed it will soon pass away. We might recall that the Catholic Church was conscious of a deep internal crisis as early as the later twelfth century; the incompetence and corruption of the clergy had already led to powerful movements of reform and schism. Yet several centuries were to pass before 'Reformation' achieved a permanent, independent base; and the Church survived that, to continue as a powerful force up to the present. Of course, the Church had coherence, wealth, and power in a way that the social institution of science does not; but if we are assessing the prospects for science over a period of a few decades to come, we must keep in mind the enormous inertia of any established institution in all but the most revolutionary of contexts.

We must also remember that the world of science is a very variegated one. Some fields are capital-intensive, and so very vulnerable to the effects of industrialization; while others can produce outstanding work with small investments for each project. Again, some are closely related to technical problems, while others can proceed in peaceful uselessness. National styles in science, which were very marked even among the successful nations before the complete domination of academic science, may emerge again so strongly as to condition the sort of work which is successfully done in different places. It is well known that the greatest strength of America lies in technical problems: both the development of physical devices and the organization of work and management. But American science is particularly prone to the dangers of entrepreneurial, shoddy, and dirty science. By contrast, British science is sufficiently small and poor (in comparison) to be led by an institution (the Royal Society) still resembling a club; and in this context the problems of contraction, and accommodation to industry, may be managed with more finesse. Again, in the Soviet Union the political pressures on intellectuals are so crude, that the leaders of science have a natural social and political role as spokesmen for an Enlightenment in classic eighteenth-century terms.[1]

The history of science provides yet another caution against over-simple predictions of the effects on science of the changes in its context. While it is relatively easy to give plausible explanations of the gross features of scientific activity in terms of its social and cultural environment, this becomes progressively more difficult as one tries to account for work of lasting quality, and the productions of genius. We have mentioned earlier that some of the greatest scientific work of all time was conducted within a metaphysical framework which would now be rejected as superstitious and anti-scientific. Yet even those who were searching for the divine harmonies of the celestial motions, or for the material location of the world-soul (as Kepler and Gilbert, for example) could, by talent and discipline, achieve results which became incorporated into the body of genuine scientific knowledge. Similarly, although men of ability will generally do better work when they are part of a vigorous community, enjoying prestige and leisure for their researches, some of the greatest advances have come from men working under difficult conditions in nearly complete isolation: such men as Copernicus, Mendel, Galois, and Lobachewski are cases in point. Hence, even if the goals of 'positive science' are totally displaced in scientific inquiry, and the major communities of science experience crises of finance and morale, there may yet be scientific achievements which will last for centuries to come. However, there is no known means of encouraging genius through adversity; and it would be dangerous in the extreme to conclude that the quality of scientific work would improve if scientists were left to starve in garrets.

Science in Society: the Problems of Morality

With all due caution in the face of the complexities of historical change, we can proceed to indicate the deepest problems that science will face as the process of industrialization develops, and to speculate on how they might be resolved. For this we will first need to analyse the inherent tensions in the relations between scientific inquiry and the society at large which supports it.

[1] See A. Vucinich, 'Science and Morality, a Soviet Dilemma', *Science*, 159 (15 March 1968), 1208–12, for a discussion of the current movements for autonomy of science, in the context of the Russian traditions. For an example of this style of work, see A. D. Sakharov, *Progress, Coexistence and Intellectual Freedom* (Norton, New York; Deutsch, London, 1968; Penguin, London, 1969).

Of itself, scientific inquiry is not a self-sustaining social institution; neither wealth nor power are derived directly from the activity. It requires support from society at large, or at least some wealthy and powerful section of it, if it is to exist. From lay supporters, then, science requires first of all 'legitimization', if its practitioners are not to be relegated to a despised or abhorred fringe of society. More than mere tolerance is required; resources must be invested in scientific work, both in providing paid research time for the practitioners, and in supplying their specialized equipment. Finally, an increasing flow of recruits, drawn from the adolescent population, is necessary if the work is not to stagnate and die. In exchange, science can offer the promise of assistance in the solution of technical and practical problems. This may be direct, or indirect; thus high-level teaching, which in recent generations has been considered to depend on an association with research, contributes to technique; and the contribution of science to national prestige, or to the strengthening of the nation's official ideology, helps in the solution of practical problems. But we notice that these return offerings of science to society are not, and cannot be, dominant components of the general goals of the work, and still less of particular research projects. Some of them may well be present, in varying degrees, in the work of particular individuals or schools, especially in the conditions of immaturity and in the endeavours of a genius. But should a large, established field, depending on the efforts of many research workers, allow its criteria of value (and hence of adequacy as well) to be dominated by such external functions, the work which results will not be science. It may have excellence of a different sort, or it may be quite corrupt, depending on conditions; but it will contribute to the advancement of knowledge only very incidentally.

Thus the social position of science is really quite precarious. Scientists are not professionals of the traditional sort, who can justify their position by the serving of the purposes of clients; nor can science be conducted on a large scale in a social context analogous to that of the fine arts, providing prestige to particular patrons. Science not merely requires very tangible support in return for quite intangible returns; but the different components of its support will in general derive from different sections of society. Each of these will need to be furnished with propaganda appropriate to its tastes; and this internal complexity, together with the great variety of social contexts within which science has operated, have produced a great variety of themes in the literature of justification and defence of science. The dominant themes from earlier times relate to the ideological functions (and dysfunctions) of science or of natural philosophy; for until the middle of the nineteenth century only a very few fields (as chemistry, itself only recently established as a science rather than a craft) could make any plausible claim to contribute to industrial production. In the later nineteenth century an accommodation with industry was recognized as necessary; and in the present period there is a strange mixture of 'images of science' purveyed to its different audiences.

In earlier times, the principal threats to the autonomy of science, or rather natural philosophy, have occurred when some fields were considered as ideologically sensitive, endangering the established religion; and in this respect they were involved in the practical problems of their age. Those with responsibility for the spiritual welfare of the lay public would use all the means at their disposal to contain or eliminate the dangerous doctrines and their perpetrators. Such measures would be more successful in places where the Church had an established machinery for handling doctrinal crimes, and the power to enforce its decisions; hence the Catholic Church has had an unfair reputation of outstanding enmity to free inquiry. As a result, there developed a belief in a close association between scientific inquiry and independent, rational, or free thinking in general. Propagandists for any of these traditions have assimilated the martyrs of science (most notably, the very complex figure of Galileo) to their cause; and some important traditions within the folk-history of science have imagined the community of science as necessarily composed of individuals who are selflessly and fearlessly devoted to Truth, against Authority and Superstition.

This identification rests on several basic fallacies. Scientific inquiry must have a subtle and complex assessment of the strength of evidence deriving from accepted authority; and in this it is similar to any other work where partly new problems are being solved. Only in sectarian religion and in teaching can the work continue successfully for any length of time without encountering the problems of the management of authority; and of course the total rejection of authority, whereby every assertion must be examined as an equal claim to truth, quickly yields chaos. Also, scientific inquiry is ideologically sensitive only accidentally and occasionally. In England, for example, the propaganda for science, purveyed from the seventeenth through the nineteenth centuries, argued that the contemplation of God's creation could not but induce to true religion. Of course, in England hardly anything gets really sensitive ideologically; and English divines could use the persecution of Galileo as evidence for the wickedness of Rome. And when English scientists were confronted with the practical problems of uncomfortable theological implications of their work, they were far more likely to devote their energies to a reconciliation than to make some specialized results a fulcrum on which they would move all heaven and earth.

Towards the end of the nineteenth century, the applications of science to technical problems were increasing in number and in power. Spokesmen for science had the delicate task of securing ever-increasing support from the community on the basis of such usefulness, while still preserving the autonomy of science itself. We recall the brilliant speech of Helmholtz (quoted on p. 39 above), where he reminded his audience of the apparent 'uselessness' of Galvani's experiments on animal electricity, which yielded the electric telegraph; but then warned them that the scientist himself must not be expected to search for anything but new knowledge about nature. This sincere plea for the social support of science, on the grounds of its accidental social benefits, is characteristic of the period of matured academic science.

With the advent of industrialization in science the claim of the technical applicability of science in general did not need to be pressed (although particular fields still need to justify their requests for support in these terms). Although there have been continuing discussions about the proportion of resources which should be devoted to 'pure' or 'basic' or 'undirected' research, there is a general recognition among policy-makers of science that it performs a variety of useful functions and so deserves some minor share of the budget. However, another audience has suddenly become crucial for the continued well-being and expansion of science: its potential recruits. Among them, there is a significant fraction who see the applications of science quite differently from the nineteenth-century optimists such as Helmholtz. Not the telegraph, but the Bomb, has become the type-example of the leading technical problems in which science is engaged. Hence a new, negative, and defensive theme has been developed in the propaganda for science: its neutrality. Of course, this will be purveyed to lay and juvenile audiences, in the hope that they are unaware of the firmly realistic terms in which 'science policy' is now cast.

Since the claim of 'neutrality' is the last defence against recognition of the political and moral problems of science consequent upon its industrialization, we can expect it to be advanced for some time to come; and it is worth closer analysis. The attempt to disclaim moral responsibility for the effects of scientific work has been made, not merely for 'pure' science, but even for work on technical problems with a military function. This extension of the domain of moral isolation of science is too implausible to survive. To be sure, one can agree that in one sense 'it is the height of folly to blame the weapon for the crime',[2] if for nothing else than that inanimate things are not appropriate objects of moral judgements. But those who are engaged in making weapons in the knowledge that their main function will be in the commission of a crime are subject to moral judgements and sometimes to legal proceedings as well. If this were not the case, then the defence of Adolf Eichmann, that he was merely engaged on a technical problem of transport, indifferent to the fact that it was a one-way transport of Jews and others to the gas-chambers, would be a valid one.

It is more plausible to assert the neutrality of science for that work which is governed by purely internal criteria of value, so that in the choice of problems the possible technical functions are either unknown or irrelevant. Even here, there is an area of ambiguity; for a particular research worker may choose to work on a problem for its functions in the advancement of the field, and for its political functions for his own career, while being aware that the investing agency is interested in the problem for its possible technical functions. If these technical functions are morally dubious or bad, can he claim immunity? The ignorance of consequences is not always a valid defence in law, and the mere absence of deliberate intent to malefaction is an even weaker defence.

However, there are severe difficulties in the way of making precise and fair moral judgements on scientists, individually or collectively. If nothing else, our experience of these problems is extremely short, and we possess neither general principles nor case-law for their interpretation, whereby responsibility and blame can be assigned in any but the most blatant cases of dirty science. Also, the division of

[2] See Sir Peter Medawar, 'On the Effecting of All Things Possible', Presidential Address to the British Association, 1969, *New Scientist* (4 September 1969), 465–7.

labour in large-scale technical problems is extremely fine; so that the scientist who publishes a result generally has no more knowledge of its possible functions than does a process-worker assembling a standard component of a device. This does not mean that the position of the agent is one of moral neutrality; rather that it is morally indeterminate. However, to the extent that his research is related to technical applications, the area of indeterminacy decreases, and the scientist's responsibility becomes defined. Once the scientist is aware of the likely consequences of his work, his sole disclaimer of responsibility can be along the lines of, 'I was only following orders.' This is no longer likely to be acceptable as a defence, in science as anywhere else.

These considerations apply strongly to scientists who are employees of a 'mission-oriented' research establishment; but for those whose experience is in a community of science still enjoying an academic style these moral problems are remote and philosophical rather than immediate and personal. There, research problems and personal achievements are evaluated by internal criteria, regardless of the motives of those supporting the work; the hard work and strict self-discipline are supported by a refined ethic; and results are shared with colleagues independently of all the boundaries which divide mankind. There, it seems, worthwhile work can be done, insulated from the moral squalor of ordinary life.

But even to the extent that the moral neutrality of academic science is real, it creates moral problems on its own; and the deep connection of science with the culture in which it is embedded involves science in its basic problems of justification and survival. The practical irrelevance of most of the results of scientific inquiry is a blessing to some, but an agony to others. What can be more selfish than to turn one's back on the sufferings of humanity, devoting one's talents to narrow tasks which will bring immediate rewards to oneself and only the most remote and unknowable benefits to one's fellow man? For a young scientist with a strong social conscience it requires an extremely strong faith in the human value of scientific knowledge to justify such a career.[3] Once this moral dilemma is recognized, it can be alleviated by various sorts of good works, but

[3] J. G. Crowther, *The Social Relations of Science* (MacMillan, London, 1941), has some sharp words on this problem: 'Young scientists who abandon science for politics often prove to be mentally unstable, and after a few years of bohemian agitation become conspicuously conservative. Conduct and opinions that appear to be based purely on moral sentiments are nearly all suspect' (p. 644).

never completely resolved. Moreover, the remoteness of academic science from human concerns is itself a result of its own traditions, deriving from its particular niche in a particular society. The openness and internationalism of science are admirable and valuable in themselves; but they are not the same thing as a sharing with all humanity. Rather, they are methods of social behaviour of a small group operating within European literate culture. Now that national boundaries within that culture are of decreasing emotional significance, the transcending of them is correspondingly less impressive. And to the extent that this culture as a whole is subjected to moral judgement, for its involvement in various sorts of colonial and class oppression, and for its creation of a runaway technology, academic science will be inescapably implicated as well.

It is quite likely that those of the present generation of elder statesmen of science who invoke its 'neutrality' are trying to reassure themselves as much as any audience of potential recruits. Consciousness always lags behind reality, and its adjustments are usually abrupt and painful. The conception of science as an essentially academic enterprise, and of 'the scientist' as an academic researcher, has persisted unchallenged in all the literature about science until very recently, in spite of the traditional industrial connections of several major fields (particularly chemistry, but also physics and biology), and in spite of the fact that the great majority of those who have ever earned their living through their scientific skills, have done so in technical work.[4] Even the interpenetration of science and industry can be traced back to the later nineteenth century, and can be seen as growing continuously since then. An awareness of a new condition of science came only when a series of dramatic events, such as nuclear weapons, and new technical problems, such as the planning of large-scale scientific research, obtruded themselves. The self-consciousness of science is still trying to cope with these changes; the purely technical problems of decision and control are difficult enough in themselves, and the deeper practical problems of responsibility and morality are only beginning to be grasped.

In the short run, the easiest response to such problems is to hope

---

[4] N. D. Ellis, in *The Scientific Worker*, has shown that the Royal Institute of Chemistry and the later Institutes of Physics and of Biology were created with the co-operation of the major employers of scientists; and their ethical principles were framed for the 'professional employee'. The academic scientist was not their concern; and these institutes are now subject to some strain because the majority of 'Q.S.E.'s' in industry have no professional status, but are rather 'scientific workers'.

that they will go away, and in the meantime to try to get the best of all worlds. Such a course of action, which was almost certainly not the result of a conscious policy, was taken by British university science teachers in the post-war period. Still believing in science as a genuinely liberal education, they expanded their departments with State funds intended to provide more units of trained manpower, and in practice taught Honours degree courses designed for that small fraction who could proceed to research. The instability of such an arrangement was revealed after only a decade or so; and whatever the outcome of the pressures for its alteration, the world of science has suffered no public discredit thereby. But the same sort of convenient wishful thinking, applied to the understanding of science as a whole, can have dangerous consequences, including a corruption of the whole work.

The sort of corruption which can occur in science has little in common, superficially, with that which is recognized in public life. Hence it is necessary for us to analyse the concept briefly, to show why it is relevant to the problem of science.[5] We can say that an activity is corrupt when the actual goals of the tasks accomplished are contrary to the professed social functions to a degree that a public trust is betrayed. Corruption is occasionally flagrant, but more commonly it exists in a penumbra of ambiguity; both the divergence between the final causes, and the awareness of that divergence, are ill-defined. Because of this, a man may work in a corrupt situation without himself being corrupted.[6] If he is ignorant of the state of

[5] The only worthwhile analytical study of corruption is *The Autobiography of Lincoln Steffens* (Harcourt, Brace, New York, 1931). Steffens was able to gain experience of corruption in American public life through his work as a journalist; but he could rise above mere 'muckraking' because he had previously spent years of (informal) study and thought on the problems of ethics. Although he never achieved a coherent solution to the problem of corruption (which is an interesting problem because, as Steffens found, the corruptors frequently have more honesty and personal integrity than muckrakers and reformers), his book is a mine of experiences and insights. My attempted formal definition derives from his interview with 'a dying boss', where Steffens told him why corruption is evil (see Part III, ch. VIII, p. 419).

The problems of corruption in post-colonial societies are discussed by S. Andreski, *The African Predicament* (Michael Joseph, London, 1968). Although the author discusses the problem with sympathy and insight, and even offers the technical term 'kleptocracy', his analysis lacks the depth of Steffens' (of which he seems unaware), and it suffers from his assumption that in advanced societies corruption is minor in its scale and effects.

[6] It is important to realize that even the practice of a legislator actively promoting measures for his direct financial benefit is not necessarily corrupt; this was a common and accepted state of affairs in Victorian England. See R. A. Lewis, *Edwin Chadwick and the Public Health Movement 1832–1854* (Longmans, Green, 1952), ch. 15, 'Reaction" which

affairs, or completely cynical, or capable of some sort of double morality, he may maintain his personal integrity. But more commonly, a shadowy awareness that things are not quite as they must be claimed to be, will force the individual agent to recognize the possibility of his complicity in something culpable. Such recognition tends to preserve and intensify the corrupted state of the activity: fear of exposure comes to dominate the purposes of the agent, and the group as a whole is held together by mutual blackmail. In such a situation, the worst elements gain power over the better, and the performance of the professed social functions is the least of the considerations affecting individual and collective decisions.

It is easy to see that this sketched analysis applies to cases of corruption in public life, where the private goals are the venal ones of personal gain. On the other hand, it is by no means necessary that every bureaucracy in which the defining functions have been displaced is corrupt; if there is no significant public which had some trust in it to begin with, there has been no betrayal. But again, it is possible for officials of a voluntary or political organization to become corrupted without desiring or achieving any personal benefit, merely by finding it impossible to achieve the purposes of their members and also finding it impossible to confess their failures. And most tragically, it is possible for a man to discover in retrospect, after years of service to an organization, that he had all along been corrupt.[7]

Cases of corruption in technical projects can be quite straightforward: a public contract for a device is sought and procured, on the basis of promised operating characteristics which the contractor has neither the ability nor the intention of achieving, but where the failure of the project will not affect his interests adversely. To protect his interests, the contractor may find it necessary to corrupt the State agencies of control, so that they will merely pretend to

describes the successful campaign in Parliament to defend the privileges of the London water companies against the needs of the population. Even in the early twentieth century Lincoln Steffens found English politicians who denied the existence of corruption in England calmly describing practices which in America are certainly considered corrupt. In trying to explain the differences between Europe and America he invoked the idea of the 'old' and the 'new' civilizations, with 'corruption' as a natural historical process. See *Autobiography*, Part IV, ch. VIII.

[7] The type-case of unwitting corruption is Steffens's Captain Schmittberger, a German immigrant who simply never knew that the policeman's job does *not* include protecting rackets and taking bribes, until the Lexow investigation exposed the truth about the system, to himself. See *Autobiography*, Part II, chs. XII and XIII, pp. 266–84.

scrutinize his operations.[8] The situation is similar in practical projects: the construction of a research empire may be publicly justified by its function in the solution of urgent practical problems, while the effective goals of the project are the provision of jobs, the securing of an academic base, and the production of titles of publications. Even in academic science, the production and publication of shoddy work involves an element of corruption, since both author and referee are participating in a deception, albeit an ambiguous one before a largely anonymous public. Entrepreneurial science is by its very nature corrupt in this sense, and an immature field in a hyper-trophic state can scarcely avoid corruption. We notice, however, that as we move away from the straightforward situations where bank-notes are passed in return for favours, the subtlety and ambiguity of corruption become more pronounced. Indeed, it is possible for one person to denounce a project as corrupt, and another indignantly and sincerely to deny it, the disagreement resting on questions of whether there is an interested public, a trust, and a betrayal.

A failure to come to terms with the new problems resulting from the industrialization of science can bring about a very subtle but none the less corrosive form of corruption within science as a whole. For science, as a part of academic scholarship, has long claimed to be discharging a public trust in the advancement and diffusion of knowledge; and it has claimed a variety of privileges and immunities for its members (not shared by other teachers or research workers) on the basis of its ethic of autonomy and integrity. These claims are different in character from those claimed by a learned profession, and in some ways more extreme: a professional is expected to use his judgement in solving the problem set by the welfare of the client, while the scientist or scholar claims the freedom to choose the problem itself. If it were possible to make a neat separation between the sections of the scientific or academic community which are devoted to scientific problems on the one hand, and technical and practical problems on the other, then each section could develop its distinct identity, with appropriate public justification of its position and appropriate methods of social behaviour. But the industrializa-tion of science brings the different sorts of problems into ever closer connection, in institutions and in the work of individuals. There is naturally a great temptation for the leaders of science to attempt to

---

[8] See Nieburg, *In the Name of Science*, for a discussion of this process.

gain the best of both worlds for as long as possible: to extol the virtues of the free search for truth to one audience, and to promise useful services to another. Except in those happy coincidences when the different criteria of value yield identical choices in a field, such a situation is liable to produce corruption. The most easily identifiable situation with such tendencies is what Americans call 'bootleg' research, where resources obtained for one project are partly, at least, diverted to another of more interest to the investigator. If this is something that 'everyone does' in a particular community, the equivalent of 'fiddling', then it is not corrupting to those involved.[9] On a large scale, however, as in big entrepreneurial science it can have serious consequences.

A more subtle but more dangerous sort of corruption can occur when the balance of real and professed final causes is tipped the other way: when scientists claim the privileges appropriate to the heirs of Helmholtz, while accumulating personal wealth and institutional power through the regular contracting of mission-oriented research. Again, the corrupting effects of such a situation may be latent, until it is exposed and challenged. The natural response is then to hide what can still be hidden and to explain away what cannot. Up to now, such exposures, and their attendant crises, have occurred only in connection with dirty science in American universities. Once the dangers of this situation were recognized, the response of many leaders of the scientific community was admirable. They disengaged from the State in military scientific establishment, doubtless at considerable personal cost, when the Vietnam war became politically and morally indefensible, and even before militant students forced the issue at leading universities. However, such a move, although welcome and heartening in itself, does not resolve the underlying dilemma of the external relations of industrialized science, with its tendency to corruption.[10]

[9] See D. S. Greenberg, '"Bootlegging": it Holds a Firm Place in Conduct of Research', *Science*, 153 (19 September 1966), 848–9. With characteristically American sophistication, some large industrial research laboratories become worried if their scientists work only on the projects formally agreed upon.

[10] Some scientific communities maintain their independence and integrity by astonishingly direct means. In Japan, physicists who associate themselves with the Japan Defence Agency are ostracized by the Japan Physical Society, not being permitted to present papers at its conferences. Military personnel sent to graduate schools by the armed forces are failed on their exams, either on entrance or on completion of their course. In these circumstances, it is not surprising that 'Japanese defence officials have also privately admitted to American colleagues that they have difficulty getting scientists

The State, and industry, need an expertise more sophisticated and prestigious than can be provided by narrowly technical institutions and personnel; it must come from the world of science based on academic institutions.[11] And if it is to continue to receive support on the present large scale, science must provide this service. To do so effectively, in the Anglo-American institutional structure, requires science to maintain a plausible semblance of its autonomy and integrity; yet the autonomy of science cannot be more than a semblance, especially as its accountability to its paymasters becomes increasingly close and obvious. The world of science will then need to half-believe itself to be still academic, free and autonomous, while half-knowing itself to be industrialized, dependent, and responsible to the State and industry. The traditional professed functions of science are internal and under its control: a means to the ultimate goal of the advancement of knowledge, as conceived and guided by itself. But the actual goals of the work are increasingly subordinate to functions that are externally defined: a means to the fostering of civil and military industry, through the provision of particularly applicable results and trained experts, on demand. When and whether this divergence will become a part of the self-consciousness of science, and by whom it could then be considered as a betrayal of a public trust, depends entirely on the complex circumstances of history and society. In this connection, one may revise Lord Acton's aphorism, 'all power corrupts, and absolute power corrupts absolutely', and substitute, 'responsibility tends to corrupt, and responsibility without power corrupts absolutely'.[12]

Critical Science: Politics and Philosophy

We can now permit ourselves some final speculations on possible trends in the future of the natural sciences. The process of industri-

to perform defence research'. See P. M. Boffey, 'Japan (I): On the Threshold of Big Science?', *Science*, 167 (1970), 31–5.

[11] 'Precisely the indiscipline of relatively free intellectual activity attracts the powerful as guaranteeing the relative disinterestedness and novelty they hope to find in the ideas of intellectual counsellors. It is one of the ironies of our time that so many intellectuals strive to identify with the perspectives of kings, whilst their rulers value them for their activity as philosophers.' See N. Birnbaum, 'On the Idea of a Political Avant-garde in Contemporary Politics: the Intellectuals and Technical Intelligentsia', *Praxis* (Zagreb, 1969), Nos. 1–2, p. 243.

[12] This aspect of corruption is discussed in my papers on 'Power, Responsibility, Answerability'. The context there was the problem of participation in university government; but it could be interesting to relate it to the position of scientists responsible to the State described by the American cliché, 'on tap but not on top'.

alization is irreversible; and the innocence of academic science cannot be regained. The resolution of the social problems of science created by its industrialization will depend very strongly on the particular circumstances and traditions of each field in each nation. Where morale and effective leadership can be maintained under the new conditions, we may see entire fields adjusting successfully to them, and producing work which is both worthwhile as science and useful as a contribution to technology. Recruits to this sort of science will see it as a career only marginally different from any other open to them; and it is not impossible for men of ability and integrity to rise to leadership in such an environment. This thoroughly industrialized science will necessarily become the major part of the scientific enterprise, sharing resources with a few high-prestige fields of 'undirected' research, and allowing some crumbs for the remnants of small-scale individual research. A frank recognition of this situation will help in the solution of the problems of decision and control. Since the criteria of assessment of quality will be heavily biased towards possible technical functions of results, they will thereby be more easily applied, and less subject to abuse, than those which are based on the imponderable 'internal' components of value.

Thus, provided that the crises in recruitment and morale do not lead to the degeneration and corruption of whole fields, we can expect emergence of a stable, thoroughly industrialized natural science, responsible to society at large through its contribution to the solution of the technical problems set by industry and the State. Scientists, and their leaders and institutions, will be 'tame': accepting their dependence and their responsibilities, they will be unlikely to engage in, or encourage, public criticisms of the policies of those institutions that support their research and employ their graduates. Such a policy of prudence is not necessarily corruption; whether it becomes so will depend on many subtle factors in the self-consciousness of this new sort of science, and the claims made to its audiences. But not all the members of any group are easily tamed, and the emergence of a 'critical science', as a self-conscious and coherent force, is one of the most significant and hopeful developments of the present period.

There have always been natural scientists concerned with the sufferings of humanity; but with very few exceptions they have faced the alternatives of doing irrelevant academic research to gain the leisure and freedom for their social campaigns, or doing applied research which could benefit humanity only if it first produced profits for their industrial employer. The results of pharmaceutical research must pass through the cash nexus of that industry before being applied, and that process may be an unsavoury one. Only in the fields related to 'social medicine' could genuine scientific research make a direct contribution to the solution of practical problems, of protecting the health and welfare of an otherwise defenceless public. Now, however, the threats to human welfare and survival made by the runaway technology of the present provide opportunities for such beneficial research in a wide range of fields; and the problems there are as difficult and challenging as any in academic science. These new problems do more than provide opportunities for scientific research with humanitarian functions. For the response to this peril is rapidly creating a new sort of science: critical science. Instead of isolated individuals sacrificing their leisure and interrupting their regular research for engagement in practical problems, we now see the emergence of scientific schools of a new sort. In them, collaborative research of the highest quality is done, as part of practical projects involving the discovery, analysis and criticism of the different sorts of damage inflicted on man and nature by runaway technology, followed by their public exposure and campaigns for their abolition. The honour of creating the first school of 'critical science' belongs to Professor Barry Commoner and his colleagues at Washington University, St. Louis, together with the Committee for Environmental Information which publishes *Environment*.[13] For some years a Society for Social Responsibility in Science, based in America but with members and branches overseas, was the main voice of conscience in science; recently a British Society for Social Responsibility in Science, with a rather broader base among the leaders of the national scientific community, has been formed. As such societies gain strength and influence, the success of the St. Louis school of critical science should soon be emulated elsewhere.

The problem-situations which critical science investigates are not the result of deliberate attempts to poison the environment. But they result from practices whose correction will involve inconvenience and money cost; and the interests involved may be those of powerful groups of firms, or agencies of the State itself. The work

[13] The first statement of 'critical science' as distinct from ecological concern is B. Commoner, *Science and Survival* (Gollancz, London, 1966).

of enquiry is largely futile unless it is followed up by exposure and campaigning; and hence critical science is inevitably and essentially political.[14] Its style of politics is not that of the modern mass movements or even that of 'pressure groups' representing a particular constituency with a distinct set of interests; it is more like the politics of the Enlightenment, where a small minority uses reason, argument, and a mixture of political tactics to arouse a public concern on matters of human welfare.[15] The opponents of critical science will usually be bureaucratic institutions which try to remain faceless, pushing their tame experts, and hired advocates and image-projectors, into the line of battle; although occasionally a very distinguished man is exposed as more irresponsible than he would care to admit.[16]

In the struggles for the exposure and correction of practices damaging the environment the role of the State is ambiguous. On the one hand, every modern government is committed in principle to the protection of the health of its people and the conservation of its natural resources. But many of the agencies committing the worst outrages are State institutions, especially the military; and in any event the powerful interests which derive profit or convenience from polluting and degrading the environment have more political and economic power than a scattering of 'conservationists'. It sometimes occurs that two State agencies will be on opposite sides of an environmental struggle; but the natural tendency of regulatory agencies to come under the control of those they are supposed to regulate can make such a struggle a one-sided affair.

The presence of an effective critical science is naturally an embarrassment to the leadership of the responsible, industrialized,

[14] The most comprehensive analysis of 'critical science' yet published is Max Nicholson. *The Environmental Revolution* (Hodder & Stoughton, London, 1970). He is mainly concerned with 'conservation', but his healthy approach to modern bureaucratic politics is developed in his earlier book, *The System* (Hodder & Stoughton, London, 1967; McGraw-Hill, 1969).

[15] Support for this new style of politics has come from the Duke of Edinburgh. Speaking on the B.B.C. programme '24 Hours', on 17 February 1970, he discussed the proposition that 'tough action against the poisoners and wreckers' was essential for the promotion of conservation and the abatement of pollution, and said that people must 'be ruder and more direct to the people in political authority'.

[16] On 22 March 1966, for example, the President of General Motors appeared before a Senate hearing to apologize to Mr. Ralph Nader, following revelations that General Motors' lawyers had hired an investigator to unearth details about Nader's private life. Nader's analysis of the defects of the 'Corvair' car was costing General Motors a lot of money.

tame scientific establishment.[17] Their natural (and sincere) reaction is to accuse the critics of being negative and irresponsible; and their defensive slogan is along the lines of 'technology creates problems, which technology can solve'. This is not strictly true in all cases, since nothing will solve the problems of the children already killed or deformed by radioactive fallout or by the drug Thalidomide.[18] Moreover, this claim carries the implication that 'technology' is an autonomous and self-correcting process. This is patent nonsense. We have already seen that a new device is produced and diffused only if it performs certain functions whereby human purposes can be served; and if the intended beneficiaries do not appreciate its use, or if those injured by its working can stop it, the device will be still-born. The distortions of technological development arise when the only effective 'purposes' in the situation are those of the people who believe themselves to derive pure benefit from the innovation. On the self-correcting tendency of technology, one might argue that no large and responsible institution would continue harmful practices once they had been recognized; but this generalization is analogous to the traditional denial of the cruelty of slavery, along the lines that no sensible man would maltreat such valuable pieces of property. And the history of the struggles for public health and against pollution, from their inception to the present, shows that the guilty institutions and groups of people will usually fight by every means available to prevent their immediate interests being sacrificed to some impalpable public benefit.[19] If the campaigns waged by critical science come to touch on some issue central to the convenience of the State or other very powerful institutions, we may experience a polarization of the community of natural science, along the same lines as has already occurred on the Vietnam issue in some of the human sciences in America. In such a situation, it will

[17] Thus *Nature* ridiculed the UNESCO conference on the biosphere of September 1968, comparing it to an earlier conference on 'communications satellites and under-developed countries'. See 'Bandwagon for UNESCO', *Nature*, 219 (7 September 1968), p. 999. There was no report on the conference itself.

[18] Up to the time of writing, *Nature* maintains a magnificent complacency, such as one hardly hopes to see in England in its epoch of imperial decline. Thus, criticizing Dr. Fraser Darling's first Reith Lecture, an editorial asserts, 'And in spite of quite proper concern for the need to make only decent use of new developments, it is hard to find contemporary illustrations of where technology has gone astray.' 'No Peace for the Wicked', *Nature*, 224 (1969), 631.

[19] The 'sanitary movement' in nineteenth-century England was involved in such struggles through its career. For a sample, see R. A. Lewis, *Edwin Chadwick and the Public Health Movement 1832-54* (Longmans, Green, 1952).

not be possible for a leader of science to be both honest and tame; and if the establishment of science chooses to serve its paymasters rather than truth, it will be recognizably corrupt.

Such extreme situations may be a long time in developing, if for nothing else than that critical science is still in its infancy. As it develops, it will be at risk of encountering many pitfalls, partly those characteristic of immature sciences applied to practical problems, and partly those of radical and reforming political movements. Perhaps the most obvious will be an accretion of cranks and congenital rebels, whose reforming zeal is not matched by their scientific skill. But there are others, arising from the contradictory relations between critical science and the relevant established institutions of society. As true intellectuals rather than a technical intelligentsia, individual members may find some 'sinecures within the interstices of bureaucratized intellectual systems';[20] but there will need to be some institutions providing a home for the nucleus of each school, and external sources of funds for research. Hence, especially as critical science grows in size and influence and society becomes more sophisticated about the problems of runaway technology, some accommodation between the critics and the criticized will inevitably develop. We can even expect to see critical research being supported, critical slogans being echoed, and leaders of critical science being rewarded, by institutions whose basic destructive policies still are unchanged.[21] Such phenomena have already occurred in America, in the politics of race; and on this issue, where the interests concerned are mainly major institutions which can hire talented and enlightened experts at will, it is even more likely. The movement of critical science would then face the pitfalls of corruption as soon as, or even before, it had skirted those of impotence.[22] But this is only a natural process, characteristic of

[20] '... the intellectuals' distance from certain kinds of material activity, their occupational repugnance for certain forms of bourgeois organizations, their attachment to abstract versions of bourgeois tradition rather than to the sub-stratum of bourgeois activity, their familiar quest for sinecures within the interstices of bureaucratized intellectual systems, combine to endow them with what was once an anticapitalist and is now an anti-bureaucratic ethos.' See N. Birnbaum, op. cit., p. 244.

[21] First prize in the 'enlightenment' stakes has been won by the Monsanto Chemical Company. The Scientific Division of the Committee for Environmental Information (which publishes *Environment*) included among its members for 1969 Mr. F. D. Wharton, Jr., St. Louis Development Manager, Life Sciences in the New Enterprise Division, Monsanto Company.

[22] On the corruption of good causes, the classic is G. B. Shaw, *Major Barbara*. The climax of the play comes when Mrs. Baines, the Salvation Army Commissioner who

all radical movements. It is easy to maintain one's integrity when one's words and actions are ineffective; but a long period of this can produce a sectarian or a crank. If one begins to achieve power, and one's policies affect the interests of many others, one must decide where one's responsibility lies. If it is to the ideal alone, then one is set on a course towards tyranny, until overthrown by the host of enemies one has raised up.[23] And if one accepts responsibility for the maintenance of a general welfare, including that of one's opponents, one is on the path to corruption and impotence. This may seem a gloomy prognosis; but a society which does not present such hazards to radical movements of every sort is not likely to retain its stability.

We can expect, then, that the future political history of critical science will be as complex and perhaps as tortured as that of any successful radical and reforming movement. But if it does survive the pitfalls of maturation, and so contributes to the survival of our species, it can also make a very important contribution to the development of science itself. For if the style of critical science, imposed by the very nature of its problems, becomes incorporated into a coherent philosophy of science, it will provide the basis for a transformation of scientific inquiry as

runs the shelter in West Ham, thanks God for the donation of £5,000 by Mr. Bodger, the distiller whose whiskey is the curse of the poor in their care. The Cockney Bill Walker, whose donation of a guinea in repentance for striking two women had just previously been indignantly rejected, utters the significant 'Wot prawce selvytion nah?' (Penguin Books, 1960; p. 106.) See also Shaw's Preface to the play.

[23] A cautionary tale that should be read by all who are embarking on political activism based on 'critical science' is the play by Ibsen, *The Enemy of the People*. Superficially, it is about an honest doctor who is hated by the corrupt forces of his town for his determination to expose the scandal of polluted waters being used in the town's profitable baths. But on closer reading, it can be seen that Dr. Stockmann's misfortunes were also due to his own naïveté and egoism. It is significant that in his own version of the play (Viking Press, New York, 1951), Arthur Miller strengthened the 'progressive' message by transferring the passage where the public meeting declares Dr. Stockmann to be 'an enemy of the people'. In his version it comes at the very beginning of the meeting, before he has spoken; while in the original it comes after the Doctor's harangue, concluding with, 'Let the whole country perish, let all these people be exterminated.' After studying the play with a class at Harvard, where this modification was discovered, it struck me that a worthwhile sequel could be written, entitled 'The People's Friend', in which the entrenched forces, if only a bit less stupid and venal than in the original, could corrupt the good Doctor without difficulty. Although Spa resorts are no longer an important focus of pollution, it is possible for an 'Enemy of the People' situation to be repeated in any of the seaside towns which dump their raw sewage into proximity to bathers. See J. A. Wakefield, 'Clean or Dirty Beaches—Which do you Prefer?' *Your Environment*, No. 1 (Winter 1969), pp. 29–31.

deep as that which occurred in early modern Europe. The problems, the methods, and the objects of inquiry of a matured and coherent critical science will be very different from those of academic science or technology as they have developed up to now; and together they can provide a practical foundation for a new conception of humanity in its relations with itself and the rest of nature.

The work of inquiry in critical science involves an awareness of craft skills at all levels, and the conscious effort of mastering new skills. The data itself is obtained in a great variety of ways, from the laboratory, from the field, and from searching through a varied literature, not all of it in the public domain. Much of it lacks soundness, and all of it requires sophisticated and imaginative treatment before it can function as information. Indeed, since the problem-situations are presented in the environment, and much of the crucial data must be produced under controlled conditions in the laboratory, work in critical science may overcome the dichotomy between field-work and lab-work which has developed in science, even in the biological fields, over the past century. In the later phases of investigations of problems, the same challenges of variety and novelty will always be present. The establishment of the strength and fit of each particular piece of evidence is a problem in itself; and the objects of inquiry (including the measures of various effects and processes, as well as conventional standards of acceptability in practice) are so patently artificial, that one is in little danger of being encased in them as a world of common sense. The establishment of criteria of adequacy for solved problems is possible, for the work will frequently be an extension and combination of established fields for new problems, and so critical science can escape the worst perils of immaturity. Also, any critical publication is bound to be scrutinized severely by experts on the other side, so high standards of adequacy are required because of the political context of the work. Indeed, a completely solved problem in critical science is more demanding than in either pure science or technology. In the former, it is usually sufficient to obtain a conclusion about those properties of the artificial objects of inquiry which can be derived from data obtained in the controlled conditions of experiment; in the latter it is sufficient for an artificial device to perform its functions without undue disturbance by its natural environment; while here the complex webs of causation between and within the artificial and natural systems must be understood sufficiently so that their harmony can be maintained.

The social aspects of inquiry in critical science are also conducive to the maintenance of its health and vitality, at least until such times as the response to its challenge becomes over-sophisticated. The ultimate purpose which governs the work is the protection of the welfare of humanity as a part of nature; and this is neither remote, nor vulgar. Critical science cannot be a permanent home for career- ists and entrepreneurs of the ordinary sort; although it may well use the services of bright young men intending eventually to serve as enlightened experts. Those who want safe, routine work for the achievement of eminence by accumulation will not find its atmos- phere congenial; for its inquiries are set by a succession of problem- situations, each presenting new challenges and difficulties. Hence although critical science will doubtless experience its periods of turbulence, political and scientific, it is well protected from stagna- tion and from the sort of creeping corruption that can easily come to afflict industrialized science.

Finally, the objects of inquiry of critical science will inevitably become different from those of traditional pure science or technology, for here the relation of the scientist to the external world is so fundamentally different. In traditional pure mathematical-experi- mental natural science, the external world is a passive object to be analysed, and only the more simple and abstract properties of the things and events are capable of study. In technology, the reactions of the uncontrolled real world on a constructed device must be taken into account, but only as perturbations of an ideal system; the task is to manipulate it or to shield the device from its effects. But when the problem is to achieve a harmonious interaction between man and nature, the real world must be treated with respect: both as a complex and subtle system in its own right, and as a heritage of which we are temporary stewards for future generations. Hence, even though studies of our interaction with the environment will necessarily use all the intellectually constructed apparatus of dis- ciplined inquiry, their status and their content will inevitably be modified. They will be more easily recognized as imperfect tools, with which we attempt to live in harmony with the real world around us; and although this attitude may seem to conduce to scepticism, it will be the healthy one which recognizes that genuine knowledge arises from lengthy social experience, and that such knowledge depends for its existence on the continued survival of our civilization. The objects of inquiry themselves will include final causes among their essential attributes, not merely the limited functions appropriate to technology, but also the judgements of fitness and success already developed in classical biology and ecology. All this is work for the future; but if it is successful, the opposition between scientific knowledge and human concerns, characteristic of the sciences derived from the dehumanized natural philosophy of the seventeenth century, will be overcome.

With a new conception of the practice of science, there will come a new conception of the history of science and of the meaning of the scientific endeavour. It is possible, and it has been natural, to reconstruct the history of scientific inquiry as a success-story leading up to the triumph of the matured academic mathematical-experimental natural science of the later nineteenth century. One can identify the historical moments, and the great men associated with them, when the very conception of the inquiry itself was significantly advanced. Thus, the origins of our sort of science are rightly located in the earlier Greek civilization, when attempts were made to account for the world of sense experience without invoking personified divine agents. The heritage of the so-called 'pre-Socratic' philosophers was further developed by Aristotle, who not merely conducted disciplined inquiries over almost all areas of human experience, but also showed that such disciplined inquiry has conceptual and methodological problems that can and should be investigated. In a parallel tradition, the idea of mathematics as a body of proved results about conceptual objects developed to full maturity in the achievements of Euclid and Archimedes. The next great advance (according to this interpretation) came many centuries later, when the pioneers of the 'mechanical philosophy' of the seventeenth century achieved a powerful synthesis of experience and reason. Galileo's slogan, 'sense experience and necessary demonstration', stood for his appreciation of the need for closely controlled experience, which could serve as evidence of the appropriate strength, in an argument cast in mathematical language. All that was required to complete the body of methods of classic academic science was the development of institutions for organized, co-operative research; and this came by natural evolution through the nineteenth century. The dominant world-view of matured academic science was atomistic in several important senses. The real world underlying our sense experience was assumed to be devoid of the characteristically human attributes of intellectual and spiritual reality, and of value; final causes were excluded from scientific explanation; and all efficient causes were to be reduced to the material cause of insensibly small brute matter in motion. Correspondingly, knowledge itself was atomic: the achievement of individual facts about the external world, isolated from any philosophical and social context, was considered possible and valuable. This approach to natural knowledge achieved magnificent successes in many fields, and was also appropriate for the development of successful large-scale research.

It was natural to suppose that this particular style of scientific enquiry could be successfully extended to all disciplines, and that it was internally stable. But both these optimistic assumptions proved incorrect. Ineffective and immature fields can be hindered rather than helped by a mechanical imitation of those whose objects and appropriate methods are very different; and the pretence of this sort of maturity only adds to the hazards of applying such fields to practical problems. On the other hand, academic natural science has been transformed by its very successes into industrialized science; and the unexpected problems and abuses of this new sort of science, ranging from shoddy science to runaway technology, present threats to the survival of science and of our whole civilization.

With the new perspective gained from our recent experience, we can look again at the long history of the human endeavour of understanding and controlling the external world. We can now see a positive significance in events and tendencies that have hitherto been considered as unfortunate aberrations. The dominant traditions in academic science have developed out of conflict with other styles of scientific work; and it can be distinguished from them by its objects of inquiry, its methods of work, and its social context. For 'our' science, the real world is devoid of human and spiritual qualities; and the scientist studies the smaller aggregations of matter, considering the most simple and mathematical properties that suffice for the successful investigation of problems. Its approach to its materials is appropriately depersonalized; any 'deeper' meaning that might be thought to inhere in its results is rigidly segregated from his reporting, and is left to amateur speculations. As a social activity, this science is necessarily élitist, presupposing a lengthy course of training and indoctrination for which only a minority have an appropriate cultural background.

To the extent that the traditional history of science has considered these aspects of scientific inquiry, it has been embarrassed by the presence of traditions and tendencies that achieved success in 'our' terms in spite of radical differences in one or more respects from the recently dominant academic style. The roots of astronomy in astrology, and of chemistry in alchemy, are cases in point. Some of the immortal ancestor-figures of the modern discipline are revealed, on unbiassed inspection, to have seen their work as contributing to what is now regarded as pseudo-science: Ptolemy and Tycho for astronomy, and Paracelsus and Glauber for chemistry. Indeed, when we look more closely at the period of the later sixteenth century, when the arts and sciences were developing quite rapidly *before* the incursion of the 'new philosophy' of dead matter, we find the very greatest scientists participating in the world-view of an animated nature: Gilbert investigating magnetism in the attempt to prove that the earth is the embodiment of the *anima mundi*, Kepler searching (with all rigour) for the divine harmonies of the celestial realm, and Harvey using 'spirit' and the macrocosm-microcosm analogy to guide his anatomical and physiological researches.[24]

It would be very misleading to imagine a simple succession of two sorts of science, each unified and coherent in itself, first that of the 'animated' world and then (since the seventeenth century) that of the 'dehumanized and disenchanted' world. History is more complex, and more interesting, than that; and within the 'old' conception of science there were many different tendencies in the interpretation of its appropriate objects, methods and social functions. I have previously referred to a 'romantic' philosophy of nature providing the vehicle for a politically radical folk-science that challenges the academic science of its time. In this tradition, the study of nature is explicitly seen as a social and also spiritual act; one dialogues rather than analyses; and there is no protective cover of belief in the 'neutrality' or 'objectivity' of the style adopted. Such a philosophy of nature will become articulated and advanced, as part of a general radical reaction against a formal, dry or bureaucratic style pervading social or cultural life. Looking back into history, we can find a similarity of doctrine or style, and sometimes a linking tradition, as

[24] Gilbert makes his programme plain; see *de Magnete* Book V, ch. 12: 'The magnetic force is animate, or imitates a soul; in many respects it surpasses the human soul while that is united to an organic body.' (tr. P. F. Mottelay; Dover, New York, 1958, p. 308.) Kepler is well-known; and for Harvey, see W. Pagel, *William Harvey's Biological Ideas* (Basel and New York, 1967).

far back as the Taoists of ancient China, through St. Francis of Assisi, to Paracelsus, William Blake, and Herbert Marcuse.[25]

Not every one of these figures would claim to be a natural scientist of any description; but as philosophers, poets or prophets, they must be recognized as participating in and shaping a tradition of a certain perception of nature and its relation to man. Granted all the variety of their messages and styles, certain themes recur. One is the 'romantic' striving for immediacy, of contact with the living things themselves rather than with book-learned descriptions. Another is 'philanthropy'; the quest is not for a private realization, but for the benefit of all men and nature. And, related to these is a radical criticism of existing institutions, their rules and their personnel. Looked at from the outside, each upward thrust of the romantic philosophy of nature is doomed to failure. Mankind will not be transfigured overnight; and the romantic style has its own destructive contradictions. Whereas the 'classic' style degenerates gradually into an ossified form and a sterile content, the 'romantic' style goes off much more quickly, through chaos of form and corruption of content. But this study of ours has shown that even in disciplined scientific inquiry, the categories of 'success' and 'failure' are neither so absolutely opposed, nor so assuredly assignable in particular cases, as the traditional ideology of science assumed. And the failure to achieve Utopian dreams, in science as well as in social reform, is not at all the same thing as futility.

The dreams of the romantic, philanthropic, radical philosopher-prophets cannot move towards realization by the accumulation of facts or of battalions. Rather, they exist through a discontinuous, perhaps erratic, series of crises and responses. Sometimes they have the good fortune of producing a creative tension in a man brave enough to attempt the synthesis of a prophet's vision with a world

---

[25] On Taoism, see J. Needham, *Science and Civilisation in China*, 2, 88–132. In his magisterial fashion, Needham provides more materials on the analogous movements in early modern Europe than is available in any general history of science. For a discussion of the limitations of his view, see note 35 on p. 394 above. See Lynn White, Jr., 'The Historical Roots of our Ecological Crisis', *Machine ex Deo* (M.I.T. Press, 1968), Chapter 5, for St. Francis.

Francis A. Yates, in 'The Hermetic Tradition in Renaissance Science' in *Art, Science and History in the Renaissance*, ed. C. S. Singleton (John Hopkins Press, Baltimore, 1968), discusses the 'Rosicrucian' style of science in considerable depth. The theme of 'philanthropy' is most clearly developed in the German alchemical philosophers in the Paracelsian tradition; and their influence on Francis Bacon is clear.

On Marcuse, see his *One-Dimensional Man* (Beacon Press, Boston, 1964).

managed by priests. He too will fail, almost certainly; some problems are insoluble. But his message, perhaps in a particular science or walk of life, perhaps of a generalized wisdom, will speak to men in later ages, coming alive whenever it has insights to offer. In this present period, we may find Francis Bacon speaking to us more than Descartes the metaphysician–geometer or Galileo the engineer–cosmologist. As deeply as any of his pietistic, alchemical forerunners, he felt the love of God's creation, the pity for the sufferings of man, and the striving for innocence, humility, and charity; and he recognized vanity as the deadliest of sins.[26] To this last he ascribed the evil state of the arts and sciences:

> For we copy the sin of our first parents while we suffer for it. They wished to be like God, but their posterity wish to be even greater. For we create worlds, we direct and domineer over nature, we will have it that all things *are* as in our folly we think they should be, not as seems fittest to the Divine wisdom, or as they are found to be in fact.[27]

The punishment for all this, as Bacon saw it, was ignorance and impotence. It might seem that the problem is different now, for we have so much scientific knowledge and merely face the task of applying it for good rather than evil. But Bacon assumed his readers to believe themselves in possession of great knowledge; and much of his writing was devoted to disabusing them of this illusion. Perhaps the daily reports of 'insufficient knowledge' of the effects of this or that aspect of the rape of the earth, and our sense of insufficient understanding of what our social and spiritual crises are all about, indicate that in spite of the magnificent edifice of genuine scientific knowledge bequeathed to us, we are only at the beginning of learning the things, and the ways, necessary for the human life.

Bacon was a shrewd man, fully sensitive to the weaknesses of the human intellect and spirit. He was aware of the superficiality of ordinary thought and discourse, at whatever educational level; and he also distrusted the extraordinary enthusiast, in religion or politics, for the damage he could cause. His life's endeavour was to overcome

[26] For a detailed interpretation of Bacon's programme for science in terms of a vision of moral and spiritual reform, see J. R. Ravetz, 'Francis Bacon and the Reform of Philosophy', in *Science, Medicine and Society in the Renaissance* (Walter Pagel Festschrift), ed. A. Debus (University of Chicago Press and Oldbourne Press, London, 1972). This is an elaboration of certain themes in Benjamin Farrington's, *The Philosophy of Francis Bacon*, and I am indebted to him for my first insights into this aspect of Bacon.

[27] See *The Natural and Experimental History for the Foundation of Philosophy* (*Works*, vol. v; translation p. 132).

this contradiction somehow, and to bring about a true and effective reformation in the arts and sciences of nature. For him, this was a holy work, a work of practical charity inseparable from spiritual redemption.[28] His audience was inevitably among the literate; and so he tried, by scattering hints and half-concealed invitations, to call together his brothers, who would gently and silently show by their example that a good and pure way into Nature is also the practically effective way. Of course he failed, in his philosophical reform as in his political career. There was no English audience for his particular message during his lifetime, and at his death he was alone and neglected.

Shortly after his death, however, there was a stirring; and Bacon's message of 'philanthropic' science began a career of its own. For a while, his followers knew what he was about; but with the passage of decades and disillusion, this was forgotten, and only the vulgar fact-finding Bacon survived. Yet when we now come back to read Bacon, perplexed and worried as we are by the sudden transformation that science has wrought upon itself as well as upon the world, we can find relevance in passages like the following:

Lastly, I would address one general admonition to all; that they consider what are the true ends of knowledge, and that they seek it not either for pleasure of mind, or for contention, or for superiority to others, or for profit, or fame, or power, or any of these inferior things; but for the benefit and use of life; and that they perfect and govern it in charity. For it was from lust of power that the angels fell, from lust of knowledge that men fell; but of charity there can be no excess, neither did angel or man ever come in danger by it.[30]

[28] See the *Meditationes Sacrae* (*Works*, vol. vii; translation pp. 243–4). Bacon contrasts the miracles of punishment wrought by the prophets of the Old Testament, with those of Jesus: 'Jesus was the Lamb of God, without wrath or judgment. All his miracles were for the benefit of the human body, his doctrine for the benefit of the human soul.' After a list of instances, Bacon comments, 'There was no miracle of judgment, but all of mercy, and all upon the human body.' Later, in the essay 'Of Hypocrites', he comments, 'The way to convict a hypocrite therefore is to send him from the works of sacrifice, to the works of mercy. Whence the text: *Pure religion and undefiled before God and the Father is this, to visit the orphans and widows in their affliction . . .*' (p. 249).

[29] On the influence of Bacon, see Charles Webster, *Samuel Hartlib and the Advancement of Learning* (Cambridge University Press, 1970).

[30] Bacon, *The Great Instauration*, Preface (*Works*, vol. iv; translation pp. 20–21).

INDEX OF NAMES

INDEX OF TOPICS

Here are listed definitions and extended discussions of basic concepts developed in this work. Compiled with the assistance of Joseph Ravetz.

GENERAL INDEX

Compiled with the assistance of Joseph Ravetz.

PENGUINEWS *AND* PENGUINS IN PRINT

Every month we issue an illustrated magazine, *Penguinews*. It's a lively guide to all the latest Penguins, Pelicans and Puffins, and always contains an article on a major Penguin author, plus other features of contemporary interest.

Penguinews is supplemented by *Penguins in Print*, a complete list of all the available Penguin titles – there are now over four thousand!

The cost is no more than the postage; so why not write for a free copy of this month's *Penguinews*? And if you'd like both publications sent for a year just send us a cheque or a postal order for 30p (if you live in the United Kingdom) or 60p (if you live elsewhere), and we'll put you on our mailing list.

Dept EP, Penguin Books Ltd,
Harmondsworth, Middlesex

Note: *Penguinews* and *Penguins in Print*
are not available in the U.S.A. or Canada

SCIENCE AND SOCIETY

Hilary and Steven Rose

'So important that no one can afford to neglect it. Who is directing research, technological development, industry, and education toward the common good? The answer, as the complex arguments in this book demonstrate, is no one' – *Guardian*

In a study which invites comparison with J. D. Bernal's *The Social Function of Science* (published before the bomb or the cracking of the genetic code), a biochemist and a sociologist attack the notion that science, like fate, is 'an unpredictable act of gods in white coats'. Since it is the product of certain men in certain societies, it can be controlled. In their opening chapters they recount the history of science in its relations with society from the founding of the Royal Society to the post-war records of Conservative and Labour governments. For comparison they add chapters on the position in America, Russia, and other countries, and on the functions of international bodies (from Pugwash to UNESCO).

This book is recommended by *New Scientist* as 'a helpful starting-point . . . for the students of the forthcoming Open University' in approaching the question of how science can be effectively harnessed for the good of all people.